SEX, GENES
AND ALL THAT

SEX, GENES AND ALL THAT

THE NEW FACTS OF LIFE

Anthony Smith

MACMILLAN

To Anne, again

First published 1997 by Macmillan

an imprint of Macmillan Publishers Ltd
25 Eccleston Place, London SW1W 9NF
and Basingstoke

Associated companies throughout the world

ISBN 0 333 64268 6

1 3 5 7 9 8 6 4 2

A CIP catalogue record for this book is available from
the British Library.

Typeset by CentraCet, Cambridge
Printed by Mackays of Chatham plc, Chatham, Kent

CONTENTS

CONTENTS

PREFACE

*There is no form of prose more difficult to understand and
more tedious to read than the average scientific paper.*
Francis Crick

Strict scientific writing requires wholesale qualification, few generalizations being acceptable. Every statement is therefore loaded with qualifying appendages. General scientific writing demands generalization; therefore many a qualification has to be omitted. Women have babies. Some women have babies, these born mostly as singletons. Women between 15 and 45 have most babies, but babies have been born to girls of 9 and women over 60. Women in developing countries have more babies when in their twenties, whereas the majority of women in developed nations are now waiting until their thirties before reproducing. Most women in developing nations have a life-long partner who is usually older, whereas an increasing number of women in developed nations, waiting until their thirties before reproducing, are opting for single parenthood and IVF is being more commonly used, which can decrease the number of singleton births ... The nightmare of such contortions can make any generalist (and reader) long for the earlier simplicity. Women do have babies; that is final.

Similarly, if someone has made a telling point, the announcement of that point can be swamped if its provenance is carefully enumerated. 'J. Luckstein, principal medical officer of the medical faculty attached to the University of the Sierra Nevada, northern Ecuador (itself affiliated to the Medical Research Development Unit of the South American Collegiate), said ... whatever he did say.' If Luckstein is famous, all well and good. If not, too bad, but at least a listed origin has not held up the prose. If other individuals, lying outside the scientific realm, are quoted and are known – good; but, if fully explained, they too can congeal the flow. 'H. G. Wells, prolific and famous author, best known for scientific fiction but also for an intriguing personal life, and who died in 1946 aged 80, said...' Such explanations will annoy (those who know) as well as stiffen the flow. Therefore named persons in this book are only briefly explained.

There are other personal prejudices, mannerisms and preferences intended to make more agreeable sense of the subject matter. (If they irritate I apologize here and now for the grief they will surely cause.) People can be professors, doctors, generals, knights, secretaries of state or even, I suppose, all of these at once, but titles can be intrusive. Besides, was he a knight/professor when he made that remark? Gregor Mendel was not yet an abbot when performing his famous experiments, and was not even alive when the world took note of them. Surely, so runs prejudice, Mendel is known, with no need to put him, as it were, in his place. In this book he has been made neither Abbot Mendel nor the late Gregor Mendel nor (it gets worse) the late Gregor Mendel, one time abbot at Brno, this city now in the Czech republic but then part of the Austro-Hungarian Empire. Similarly professors and the like have (usually) been stripped to the bare bones of their names.

The same goes for words now equally well known. Ovular, amniocentesis, menarche, Parkinson's, implantation, prostate, in vitro, hypertension – all these, and countless more, have entered common speech increasingly of late. They do not demand lengthy explanation; the briefest kind will do. Words from Nicholas Culpeper, brilliant but sad author of *Genitals of Men*, can serve as precedent. He wrote (in the seventeenth century) that 'Latins have invented very many names for the *yard* ... I intend not to spend time in rehearsing the names, and as little about its form and situation, which are both well known, it being the least part of my intent to tell people what they know, but teach them what they know not...' The sentiment is sound, even if that opening phrase is now tinged with regret: we no longer use the yard as a term for each male's most distinct possession.

Having raised masculinity it should be added that this book is written by a male. Hopefully the fact does not obtrude but, in discussing sex appeal (for example), it seemed easier never to view the matter as if from a female. Of course appeal works both ways, but wrong notes would be struck if pretence were involved. Hopefully, yet again, there is balance, and a kind of impartiality, but this is never total. By way of making this point, and serving as belligerent apologia, the opening chapter is 'Male' and its successor – its *longer* successor – is 'Female'.

There is also trouble, particularly in a book on reproduction, with the word Sex. It ought to be solely an abbreviation of Sexual, but it has come to mean intercourse, mating, copulation. Once upon a time, as with Jane Austen, intercourse was a synonym for communication, but the word has been usurped. So too sleeping (with), going to bed (with) or, as comedians know full well, practically every other turn of phrase, given the range of innuendo possible these days (and perhaps, less obtrusively, in other days as well). It is ridiculous that Sex should be a problem word, but in a book dealing with sex (as a system) and sex (as an activity) there are sentences which tie themselves in knots to make clear their reference.

Science is sensible in using the metric system rather than 8 pints to a gallon, 12 inches to a foot, 1,760 yards to a mile, or any other of the splendidly archaic units we have inherited. The problem, as of now, is that many of us prefer the archaic to the metric, and this book has tended to follow suit. Scientists are quoted with their preference but, given half an inch of opportunity, the remaining text is more traditional.

Science, unfortunately, is less than helpful in what it calls cohorts. One treatise may tell of 'infants' dying in their first six months, another of 'neonates'. One text may be most revealing about the elderly (over 60) and another about the over-65s. How therefore to compare and contrast, as school used to say, these two sets of information? Similarly the United Kingdom prefers being disunited, in that its facts are published under England and Wales, or Scotland, or Northern Ireland, or even (it has been known) under the UK as a whole. The United States competes by being as little federal as possible with, on occasion, thirty-six states promoting some measure, twelve against, and two with quite different edicts of their own.

Science loves averages, whether mean, mode or median. People love averages too, even if less aware of their contrivance. Statistics has been described as 'the science which tells you that if you lie with your head in the oven and your feet in the fridge, on average you'll be comfortably warm'. People, in general, prefer average to mean normal, or perhaps the normal range. This book, it is hoped, has steered a path between the various preferences.

Finally, in a world addicted to initials, these too can irritate (if poorly known) but are convenient. Some are recent, such as AIDS, HIV and HRT, while others have longer histories – IQ, WHO, USA. Therefore beware, as there are many more to be encountered in the following pages, such as CSA and FGM or, sounding a touch more friendly (if misleadingly), TURP, BEDS, NORM, RECAP and BUFF In short, welcome to all of them, and to the pages now ahead.

A.S.

INTRODUCTION

Few topics are more thoroughly obscured by unsound
information, contradictory religious and cultural beliefs, and
illogical thinking than human reproduction.

Caroline Pond, *New Scientist*

Disinterested intellectual curiosity is the life-blood of real
civilization.

G. M. Trevelyan, in *English Social History*

'The division of sexes is a biological fact, not an event in human
history,' wrote Simone de Beauvoir in *The Second Sex*. It is so much a
piece of us that we rarely contemplate the notion of some alternative
method of reproduction. No fundamental law states that living matter
can only create additional living matter by having male and female,
one to fertilize the other, one to act as incubator and the other as
instigator. Many plants and animals do very well, by way of procreation, without a pair of sexes; but we humans cannot do so any more
than we can grow six legs or eyes at the back of our head. All
mammals are emphatically male and female. They do need sperm
and eggs for replication. Each human then needs many years before
achieving first sexual maturity, then physical maturity. We next grow
old, and then we die.

As Jeanne Moreau said with awesome clarity, 'There's a certain
moment in life when you realize you're born with a deadly disease
which is life.' Or, as biologist Michael Brambell put it, 'Whenever a
life begins, another death becomes inevitable.' The only immortality
is leaving genes behind. We can continue to live only through our
children, and then through their children. We are our genes and can
take comfort that our portions of DNA will last as long as reproduction lasts in the family tree we have helped along its way. For males

there is the extra fact, should we either contemplate our inheritance or wonder at our future, that our single Y chromosome was possessed identically by our father, and by his father, and that man's father, and so on back in time. Should we have male offspring they possess the very same Y, and will donate it to any males they might sire, and into future centuries. (Females can think similarly, save that their X chromosomes are a pair, and there is never the clarity of inheritance as with that solitary Y.)

This book embraces sex in the broadest biological sense. It therefore starts with male and female before leading on to sexual reproduction, to pregnancy and birth. It then, as is the nature of the sexual cycle, carries on through growth and puberty to maturity. Along the way, as incidentals (or barriers) to this development, lie infertility, contraception and abortion. Never irrelevant, but more unravelled than ever before, are genetics and the gene, the coding system from DNA to protein, and the genome which is the blueprint of us all. Eugenics also finds a place, this theoretically desirable intent with some awful consequences. Decay and eventual disappearance are as much part of sex as union and appearance; so there is ageing and death, along with disease, and suicide, and euthanasia. Biologically anything and everything which interferes with the passage of genes from one generation to the next is entirely pertinent. Miscarriage, still-birth, infant mortality, early death, impotence, sterility – they are all the same. Sets of genes have been prevented from their own form of fulfilment.

Not everyone today accepts Genesis with the fervour of their forebears, but belief lingers that humanity is in a different class from all the animals. Its intelligence is undeniable, so too its ability to transform; but, in the matter of reproduction, it is as sexual as the rest. What goes on in testes and ovary, or in mating, or within the uterus, or with lactation, is identical save for minor variations. Adolescent infertility, a feature of children sexually mature but not ready for a family, is best examined by comparison with near neighbours, such as gorillas and chimpanzees. We are never equal to animals, not even to the apes, but we should see ourselves as one more fragment of the natural world, however unnatural we are making it.

Human beings are distinguishable, it has been said, because of their enthusiasm for taking medicine. They have always done so, so far as can be judged, and have certainly not given up the practice. More medication is available now than ever before (and more is being taken). There is also every possible form of amendment. Nothing, it would seem, has escaped our longing to alter. We can conceive in test tubes. We can use the genes of others. We are making embryos externally, and using surrogate mothers. We may soon take out embryos conceived in the old-fashioned way, have a look to check on their credentials, and then replace them. We still have to wait for the day of birth but we inspect beforehand however possible. We hate fertility – often, and may soon reach a time when infertility is the norm, with frozen gametes waiting until procreation is convenient. There is also the procedure, perhaps the ultimate amazement for our ancestors should they return today, for making women out of men.

So much change, leading to so many rights, to pressure groups of every kind, can make us forget the astonishments of reproduction. All the DNA that initially forged everybody on earth since the time of Christ has weighed a fraction of a gram. A foetus's heart starts beating twenty-two days after sperm meets egg, and does so unceasingly until finally shutting down. A human baby is so incapable, not even standing surely for a year, and yet learns language long before its sphincters are properly controlled. To see mice or lambs fresh-born is one amazement. To see our own kind, or – more particularly – our very own kind which we have helped to make, is a thousand or a million times more amazing still. It happens every day, in fact 77,000 times every single day, and is none the less astonishing. So too growth. And comprehending. And developing. And seeing one's own life span within this aged tale. 'It has all been very interesting,' said an old lady as she lay dying. It has indeed, and is.

There is also need for change. Population increase is daunting. Abortion, as ever, is disconcerting. But without abortion the population of China (over a billion) would have been added to the world's total number even in the past thirty years. Contraception is not always pleasing, but anything which keeps the quantity of newborns down to the current nine a second must have virtues. Its blessings should be available to the hundreds of millions who could also benefit. Ageing and the aged are on the increase, notably in the developing world. The elderly have no wish to be a burden, but are too frequently. Advances which enabled them to survive must be partnered by other advances permitting their extra years to be a blessing, for them and their relations, for their community and the state.

All progress confuses the old simplicity. It is good that extremely premature babies do not die, but bad if they survive mentally impaired. It is good that previously infertile couples can conceive, but less satisfactory when they wish only for a certain sex, or for some superior genetic material, or for a range of guarantees which never formerly existed. Surrogacy has obvious advantages for some, but is less appealing when the surrogate mother subsequently disputes the rights of those who gave the genes. To know of some dread disease like Tay–Sachs sounds beneficial, but how many of us wish to know of our actual catalogue, of the recessives we each carry, of the likely consequences for offspring should we choose to mate with the one we most desire? It *is* a brand new world we are creating but, as partner, there is a brand new set of ethical upsets which we must now address. To take eggs from aborted foetuses may seem abhorrent or sensible, a nauseating practice or a pragmatic solution to the need for donor material. Such choices will pepper the years ahead.

They are also raised among the pages of this book. A knee-jerk reaction is not necessarily the wisest, either in the long term or the short. A considered response, founded upon much relevant information, is more likely to be profitable. Therefore countless facts have been included which should help with decisions, and anxieties, and with desperate concerns about future possibilities, some of which are waiting in the wings, and some of which, however futuristic, have already been tried by and on and with a few of us.

It is not quite the brand new world of the twenty-first century, but we now have a telescope to anticipate a large part of it. The following pages are as up-to-date as could be contrived to bring that future sharply into focus. Whether we like much of what we read is not the point. It is what we do with the range of possibilities on offer in the realm of human reproduction that is paramount. And what we do is up to all of us, the bewildered, uneven assortment of individuals who, as *Homo sapiens*, have – and will have – so many more choices on their hands, for good or ill. The old and entirely natural ways are as embedded in the past as all the battles, and kings, and empires in our lengthy history. Much even of the twentieth century appears almost medieval when viewed from now. As for the twenty-first century, that can seem daunting, but it is about to be upon us and we, in theory, are in charge.

MALE

SPERMATOZOA   CASTRATION   THE PENIS   EJACULATION

SIN OF ONAN   IMPOTENCE   DETUMESCENCE

THE PROSTATE   PREPUCE   PATERNITY   HOMOSEXUALITY

SPERMATOZOA The world is aware of shortages, many important for maintenance of life, but the output of human spermatozoa is not one of them. Assuming 1,300 ejaculations per second (a World Health Organization estimate) and 3.5 cc of ejaculate (an average amount) possessing 50 million sperm per cubic centimetre (frequently quoted), a couple of hundred *billion* are therefore expended every *second* of each day. Only nine or so of this formidable total fulfil their promise by providing their half of the genetic material necessary for another human being. The extravagance is therefore a prime astonishment, with the massive army succeeding in fertilizing only nine ova.

Each spermatozoon is extremely small, at least as compared to the (just) visible ovum. Its head is 0.004 mm long and 0.003 mm wide at its widest. The tail, which contains none of the genetic material, is far thinner (0.0005 mm) than the head, but far longer (0.04 mm), so much so that its total volume is greater than that of the head. The tail's length is perplexing, in that a shorter flagellant might be as effective, but the human measurement is quite outclassed by, for instance, the fruit-fly. *Drosophila melanogaster*, known to every student of chromosomes (and to others who observe these midge-sized insects hovering by garbage), has sperm tails of 1.76 mm – or 300 times longer than the human appendage. *Drosophila bifurca* is even more bizarre. Its tails are 60 mm, or twenty times longer than the flies manufacturing them. No one – at present – can satisfactorily explain the evolutionary advantage(s) of such exuberance.

Only when human sperm have been ejaculated do they 'display their full pattern of motility', as one textbook phrases it. In short, after idling casually (perhaps for weeks), they abruptly achieve full speed following their release. This velocity is considerable, being 1.5 to 3 mm a minute (or 3.5 to 7 inches an hour). Movement of this kind can be maintained for many hours, or even days (up to seven in suitable in vitro solutions), but the ability to fertilize is lost much earlier. Although spermatozoan motility is important these minute swimmers reach their destination – the uterine tube – after little more than an hour, and far more swiftly than their own flagellation could achieve. It seems that contraction of the uterine and tubal musculature helps them to get there, much as swimmers can surf on waves to shore.

Abnormal waves, or the lack of them, explain the failure to conceive of nearly half the women with hitherto unexplained infertility, according to researchers at the University of Manitoba. The waves in healthy women move upwards towards the tubes for 80% of the time at about three waves per minute. They also occur at their maximum frequency in the middle of the menstrual cycle when a woman is ovulating and at her most fertile. Speed is necessary to hasten the sperm towards the egg, and it is because the environment of the upper vagina is so acidic that few sperm survive it. Unfortunately – for fertility – the waves are not always helpful, being sometimes weak, occasionally moving in the wrong direction, or even unco-ordinated, spreading from a central point in opposite directions. Hence one group of reasons for infertility.

More problematical for the sperm than mere distance is the range of environments through which they have to pass. From their home base (within the originating male) to the female's vagina, to her uterus and then her right or left tube, the surrounding fluids encountered by the invading sperm vary considerably. Successful sperm have to survive them all. Not only do these environments create a varied obstacle course, but they are crucial to capacitation, the process – poorly understood – whereby sperm achieve their ability to fertilize. This is not possible until the sperm have been in the female's genital tract for a period of time, a matter of hours. Spermatozoa are, after all, foreign bodies. In theory they should be resented by the recipient, the female. In practice this cannot be allowed to occur, and capacitation must involve an immunological reaction – of some sort.

A male's inability to reproduce can have many causes – impotence, for example – but faulty sperm are more frequently to blame. This form of fault is reputedly – and surprisingly – getting worse. At least one man in twenty is now infertile. Not only do there have to be sufficient sperm – fertility decreases if there are fewer than 50 million per ejaculation – but the proportion able to move should be at least 50%. Some studies have suggested that 'sperm concentration per unit volume' has fallen by 40% over the past fifty years. The alleged diminution is all the more worrying (for potency) because 'normal' human semen is already poor when compared with that of other mammals. More than half of human sperm usually have morphological abnormalities as against 5% in a general mammal assortment. Any worsening of the human situation will lead to further unhappiness among couples wanting children.

So why the fall? Some say smoking is to blame. (Is there anything not allegedly aggravated by the cigarette?) Others argue that soap and washing powder may be guilty, their 'surfactants' becoming oestrogen-like as and when they are destroyed by bacteria. Even food has been blamed – unless it has been grown organically. So too has water, its current impurities serving as culprit. And so has DDT. Finally, and almost inevitably with any novel finding, the finding itself is questioned. Is it really true that sperm count is decreasing? The answer is perhaps. (Or even No. Two new studies, reported in the *Lancet* during May 1996, suggested that there may not be a decline in sperm counts after all. 'All the studies before 1970 were from the USA and 80% were from New York ... areas with the highest sperm counts ... After 1970, 80% of the studies were from locations not represented earlier, including five studies from third world countries, where sperm counts are low.')

A survey undertaken in seven French maternity hospitals (and published in January 1996) asked a basic question of women who had recently been delivered of a child: what kind of work was the father doing around the time of conception? The researchers wanted to know about 'heat exposure as a hazard to fertility'. They were able to conclude, after their extensive questionnaire had been completed by 577 women, that pregnancies had been (slightly) delayed if their partners were exposed to heat at work, if these men were seated in a vehicle for more than 3 hours a day during work, or if the women themselves were smokers. It can seem odd that many mammalian testes, including man's, are located externally, but there is a reason. As the textbooks phrase it: spermatogenesis requires physiological scrotal hypothermia. Sperm production is improved (for whatever reason) if the testes are cooler than the body – 35 degrees rather than 37. Exposing them to heat, or preventing heat loss (by confining them within trousers in a restricted position), diminishes that necessary coolness – and therefore lengthens the time before fertilization and pregnancy are achieved.

Whether or not there is a general decrease in sperm production within modern society as a whole the relative sperm lack is not immediately troubling. There are ample to maintain a disturbing population pressure. Jack Cohen, British biologist and frequent sceptic of conventional opinion, has reported that many of the sperm reduction figures 'are so smooth that they shout "artefact" rather than biology'. He considers the major problem is that 'we don't understand infertility ... Bulls produce a billion at a time, for one calf. Until we know why so many sperms are offered, we can't assess the seriousness of changes – if indeed they've happened.'

Nature spoke forth on the subject (in its leading article of 7 March 1996), admitting 'the insidious threat posed by environmental pollutants ... not that pollutants are necessarily the source of the falling sperm counts reported over recent years'. It considered 'the statistics of sperm decline are themselves controversial'. Moreover, 'changes in women's lifestyles may have affected the development of male fetuses, while men's smoking and prolonged driving in tight trousers' may be relevant. Nevertheless 'endocrine-disrupting chemicals affecting fetal testes are prime suspects'.

Pessimists (who worry about falling sperm counts) can have their fears enlarged by reading *The Children of Men*, a best-selling thriller by P. D. James. Set in the twenty-first century, it describes what happens when sperm counts, for men *and* animals, decline inexorably, the trend having first been detected (but not amended) in the 1990s. Optimists can be encouraged by recent work, as from the Centre for Reproductive Biology, Edinburgh, which has greatly improved male fertility. 'Immature spermatozoa taken from testes of men with the severest forms of reproductive dysfunction can achieve both fertilization and normal development of the embryo,' wrote R. John Aitken and D. Steward Irvine in February 1996. They added that 'even precursor germ cells ... that have not yet differentiated into spermatozoa have been used to accomplish the same end'. There may come a time when a male need only produce a single sperm, rather than millions, to effect fertilization, given a little assistance from some laboratory.

As a digression, the French authorities had no wish to diminish sperm numbers by careless driving. Early in 1994 it became necessary to transfer 300,000 sperm samples from the Cochin Hospital, Paris, to a new site on the city's other side. A special truck made four trips at dead of night, and at slow speed along smooth roads, to effect the transfer. Liquid nitrogen, in which the sperm were stored, becomes gaseous when actively shaken, thereby rendering the samples useless. Fortunately, according to a subsequent report, there was no sperm diminution following this highly abnormal and gentle form of Parisian driving. (A different form of twentieth-century hazard hit Parisian sperm when a sperm bank became bankrupt – in 1996. Couples who had donated semen were unable to retrieve their deposits until the bank's precious assets had been renegotiated to another bank.)

Wrongful driving in the United States led (in January 1995) to the preservation of some sperm which would otherwise have perished. Their owner was killed in a road accident. His widow, with speedy presence of mind, requested the preservation of some of her husband's spermatozoa. Peter Schlegel, urologist at the New York Hospital, cut the cord which carries sperm from testis to penis and acquired the necessary sample. This was immediately stored in liquid nitrogen. The widow's point of view is understandable – she may want a child in the future and may prefer that child's father to be her former husband, but conflict is inevitable if the procedure becomes commonplace: paternity is being conferred upon a man without his consent. What of inheritance if his estate is to be shared between descendants? And for how long can he continue to sire offspring *in absentia*? Is this the first form of immortality likely to come our way, with frozen sperm creating future generations so long as they remain viable within their nitrogen? If the procedure is good enough for sperm it will doubtless become good enough for ova. The original couple may both have died but, via their frozen gametes, may each continue to live – and breed – in perpetuity.

CASTRATION (Greek, to guard the bed) is no longer prevalent. When testicle severance is performed after puberty there can be both desire and performance (making one wonder about harems believed safe from male interference). If the operation occurs before puberty the man will not be capable in sex, will not grow hair in masculine places, and will not grow bald, but will have pubic hair like a woman's and, most importantly – for some – will retain an unbroken voice. Castrati could span three and a half octaves. Their power to sing was also exceptional, enabling them to hold a note for more than a minute. The very talented could sing passages of 200 notes without taking a breath. Such castration was outlawed during the nineteenth century and Alessandro Moreschi, last of his kind, died in 1922. The recent film *Farinelli, Il Castrato* needed a good voice of the correct range and intensity to mimic that of its eponymous hero (who had lived from 1705 to 1782). There being no suitable candidate, the singing was performed by a clever computer which created a single voice from the combination of a male contralto with a female coloratura soprano.

THE PENIS was not normally severed during castration, save on occasion in China. This extra act not only left the victim with a mass of scar tissue in lieu of external genitalia but removed all possibility of sexual performance (making such total eunuchs favoured for the Imperial Court). It was all very well – to speak loosely about evolution – for the system of sexual reproduction to be developed, and for two forms of gamete to have to meet after their creation within two forms of individual, but this necessity posed problems, particularly after further development demanded that fertilization should occur

internally. Somehow male and female had to unite. Sperm and ova
have to meet.

Initially, before internal conception, the system was casual, with
sperm and eggs liberated into the surrounding medium. Most fishes
perform in this fashion, usually with some form of synchrony of
action increasing the likelihood of union between sperm and egg.
Certain fish, such as sharks and rays, have a more advanced arrangement. The males possess claspers that are inserted within the female
and along which the sperm travel to meet the female's eggs. Birds
have no specific organ of copulation, save for ducks, geese and
ostriches. The general avian system, with males releasing their sperm
when on the backs of females, and those sperm having to reach the
cloaca, plainly works, but it is thought that ancestral birds did have a
copulatory organ, if only because reptiles – the group from which
birds sprang – possess one (or even two). Snakes and lizards own
what are called hemipenes, organs turned inside-out at the appropriate moment before insertion into the female. Crocodiles and turtles
have an organ so similar to the mammalian penis that it merits the
same name.

Mere possession of a penis is not inevitably partnered by an ability
to inseminate. The organ has to be made capable of insertion. With
humans this is a purely vascular phenomenon, transforming a flaccid
object into one capable of penetration. The pressure involved is blood
pressure, and is therefore caused by the heart. A youthful male's
blood pressure oscillates between 120 and 80 mm of mercury (or
about one-seventh of atmospheric pressure). Rigidity is a function of
both pressure and diameter; the greater either the diameter or the
pressure the greater the firmness, and capability, of the penis. (As
elderly males often have higher blood pressure they may, in consequence, be able to create more rigid erections, not that there seem to
be any scientific reports to this effect.)

To build up this firmness the invading blood must have its outflow
restricted. As *Gray's Anatomy* phrases it: 'Rapid inflow from the
helicine arteries fills the cavernous spaces and the resulting distension
of the corpora cavernosa acts as a contributory factor by pressing on

the veins which drain the erectile tissue.' The more these spaces are filled the harder it is for blood to escape via the veins. The penis is therefore able to expand from its modest size (yet more modest when cold or fear is paramount) to something both capable of insertion into the vagina and of arousing that partner appropriately. Human males can take quaint comfort that they are the only primates without a bone in their penis to assist in creating the necessary rigidity, this lack of an os penis shared with hoofed mammals, most marsupials, and whales. In humans the system becomes less competent with age. A. C. Kinsey's famous survey of human sexuality discovered that 90% of men still had 'erectile potency' at the age of 60, 70% at 70, and 25% at 80. 'Making love at my time of life,' said an elderly Yorkshire comedian recently, 'is like using a length of rope.'

Engorgement can be extremely rapid, perhaps no more than three to five seconds from the start of stimulation. Average final length is 6.25 inches and average basal diameter 1.5 inches. According to *Human Sex Anatomy* (by R. L. Dickinson) there is relatively little difference in penis size as compared with height and weight differences among humans. Jack McAninch, and others, of the University of California injected sixty men with an erection-creating drug before concluding that an average erect penis measured slightly more than 5 inches. (Outsiders can wonder if the injection, however effective, provided stimulus with the same merit as normal inducement.) Penis-lengthening operations are performed in the US by some surgeons who charge up to $6,000. A severed ligament allegedly permits the increase. Extra girth is achieved by implanting fat taken from thigh or abdomen. McAninch is not wholly supportive of such surgery, stating that uneven swelling, bleeding, loss of sensation, infections (themselves causing skin loss or deformity) and psychological problems can result, causing a subsequent inability to achieve an erection of any girth or length. It could be that other ways of spending $6,000 might be deemed more pleasurable after hearing of such possible prognoses.

Of course size difference among mammal species is associated with considerable differences in the size of their copulatory organ. In general the bigger the animal the bigger its penis, being 18 inches in the boar, 30 inches in the stallion, 3 feet in bulls, 5 feet in elephants, and 7–8 feet in the blue whale. The problem of insertion is not solely of rigidity and size. There is also the unique circumstance of each species, and thoughts should be spared for fellow mammals with major problems, such as all the whales (despite that length), the porcupines, the hedgehogs, the armadillos, and the aardvark. At least humans have hands, useful aids for both insertion and arousal.

EJACULATION can occur even before penetration (particularly among the young). It always begins – following suitable stimulation of neuronal centres within the spinal cord and then the transmission of impulses via nerves to the genital organs – with rhythmic peristalsis of the epididymis. This convoluted store-house of sperm, named after the Greek for 'on the testicles', is a small oblong lump of tissue where the sperm both mature and wait. Peristalsis (a co-ordinated squeezing which occurs elsewhere in the body, as in the gut) first passes the sperm upwards along the vas deferens. This is no straightforward pathway. As one medical dictionary defines it: the duct carries spermatozoa by a tortuous route up the neck of the scrotum, obliquely across the groin, through a gap in the abdominal muscles, behind the bladder and through the prostate gland to the urethra.

The total distance is more than a foot, and there is a suddenness about the event of ejaculation out of all proportion to the relaxed time spent within the epididymis. There is not only the abrupt journey from store-house to penis, but the collection of mucus-like fluid from the seminal vesicles, the further collection of a more watery substance from the prostate, the expulsion into the vagina, the abrupt acquisition of motility, the arrival of capacitation (a form of maturation already mentioned), a hastening of speed by uterine contractions and then – or not, as the case may be – an encounter with an egg. Orgasm, defined by *Chambers* as 'immoderate excitement', is from the Greek for swelling. To many this single word does not properly convey the headlong, precipitate, triumphant and impetuous event that, at some time in the past, was the start of every one of us, when perhaps two-fifths of a billion sperm headed towards a single goal and one of them succeeded.

SIN OF ONAN A good many more billion sperm are even less likely of fulfilment following the 'heinous sin of onania', as an eighteenth-century best-seller called it. This practice has not been blessed with official sanction over the centuries. One student's manual, first published in 1835, did well (achieving twenty-four editions) but the relevant onanian section was written in Latin to prevent youthful minds being corrupted. Kellogg, of the flakes and much else, even developed a cereal whose express purpose was to lessen enthusiasm for, as another phrase defines it, self-pollution. Ordinary dictionaries can also exude a reprimanding tone – 'self-defilement, self-abuse'. The US Surgeon General was recently dismissed following her speech on a World AIDS Day Conference affirming that the act 'is part of human sexuality and it's part of something that perhaps should be taught'. President Clinton was most abrupt in expelling Jocelyn Elders from her post.

Yet masturbation is common, with many countries considering it the most frequent form of sexual activity after heterosexual vaginal intercourse. A. C. Kinsey, whose survey is half a century old, estimated that 'at least' 92% of all American males and 58% of all females had indulged in 'masturbation to orgasm' at some time in their lives. A more recent British survey, after tying one question in a knot about 'genital contact with a man (woman) NOT involving intercourse (for example, stimulating sex organs by hand but not leading to vaginal, oral or anal intercourse)', concluded that 82% of men and 75% of women had so indulged, with 25% having done so in the previous week. Even *Encyclopaedia Britannica*, now as American-based as the president, states that 'the stigma against masturbation is decreasing, and many students of sexual behaviour extol its virtues as being healthy, pleasurable, sedative, and a release of tension'.

The Holy Bible can be exceptionally straightforward in expressing opinion, as well as describing events. After Judah met Shuah, daughter of a Canaanite, 'he took her, and went in unto her'. There were two fruits of this union, Er and Onan. Er was then wicked (details are lacking) 'and the Lord slew him'. Judah then told Onan to go into his brother's wife, Er's unfortunate widow. Now Onan knew 'that the seed should not be his'. Consequently, 'after he went in unto his brother's wife', he 'spilled *it* on the ground, lest that he should give seed to his brother'. This caused further displeasure; so Onan too was slain. Hence 'onania' for the sin of self-pollution and also for the supposition, frequently quoted, that Genesis 38 and Onan's story lie at the root of masturbation's difficult history. On the other hand, say others, an act reported as healthy, pleasurable, sedative and a release of tension is likely to have been condemned, even without biblical authority. The pleasure alone, whether singly or in unison, was quite sufficient to earn opprobrium.

IMPOTENCE, the inability to perform coitus, is normally a male disorder. In women it can be caused by structural anomalies or (as Peter Wingate's dictionary phrases it) 'painful inflammation with spasm of the muscles around the vagina'. With men it is 'the persistent failure to develop erections of sufficient rigidity for penetrative sexual intercourse'. It is age-related, affecting 2% of males at age 40, 25–30% at 65, and 'probably' (according to one survey) over 50% of men over 75. There is also a relationship with diabetes, itself age-related. Impotence does not affect life expectancy, but can alter well-being, the quality of life and relationships – with partners often suspecting that infidelity may be at the root.

The causes can be psychological, neurological, endocrinological, vascular (venous or arterial), traumatic, or even iatrogenic (with medicine, perhaps surgery, perhaps drugs, being to blame). The result can be similar in each case – the corkscrew-shaped helicine (hence their name) arteries, normally contracted, do not relax sufficiently to permit arterial blood to flow into all the lacunar spaces. Or, conversely, venous outflow is not restricted and blood therefore flows from these spaces as fast as it flows in. Erectile dysfunction (as impotence is more formally known) is, on occasion, self-perpetuating.

Each failure increases concern about subsequent attempts. So called psychogenic impotence is the commonest cause of a failure to achieve an erection by the young. It probably begins suddenly, occurs only in certain situations, and is accompanied by normal erections at other times, such as early morning.

With older individuals there is more likely to be some organic cause, such as a drop in testosterone levels. There can be neurological disorders, with perhaps diabetes or alcoholism influential. Most common cause of all is vascular impotence. Either the flow inward is inadequate or the outward flow is not properly restricted, or both. Arteries of the elderly suffer constriction – atherosclerosis being a thickening and hardening of arterial walls – and the blood-vessels supplying the penis are not exempt from such deterioration. Normal inflow to the penis partnered by excessive outflow is a much rarer condition. Drugs (given for other reasons) can also cause impotence as side effect, with no one wise why this should happen. Drugs for banishing impotence are sold in tremendous numbers, with rhino-ceroses suffering famously to supply a portion of this demand, but practitioners of medicine look askance at all the homoeopathic remedies. In a major article on impotence (written for the *British Medical Journal* in 1994) Roger S. Kirby, consultant urologist at St Bartholomew's Hospital, wrote that 'few drugs have had rigorous scientific testing and it is doubtful whether many would, in such circumstances, perform significantly better than placebo'.

Kirby also discussed treatments of greater worth. Psychosexual counselling can be effective, notably if aimed at decreasing perform-ance anxiety with 'close cooperation of the sexual partner'. Giving more male hormone is another attempted remedy, but the procedure is only beneficial when hormone levels are well below normal. What is known as intracavernosal pharmacotherapy resulted from the discovery that certain muscle relaxants could produce an erection when injected within the penis's cavernous spaces that fill with blood. The procedure cannot be entirely pleasant, but 'most patients are able to learn how to inject the drugs quite quickly if they have reasonable manual dexterity and are given a detailed instruction and information sheet'. The subsequent procedure, that of return to flaccidity, is also not without its drawbacks, the principal side-effect of the earlier injection being a prolonged erection. If this lasts longer than four hours, warns Kirby, there is need for 'pharmacological detumescence', namely further drugs to diminish what had, via the injection, been achieved. Once again a needle is involved. Some 20–40 millilitres of blood are first extracted before being replaced with a suitable drug. This method of detumescence 'rarely fails' except when the erection has been present 'for more than 12–24 hours'.

Not everyone warms to the notion of self injection, particularly in such a location and when disturbing after-effects are possible. Surgery is another alternative, even though – according to Kirby – 'only fair results' have been obtained by tying off the relevant vein (for those with venous leakage). The converse procedure – of improving the blood supply – has been attempted, with a reported 50–60% success rate among 'younger patients in the short term'. Better long-term results have been achieved with penile prostheses, these being either semi-rigid or inflatable. The former are 'difficult to conceal' whereas the latter 'provide good patient and partner satisfaction'. Unfortunately – for most patients and their partners – these extras can cost £3,000. Even if money is not a problem they are not suitable for everyone, and 'expectations that penile prostheses will reinvigorate an already failing relationship nearly always prove unrealistic,' concludes Kirby.

DETUMESCENCE is achieved by a reversal of the processes that led to an erect penis. The cavernosal and helicine arteries contract, thus restricting blood flow to the cavernous spaces. The consequent fall in pressure permits increasing outflow to the veins. Lack of input and increase of output simultaneously cause the erect member to become flaccid once again. This normally happens after ejaculation has been achieved. 'Erection and Demolition Experts', stated an advertisement in London's *Daily Telegraph* during the 1960s shortly after new legislation had all but banished prostitutes from the streets. The telephone number for such expertise was located in Belgravia, which might have alerted the *Telegraph*'s authorities, it being an unlikely headquarters region for destruction and construction, but words like engorgement and detumescence would have given the game away – even if the undoubted experts ever used the terms.

These are If-there's-a-problem-start-an-organization days, and impotence is not exempt. BEDS, the British Erectile Dysfunction Society, was initiated in 1995. So was the Sexual Disorder Foundation. Both have arisen in response to what they call 'a growing problem', claiming that 10% of British men are affected. One reason is an ageing population, coupled with sex now more legitimate among older people. (Two British comedians, sitting – on TV – in desultory fashion while mounting sounds of orgasm came from the room above, looked at each other solemnly when silence reigned again. 'Just how long are your parents staying?' said one to the other.)

Many older men are 'scared' by sexual demands made by women of the same age, said a psychologist at a conference in July 1995. 'Not only are women longer lived and generally more robust but they retain their sexual capacity to a greater age,' reported Hamilton Gibson. He thought sex education should be available for older people, not just children.

Impotence is not exempt from alleged and speedy remedies (and presumably never has been). 'There is *only one* clinically proven product available,' claimed an advertisement in the *Oldie* magazine. The ad was not alone, with 'rapid recovery' being promised by another. A 1994 article in the *Journal of Urology* was also optimistic. The most common cause of organic erectile dysfunction 'is the venous leakage phenomenon', with surgical treatment beneficial in half of patients provided they are carefully selected. Impotence may be increasing; so is care and possible cure for its condition.

THE PROSTATE is less than half an ounce in weight and, notably in a male's later years, can provide half a ton of trouble. It can disturb his nights by sending him off to urinate once or twice or more. It can block off urine flow, perhaps suddenly and absolutely, causing a most miserable form of pain. And it can become cancerous, being the most common site for male cancer and the second most common cause of male death from this disease (with bronchial cancer as number one). A male may wonder why he has to have this minute organ in the first place, and why it is the cause of so much grief when only one seven-thousandth or so of his total bulk.

Initially it is an even smaller proportion. For the first nine years of a boy's life the prostate gland undergoes little change. It then starts to grow. At puberty there is a sudden spurt, lasting some 6–12 months, with the organ becoming twice as big as it had been before puberty began. A greater production of testosterone (from the testes) is thought to be responsible for various changes in the structure of this diminutive gland. Its size thereafter remains fairly constant until age 45–50. Then, as blessing or its converse for the possessor, it either atrophies or increases its bulk steadfastly until death.

The prostate's standard size, before it begins its middle-aged hypertrophy, is about one inch across; but, by being more of a cone than a cube, some widths are more than an inch and some less so. This organ is part glandular, part fibrous, and part muscular. Its function is to add seminal fluid to the passing sperm (just like the seminal vesicles and the bulbo-urethral glands). The prostate's addition gives an ejaculate its characteristic smell, a memorable odour but no more describable than any other distinctive smell. (Oddly there is a bush growing near Cape Town mimicking this smell precisely.) The prostate's location is most relevant to its subsequent interruption (for some) of urine flow. Not only is it immediately below the urinary bladder but it surrounds the urethra, the outlet from that bladder. Hence any swelling of the prostate (and no one knows why this swelling occurs) impedes the urine flow or blocks it off completely.

In pre-surgical times such a blockage must have been fatal. The first successful prostatectomy was performed almost a century ago in Baltimore. It is now both a common and a simple operation, albeit with a consequent risk of erectile impotence (some say as high as 50%) and a far lesser risk of incontinence. A different operation, known as TURP, attacks the problem via the urethra. Essentially this is akin to apple-coring, attacking not so much the prostate as the urethral blockage. The inserted corer enlarges the urethra in its constricted region near the bladder and also destroys one of the two sphincters between bladder and exterior. Some incontinence therefore becomes more of a possibility. So too, and to a greater extent, does retrograde ejaculation. Instead of the sperm shooting/emerging from the penis they proceed conversely and enter the bladder, thus destroying the possibility of fertilization and diminishing sexual pleasure.

Prostate cancer is of far more concern than possible impotence or infertility or a lesser orgasm; it can kill. Within the United States this disease is detected annually in 140,000 men and 34,000 die from it. European figures are proportionately similar. Worse still, its incidence is increasing, having risen – in America – by 30% between 1980 and 1988, with deaths increasing by 2.5% despite greater awareness of the problem and, in general, an earlier detection rate. The Prostate Cancer Cure Foundation Ltd, with its secretarial office in Switzerland, has even offered a $5 million award 'to recognize the contribution of the dedicated team of people who find a permanent cure for Cancer of the Prostate and surrounding Lymph Nodes'.

Currently there is doubt whether prostate cancer is worth diagnosing. Cancer screening consists of a rectal examination and measurement of prostate-specific antigen (PSA) concentrations. Unfortunately PSA is not a firm indicator (it can produce false-positive results if the man being tested has ejaculated within the previous two days) and screening is expensive (costing £2,477 per case detected according to a Swedish survey). An article in the *Lancet* (of 4 November 1995) expressed most optimism in future genetic work which could improve diagnosis and (possibly) 'specify radical therapies'.

The prostate may be diminutive, but its ability to cause all manner of trouble, not least death, is giving the gland increasing renown, particularly as elderly males become more numerous. It managed to make a misery of Groucho Marx's final years, as with many tens of thousands of others who probably did not even know about the prostate until, night after night, it made its presence felt. Only then does a urologist explain about this inch-long lump of tissue and, following investigation, has either grave news to report (of cancer) or news less grave but of one more annoying extra among all the other inconveniences of growing old.

PREPUCE The medical profession, in general, has been performing one particular operation for a century. It is the most common neonatal surgical procedure. It increases heart rate, decreases oxygenation, increases blood pressure, induces sweaty palms, and alters behaviour. It is usually carried out without anaesthetic, either local or regional. Without doubt it is painful, but most patients are not even asked for their consent. One Los Angeles doctor reported (to the *Lancet*) that the effects of this pain, adjusting so many metabolic events, are 'consistent with torture'. In many nations it is extremely prevalent, such as Canada, United States, Algeria, Israel, and Nigeria, their rates varying from 50% to 100%. In others, such as the United Kingdom, France, Denmark, Russia, it is relatively rare. It has been called unnecessary, intrusive, mutilating and barbaric. It has an extremely ancient history, and is still performed upon tens of millions of youngsters. A dictionary defines it as removal of the prepuce. The general term is circumcision.

There are reasons for this 'rape of the phallus', as it has also been called. It is lucrative. It is traditional. All male Egyptian mummies had been circumcised. Various religions advocate it, such as Islam and Judaism. The Bible frequently harangues the uncircumcised (although St Paul, a circumcised Jew, decreed that it was unnecessary for Christian converts). Herodotus reported that Egyptians were the first to circumcise, and there are illustrative wall carvings in the temple of Karnak. Moses allegedly asked his wife to operate with a stone implement. Pythagoras, no less allegedly, had to be circumcised before receiving permission to study temples in Egypt. Saul told David to bring him 100 Philistine foreskins as proof of their owners' demise. David, in exercising this early form of scalping, then brought 200 to show himself more worthy (of being a son-in-law).

Circumcision was once said to be a cure for masturbation. Nowadays it is frequently called a hygienic measure that diminishes the risk of penile cancer (a rare cancer in any case). There is great momentum to the procedure, with certain hospitals considering it as much part of the routine as providing food and care. Any mother wishing her child to remain intact 'would be well advised to maintain permanent guard over it until such time as they both leave the hospital', wrote a British doctor in the 1960s when practising in an American hospital. She should still be watchful. The majority of American male babies – roughly 60% – lose their foreskin very speedily.

Somewhat inevitably, particularly today, there is backlash. 'Foreskin fundamentalists', so-called, have created – for example – NORM (the National Organization of Restoring Men) and BUFF (Brothers United for Future Foreskins). The former used to be called RECAP (Recovering A Penis), with its members principally interested in non-surgical methods for restoring their lost portion. *Tugger*, for instance, is used for stretching skin over the penis (and, in 1995, cost $115). Slogans promote the cause of prepuce retention, such as 'Are ears next?' and 'Boys deserve to have it all.' Even before fundamentalists started upon their campaign a lessening of enthusiasm for circumcision had already occurred. In the 1960s the proportion of American boys experiencing the operation was over 90%. Today's figure of 60% therefore shows a drop in incidence of 1% a year. With so many Jews and Moslems adamant, along with various other traditionalists, this form of mutilation is unlikely to vanish soon. Besides, it is one more item readily added to the obstetric bill.

In Britain it used to be commonplace but, following the introduction of the National Health Service, questions were inevitably asked whether a mutilation procedure of doubtful medical value should be conducted and paid for by the state. It speedily dropped from favour, with many parents still requesting circumcision almost out of habit but finding their request flatly denied by hospitals. Paediatrician Sir James Spence was notably outspoken: 'If you can show good reason why a ritual, designed to ease the penalties of concupiscence amongst the sand and flies of the Syrian deserts, should be continued in this England of clean bed linen and lesser opportunity, I shall listen to your argument...' Individuals whose ethnic origins lie nearer the

Syrian desert than the River Thames are still enthusiastic. They pay for it, and there are many who like to be paid.

There is no need for a medical practitioner to perform the operation. Specially trained rabbis do the job for Jewish boys at eight days. There is a similar arrangement for Muslim boys, but later and with no fixed date. Problems can arise in mixed marriages. A London woman whose 18-month-old son was circumcised on the instruction of his Nigerian father won compensation (in 1994) from the Criminal Injuries Compensation Board. The parents were neither married nor living together. The father, himself circumcised, wanted a similar son 'for cultural reasons'. The mother would only accept medical arguments in favour of the operation. She received compensation of 'more than £1,000' because a 'crime of violence' had taken place. Contrarily both cultural and medical reasons have been successfully exploited by certain explorers, notably those travelling in Arabic areas. Should the foreigner's instruments and his skill be better than the local variety, and if less infection or death results among male offspring, the visitor will be assured of hospitality and favour.

Discussion about circumcision, and whether or not to cut, has recently shifted its ground. Relevant pain is now a bigger issue. Two doctors from a Canadian hospital in Alberta, writing in *Pediatrics* (in 1994), stated that neonatal circumcision was 'performed without any form of analgesia in most institutions', including their own. At their hospital only one of the twenty physicians, who regularly performed such operations, took the extra step of using local anaesthesia. As over 1,000 babies a year were circumcised (50% of all born there), this represented, in their medical opinion, a considerable quantity of suffering. 'There is no doubt that the newborn infant experiences pain during circumcision as reflected by physiological and behavioural changes.' These include alterations to the heart rate, to blood pressure, to sweating, and to the take-up of oxygen. Another Canadian hospital has reported that circumcised babies have 'short-term alterations in behaviour, sleep patterns, frequency of feeding, crying, fussiness, and heart rate'.

Physicians from this second hospital (in Toronto) went further in their assessment of this pain; they considered its influence might go beyond the trauma of the circumcision operation. In a trial of relative pain scores between boys and girls during vaccination at 4–6 months they found that boys scored higher than girls. They postulated that 'if this sex difference is a real effect, it may be partly related to previous experience with acute pain, such as circumcision'. When vaccinated, the circumcised boys 'had significantly longer crying bouts' as well as the higher pain scores. Only speculative conclusions were possible, mainly because numbers were small, but the paper's authors did conclude 'that analgesia should be routine for circumcision to avoid possible long-term effects in infant boys' pain responses'. At the Alberta hospital, where only 5% of physicians had used analgesia before this topic was investigated, that proportion rose to 75% within a year. There was no indication that the number of boys experiencing this mutilation, with or without pain, had been reduced during that period of reassessment. In Britain, by contrast, the quantity is currently around 6% as against more than half for North America.

Unfortunately – for those who rail against circumcision – individuals still in possession of their foreskins are reported, in several studies but not all, to be more susceptible to HIV. Even if the association was more clear-cut there are confusing extras in the equation. As some Swedish workers wrote (in the *Lancet*) after their experience in Kenya 'nobody is circumcised at random'. Among Kikuyus of Central Kenya 'almost all boys aged 12–15 years undergo circumcision before they start secondary school'. The practice is not as it used to be, with parents heavily involved and choosing an appropriate *mutiri* to perform the task, for there is now a youth subculture dictating that, 'as proof of manhood, intercourse must follow soon after circumcision'.

So what is happening? Are the youngsters achieving their intercourse in the commercial sex market? Or, being young (and poor), are they more likely to use equally immature and fellow students, less likely to be HIV positive, to gain their manhood? The value of circumcision, if any, is therefore no less obscure than in other societies. But, as Marilyn Monroe said about sex, the act 'is plainly here to stay'. It is unlikely to go away, just yet, with money, religion, custom and momentum *all* heavily involved.

PATERNITY The single certainty, in former times, focused upon the mother. The baby being born was, without doubt, her offspring. There is argument that matriarchy used to be commonplace. If power and privilege were to be inherited, they should pass down the female line because only then was there assurance of legitimacy. Whatever a man might say, or the mother might affirm, absolute truth lay only with her delivery. She was the certain parent. Anyone else was conjecture.

There is still doubt. Science can deny an alleged father but cannot prove fatherhood. As a paper in *Trends in Genetics* (June 1994) phrased this generality: 'Exclusions of paternity are logically irrefutable. Proof of paternity, on the other hand, depends on statistical inference based on failure to exclude.' The need to establish paternity is not a matter of idle curiosity. Money can be involved, whether for upkeep of the child or from inheritance. So can immigration, and whether or not an offspring is a true descendant. So can medicine, with genetic counselling needing to know the true parents (and, quite probably, *their* true parents as well).

At a Ciba symposium, held during the 1960s, one delegate quietly affirmed, following his work 'in a town of south-east England', that 30% of the children he examined could not have been the product of their alleged parents. There was no initial comment following this assertion. Only later was the gynaecologist, Elliot E. Philipp, questioned about his remark. He explained that parents were being examined concerning the rhesus factor (and possible maternal reaction to second conceptions if the partners were rhesus-negative and rhesus-positive, thus damaging the chances of healthy babies). His patients were all worried parents, eager for further children. There was no reason to suspect infidelity among them, and yet that 30% – discovered by examination of several blood groups, including rhesus – was proof that relationships were not always so straightforward as either declared or believed. Many a genetic conclusion, about the likelihood or otherwise of some characteristic appearing in offspring, should therefore be viewed in the light of that 30%.

Or in a light even more revealing. Other gynaecologists, initially doubtful of Philipp's contention, attempted to reproduce his work by similar examination. Some discovered that 50% was nearer the truth. There is talk of shotgun marriages – where couples are forced to formalize liaisons – but many a shotgun has two barrels. The lady in question may earlier have had two partners. (Or more, it has been known.) When a pregnancy arrives she will then decide in which direction to point her finger, one partner being judged more suitable for parenthood than the others. This male, whether pleased or appalled by her decision, may then accept the situation. And may prove to be an excellent father, such as many (or most) of those who sat in Philipp's surgery.

Over thirty years later he revealed that Romford, in Essex, had been the 'south-east town'. His patients had largely been the wives of workers in Ford's nearby factory at Dagenham. Many industrial strikes were occurring at the time, giving greater opportunity for both boredom and promiscuity, but Philipp does not believe 30% is in any way 'unusual'. 'When we confronted some of the patients with the possibility of the putative father not being the father of the child they admitted that it could have been another "Tom, Dick or Harry".' It is therefore better to believe that the presumed fidelity of all our ancestors – 'yes, that photo is of my grandfather, and she was my great-grandmother' – is not the total truth. Tom, Dick, Harry and many others have been alive, and well and active, long before Ford's Dagenham workers had plenty of time to spare.

Recent events have not altered this general statement, but have given it extra weight. Firstly, notably in the UK and Australia, absentee parents – usually fathers – have been vigorously pursued for maintenance money. Secondly, DNA fingerprinting (so-called) is a more certain procedure than mere resemblance (of child to its parent) or the subsequent blood-group testing (which could exclude many alleged parents rather than confirm them). DNA testing also works on exclusion but with massive probability odds against error. Or, as two Brazilian scientists phrased it – 'By using multiallelic variable number of tandem repeat polymorphisms, paternity disputes can be resolved with certainty in virtually every case.' There is perhaps a 50 million to one chance in favour of a particular man, for example, being the father of a particular child.

The Child Support Agency, established in Britain (in the early 1990s), had much of its heart in the right place but little of its head. It was rightly intent upon extracting money from absent parents but wrongly cavalier in its method. It did not care if parting parents had made a clean-break settlement, with perhaps the mother keeping the house. It did not worry about the cost of access between absent father and distant children. It did not even care if the father looked after the children for a major portion of each year. In short, it often managed to create hate in place of mere discordance. Much to the surprise (and horror) of many males, the agency did not differentiate between one-night affairs and long-term marriages. Paternity was paternity, whatever the time-scale, the intent, or pre-intercourse assertions by the female of being 'safe'.

Paternity testing is huge business. The worldwide market is now estimated to be 1 billion dollars. Even in 1990, and in the USA alone, 120,000 paternity tests were carried out. Many fathers must be discovering, to their delight or devastation, that sperm to create a certain child could never have come from them. Whereas proof of paternity depends on 'statistical inference based on failure to exclude' (another definition) the exclusions of paternity are irrefutable. With the CSA and other organizations demanding maintenance (even directly from wage packets), with family inheritance (of funds, possessions) always such a major issue, and with genetic involvement in both disease and prowess better and better understood, it is certain that parental testing will become bigger business in the years ahead. Even now the CSA (as from late 1995) offers men free DNA tests in cases of disputed paternity, but there is a sting in the tail of this generosity. If the disputing father is shown to be the true father he has to pick up the bill – of approximately £400.

Three recent cases, described by the two Brazilians mentioned above, indicate the widespread nature of current ability to prove (as near as dammit) and disprove parenthood. Case 1 involved a married, fair-skinned couple. Their daughter was strangely dark. Early blood-group testing rejected the father as possible parent. A marital storm then ensued, with both parents swearing absolute fidelity. Subsequent DNA testing showed that neither parent was responsible for the child. A hospital mix-up was therefore to blame. Case 2 concerned a 4-month-old skeleton. When discovered, this child was thought to have died in a satanic ritual. A couple whose 6-year-old child had disappeared at about the same time were suspected, and then proven – by DNA testing – to be the parents of the skeletal remains.

Case 3, markedly dissimilar to the other two, involved inheritance. A wealthy man had died, and a younger man claimed to have been his son. Nothing of the alleged father was available for testing, but there were two sons of his two brothers, those brothers having already died. The Y chromosomes in each of these six males should have been identical, namely those of the dead father, his two dead brothers, their two sons and the individual who had filed the paternity suit. Following DNA testing it was clear that the Y chromosome of the claimant was quite different from the two Y chromosomes of those who should, if his assertion of paternity was correct, have been his cousins. No wealth therefore came his way.

A final paternity point. Fathers, certainly in Britain, have been experiencing a bad press recently. Half have been said to prefer DIY to playing with their children. One in six of their offspring is said to have been abused, and marriage is going down the drain as male partners leave home. A lobbying group named Care produced such disturbing information, and prompted Libby Purves, a determined journalist, to investigate the tidings. She discovered that only 920 men had been interviewed, of whom 504 were fathers, and no more than 218 had children under 15. They were not asked if their children were asleep when they reached home, or if older children said 'Hi, Dad!' before vanishing for the evening. As for abuse, that included indecent exposure and uncomfortable sexual remarks. In which case, added Purves, she herself had been abused – by two park flashers and 'a French exchange teenager with a dirty mind'.

As she concluded, by examining more careful sets of figures, '70 per cent of first marriages succeed, most divorces do not involve dependent children, and between 78 and 83 per cent of children under 16 live under the same roof as both natural parents'. She longs to see the headline, somewhere, sometime, saying 'Most Families Doing OK.' There might even be another: 'Most fathers satisfactory.'

HOMOSEXUALITY On the one hand is Leviticus: 'Thou shalt not lie with mankind, as with womankind: it *is* abomination.' On the other is the 'rise and rise of the gay global village', as one newspaper phrased it, with the number of homosexual organizations leaping from twenty to 1,200 (in three years in America alone) during the early 1970s – and no kind of diminution since then. On yet another hand, the biological one, there is curiosity that such a fundamental drive as sexuality should ever be diverted into a similar drive within a gender. Heterosexuality leads to offspring and future generations. Homosexuality, if maintained exclusively, does not. The biological query becomes stranger still if there is a genetic basis for this deviation.

In 1993 Dean Hamer, of Bethesda's National Cancer Institute, announced that a portion near one end of the X chromosome probably contained a gene or genes for homosexuality. Precisely thirty-three out of forty pairs of homosexual brothers had inherited an identical version of this chromosome region (each X chromosome being the length of genetic material inherited doubly by women, singly by men). In 1994 Hamer then reported confirmatory results after studying another similar series of males. Headlines around the world made great play of his conclusions, causing Hamer to express concern. 'We have never thought that finding a genetic link makes sexual orientation a simple genetic trait like eye colour; it's much more complex than that.' For example, there were the original seven (of forty) who did not conform. Other work has shown that even the one-egg twin of a homosexual man is only 50% likely to be homosexual.

'My guess,' added Pepper Schwartz of the University of Washington, 'is that there are different kinds of homosexuality, some . . . more hardwired than others.' Homosexuals have tended to agree. A proportion have always considered themselves homosexual from their early years, others have gradually adopted such a stance, and still others can oscillate between homo- and heterosexuality. Even if genetics forms the basis its rulings are not always straightforward, not even with simple genetic traits. Blue eye colour is generally said to arise with certainty if both parents have blue eyes – but exceptions occur even within this allegedly single-gene phenomenon. 'There will never be a test that will say for certain whether a child will be gay; we know that for certain,' concluded Hamer. Appreciating the ethical questions which would inevitably follow such a procedure, he has promised a simple solution. If his laboratory does discover a gene for homosexuality, and is then able to win patent rights over it, he will use that power to prevent others from developing such a test.

Laws, prejudices, advances, retreats and stated percentages of homosexuals are now changing faster than ever before. Some recent shifts, in opinion and consequences, help to affirm the speed of alteration. Dallas, Texas, known as the buckle of the American bible belt, has three homosexual churches. (One is purpose-built of stone and steel, complete with bullet-proof stained-glass windows.) Virtually every US city now has a homosexual district – Castro in San Francisco, Boystown in Chicago, and West Hollywood in Los Angeles. So do other world cities – Oxford Street in Sydney, Ghetto in Toronto, Marais in Paris. Australia has recognized property rights for same-sex relationships. Ontario has debated the possibility of homosexual couples adopting children.

Montana seemingly set the clock back in March 1995 when it passed a law (first and only state in the US to do so) requiring homosexuals to register with their local police, along with convicted felons and sex criminals. It already had laws which carried up to ten years in jail and a $50,000 fine for homosexual activity, and a slogan 'Register gays, not guns' had been actively promoted, but the new registration law was dropped two days after its introduction. Homosexuals were then no longer covered by the state's Criminal Sex Offenders Registration Bill as were those guilty of murder, rape, incest, and sexual assault. In Britain the Criminal Justice Bill of early 1994 lowered the age 'of homosexual consent' from 21 to 18 – but not to 16 which would have brought it into line with heterosexual consent.

Later that year two British judges ruled that 'only women' could be charged with 'loitering for the purposes of prostitution' under the 1959 Street Offences Act. This inequality had arisen as a consequence of the homosexual reform laws of 1967. These, it would seem, had not anticipated the possibility of male prostitutes becoming a nuisance. Hamilton Gibson, the psychologist already quoted (on elderly male timidity), reported (in July 1995) on a survey which showed that more women than men had admitted feeling homosexual urges, but more men than women had had homosexual experiences.

Alterations in homosexual laws have been partnered by vigorous statements from those opposed to a more relaxed attitude. The Catholic Church admitted in 1992 – 359 years after condemning Galileo as a heretic – that it had been wrong to refute his Copernican belief about the Earth's status as a planet, but in 1992 it reaffirmed (in a proclamation) its position on homosexuality. This was 'an objective disorder'. There is 'no civil right' to homosexuality, it being a 'tendency ordered towards an intrinsic moral evil'. If rights were requested 'neither the Church nor society should be surprised when irrational and violent reactions increase'. Cardinal John O'Connor, of the US, has stated that homosexual Catholics dying of Aids should not be given the last rites unless they repented appropriately. In Nicaragua a new criminal code pledges three years' imprisonment for anyone who 'induces, promotes, propagandizes, or practises in scandalous form concubinage between two people of the same sex'. Cardinal Hume, leader of the Catholic Church in Britain, has stated that 'homosexual genital acts, even between consenting adults, are morally wrong'. Parliament should therefore be 'cautious'. Leviticus, it has been pointed out, deplores homosexuality, not homosexuals. The world, in general, is still undecided.

(That Leviticus comment, so frequently quoted, should best be encountered in its context, for this makes powerful reading. Whereas it does say that lying with another man as with a woman is an abomination 'and both should be put to death', it adds that anyone who curses his father or mother should also 'be put to death'. As for adulterers and adulteresses they too should 'be put to death'. Leviticus also did not tolerate the notion of sleeping with a menstruating woman, however much a wife: 'Both of them shall be cut off from among their people.' There was nothing undecided about this biblical indictment. Everything was forthright and never outstandingly lenient in the matter of punishment, whether for homosexuals or any other deviant from the rigid code.)

Finally, as one doughty lady said to another (in a cartoon published by *Nature*): 'If homosexuality is inherited, shouldn't it have died out by now?'

FEMALE

VAGINA   UTERINE TUBES AND OVARIES

MUTILATION   MENSES   MENOPAUSE

HORMONE REPLACEMENT   LACTATION

HUMAN MILK   BREAST-FEEDING   POLYTHELIA

BREAST IMPLANTS   CANCER   SCREENING

Gentlemen, woman is an animal that micturates once a day,
defecates once a week, parturates once a year, and
copulates whenever she has the opportunity.

A professor of gynaecology, quoted by Somerset Maugham who
thought it 'a prettily balanced sentence'

Every luxury was lavished on you – atheism, breast-feeding,
circumcision. I had to make my own way.

Joe Orton, in *Loot*

VAGINA It 'is lined by a non-secretory, stratified squamous
epithelium and bathed in fluids from Bartholin's and Skene's glands,
from the mucus secreting cells of the cervix and from oviductal and
uterine sites'. Men may not recognize this object of much desire, and
even many women may not immediately appreciate this textbook
definition referring to their passage between uterus and exterior. The
vagina is not homologous with the masculine penis – her biological
equivalent being the clitoris – but is its correspondent form. Engineers, finding the analogy helpful, refer to male and female ends of
pipework, one inserting and one receiving.

As a female's eggs are contained within her, and a male's sperm
must reach them, there has to be an access, and the vagina fulfils that
role. It is both starting point and destination in the creation of
another separate human form. Not only does the male have erectile
tissue which becomes active during intercourse, but a female has
similar tissue in abundance around the opening of the vagina. This
swells following psychic excitation and stimulation of her genital
organs. Nervous impulses passing from the spinal cord cause her
erectile tissue to expand, thus providing a tight but stretchable
opening. Simultaneously the two Bartholin's glands, immediately

within the vagina on either side, secrete much mucus, this being of considerable benefit to the movement of the penis. The massaging of the clitoris by the penis provides most of the female's sexual stimulation, a case of two homologues encountering each other. The clitoris lies outside the vagina at the front edge of the vulva. When the female climax occurs (or if it occurs) both the uterus and the uterine tubes begin rhythmic, upward, peristaltic contractions. These are presumed to assist, or propel, the sperm to reach the tubes.

The vagina's length is 3 inches (at the front) and slightly longer at the rear. In position it lies between the bladder (at the front) and the rectum and anal canal (at the rear). Its fibromuscular walls are normally in contact with each other, but they can part to accept not only the penis but also – and eventually – the far larger bulk of a neonate. Its upper end surrounds the cervix of the uterus, but there is a 90-degree bend between the vagina (which slopes backwards and upwards) and the uterus (which slopes forwards and upwards).

UTERINE TUBES AND OVARIES The uterus is also a thick-walled, muscular organ, about 3 inches long, 2 inches wide at its widest point, and 1 inch thick. It normally weighs little more than 1 ounce. The organ will expand tremendously to accommodate the developing foetus, and will weigh about thirty times as much when the enclosed infant is ready to be born. After birth the uterus shrinks almost to its original dimensions, weighing perhaps 1.5 ounces by the end. Its normal position varies, being shoved this way and that by altering dimensions of bladder and rectum. During menstruation the organ is slightly enlarged, with its lining membrane a darker colour. In old age the uterus atrophies, becoming paler and denser, its role as possible incubator having been firmly concluded.

The two uterine (or Fallopian) tubes form the link between the two ovaries and the single uterus. Each is about 4 inches long, and particularly thin at the uterine end where the tube's opening, as *Gray's Anatomy* phrases it, 'admits only a fine bristle'. At its other end, the one near an ovary, there is a trumpet-shaped expansion, the infundibulum. Each uterine tube varies its form along its length, being thin-walled and tortuous in the half nearer the ovary. Fertilization, the union of egg and sperm, is less likely to occur either in the isthmus, the tube's central portion, or the remaining and smaller portion that links with the uterus.

The two ovaries are homologous with the male's testes, in that they arise from the same portion of embryological tissue, but they end up in quite different locations, lying at the conclusion of each uterine tube (and not externally in a scrotal sac). Each ovary is amygdaloid (Latin and Greek for almond-shaped), slightly longer than an inch, about half as wide, and less than half an inch thick. Each is grey-pink in colour, and initially smooth. They too are mobile, being displaced by neighbouring organs when these alter shape and volume, and are also shifted by the first pregnancy, so much so that neither ovary subsequently returns to its original site. The tremendous quantity of eggs (or, more precisely, primary ovarian follicles) are located within the ovaries, the 40,000 (give or take several thousand) at puberty. At puberty a few of these develop each month. Each then consists of a central cell surrounded by a layer of flattened bodies, the follicular cells. Known as ovarian (or Graafian) follicles, these grow until one of them ruptures to discharge an ovum, or egg, the stage known as ovulation. By then the egg is about a thousand times bigger than its earlier, undeveloped form. Each rupturing halts the other developing follicles in their tracks, and they swiftly diminish without further development.

As for the single ruptured follicle, its lining of cells multiplies to form a small mass known as the corpus luteum, the yellow body. This, in essence, is a small ductless gland, producing the hormone progesterone (or progestin), a substance that prepares the uterus for the task of receiving and growing one embryo. If no such fertilization occurs the egg begins to degenerate. So does the corpus luteum and, following its disintegration, the flow of progesterone ceases. Without a supply of progesterone the uterus begins to revert to its earlier condition and, in time, the flow of menstruation indicates that another month has passed without conception. About eighteen days later there will be yet another rupturing of a follicle, another ovulation, another release of one more egg into a Fallopian tube, and therefore the further possibility of fertilization should sperm and egg meet before either has, in modern terminology, passed its sell-by date.

If the development of sexual reproduction can be addressed straightforwardly, it was all very well for this process to evolve, but the subsequent complications are a splendid confusion of needs, priorities, instincts and compulsions. The fundamental facts are that male sperm have to meet female eggs, these eggs have to develop within the female, and she then has to give birth to the product of this conception at some later date. Equally basic is the need for sperm and egg to meet when both are ripe, and when the female is ready for her role of incubation. Courtship is therefore involved, the selection of mates, and then the act of mating. With many an animal species the timing of this procedure is most accurately controlled, with mating only possible and/or permitted when most likely to serve its function. With humans there is no close correlation, although intercourse during menstruation (a bad time for conception) is generally less favoured. The preparation of the uterus each and every month is a most steadfast procedure, and the 'weeping of a disappointed womb' (as menstruation has been called) equally determined as the female organs accept one unfulfilment and start preparation for the next opportunity another month ahead.

*

We humans are no odder than the rest of sexual life, whether oak tree or octopus, dandelion or mountain lion, and most of us care but little for the extraordinary physiology involved, the astonishing event that may or not take place, as we mate in intercourse. Our oddness lies in attempting to amend or reverse the procedure in every way we can. We try not to conceive or, as total converse, we try enthusiastically to do so. We abort conceptions and we also stimulate others that seem to be failing. We have in vitro fertilization and, at the other end of pregnancy, we keep prematures alive who would otherwise have died. We inspect before birth and reject if so inclined. With each amendment to the ancient system we encounter different choices, and different ethics to be agreed. Most amazing still, we are only at the beginning of all this chop and change (and see the chapter on Infertility, page 161, for a full account).

MUTILATION Strictly it is not circumcision, in that there is no 'cutting around'. Many a medical dictionary omits it altogether, defining circumcision solely in a male context. Interested and relevant authorities, such as the World Health Organization, refer to FGM (female genital mutilation), but neither the initials nor the full name are as well known as they ought to be, bearing in mind that (so stated the WHO in 1995) between 85 and 115 million girls and women have been subjected to FGM. More worryingly an earlier WHO assessment (1993) was '80 million female infants, adolescents and women'. The act may therefore be increasing. For those who consider this form of mutilation an anachronistic obstacle to female welfare, it is depressing that numbers are increasing instead of vanishing. Currently 2 million young girls a year are believed to experience FGM.

The initials cover a range of procedures:

Clitoridectomy – removal of all or part of the clitoris.
Excision – removal of the clitoris and the labia minora.
Infibulation – removal of clitoris, labia minora and inner surface of
  the labia majora. This most extreme form of FGM is followed by a
  stitching together of the two sides of the vulva, leaving only a small
  opening.

Such operations, often carried out in conditions of poor hygiene without anaesthesia, are usually performed on females between the ages of 4 and 10, but may involve younger infants or older adolescents, or those about to be married or having their first pregnancy. Unlike male circumcision there is no religious authority to justify the mutilation. No reference to it is made in the Koran, for example. Certain sayings of the prophet Mohammed have been implicated, but these are 'weak, unauthenticated, dubious and hence unreliable', according to a high level committee chaired by Egypt's Minister of Health which included Egypt's Mufti.

The consequences of FGM are considerable. Short-term effects include severe haemorrhage, shock and infection, any one of which may lead to death. Long-term effects include urinary tract infections, infertility, sexual dysfunction, and obstructed labour – which itself can lead to maternal and infant mortality. There can also, most understandably, be psychological trauma. During birth there is need to de-fibulate the mother, if she has been victim to the most extreme mutilation, and the woman is then often re-infibulated after the delivery. In modern times HIV infection can also be contracted from the instruments used, an unwelcome extra to the age-old risks.

FGM is mainly an African business, notably in the east and west. In Djibouti, Mali, Sierra Leone, Somalia, large areas of Ethiopia and of Sudan 'nearly all women are affected', according to the WHO. In Kenya, Guinea Bissau and Burkina Faso the proportion varies from 25 to 50%. In Mauritania, Togo, Uganda and Tanzania it is only 5%. It also occurs in Europe, in North America and in Australia, mainly because immigrants have imported the custom with them. Some European countries, like the UK, Sweden and Switzerland, have specific legislation against such excision, but few excisionists (most have no medical training) have ever been brought to trial. France has been the most determined European nation in this regard. More than twenty cases have been brought to court, and one professional in this form of mutilation was sentenced to five years' imprisonment. Doctors must frequently notice signs of FGM, but are not over-zealous in reporting their observations. It has been argued that failure to report is a breach of ethics.

Britain has a considerable ethnic community originating from Africa, and FGM is undoubtedly practised, but there has been no prosecution in the ten years since legislation was introduced (providing for five years' imprisonment). An Egyptian doctor, accused of female circumcision, was found guilty of serious professional misconduct, and struck from the British medical register, but he was never prosecuted. More than 10,000 children living in Britain are thought to be 'at risk' from experiencing FGM. 'Cultural asylum' was offered to a Nigerian woman who argued that her daughters would be circumcised if forced to return to Africa.

Egypt has been both traditional and radical in recent years. The operation is legal if carried out by a doctor, and is common (with 3,600 operations a day, according to the Egyptian Organization of Human Rights), but a nationwide health education programme has been launched, its aim to curtail and then eliminate the practice. The high-powered Egyptian committee, already mentioned, condemned the act as physically and psychologically harmful. They called on mosques, schools and the media to advocate banning the custom, and for health centres and hospitals to warn their female patients of its dangers. Egyptian practitioners without medical training are liable to prosecution for carrying out any surgical procedure without a licence. The same authoritative committee even thought it appropriate to take steps for the regulation of masculine circumcision. As the cutting of males is mandatory in the Islamic religion, and female mutilation is so widespread – 95% of all girls in the rural areas and 73% in Cairo – Egypt's Ministry of Health has a considerable struggle ahead if it is to reduce either form of ritual injury.

The most widespread existing violation of human rights in the world, as it has been called, received a publicity boost (in 1994 in Britain) when *The Story of a Somali Girl* was published. Aman, its author, had been circumcised at the age of 8. 'Afterwards I was sick and had a fever. And when I peed, it felt as though it would kill me. It felt like fire. Or like alcohol when you put it on an open wound. Anyway, I got a fever and they had to cover me up, and my teeth were chattering and I was shaking over my whole body when my mother came back.' Her mother had opposed the operation, being unable to pay for a circumcision feast. The remaining family had disregarded her wishes and Aman herself then attempted to comfort her mother: 'It's already done. There's nothing you can do. They did it because they love me.'

Others, such as human rights groups, consider that FGM should be branded as torture. The guilty parents, and their children, can believe they are behaving correctly, being reluctant to accept that the act is anything other than part of their culture.

MENSES A century ago many a young woman was taken by surprise at the sudden discharge manifesting menstruation. Even thirty years ago, and in Britain, it was estimated that a third of girls received no pre-menstrual counselling and were also much alarmed by the bloody flow. Such alarm is now less widespread, but there are millions of women for whom menstruation is a rarity, or virtually non-existent, with pregnancy so dominant. Conversely, a woman who never conceives can expect to experience roughly 440 so-called periods, as the curse – once called a blessing (for not being pregnant) – repeats its cycle.

Not only women were ignorant about menstruation in the past. Students of medicine assumed it was related to and even caused by the lunar cycle – hence menses after the Latin for month (calendar months being already related to the moon). Not until 1865 was it ever suggested that menstruation was due to the development of ovarian follicles. The discovery of the mammalian egg a few years beforehand had prompted this line of thinking. Not until the first year of the twentieth century was the ovary's role in producing not just an egg but hormones first shown. Then, in 1903, the corpus luteum was identified as hormone-producer, and in 1908 changes to the uterine wall, the endometrium, during menstruation were properly described. It quickly became possible for the ovarian cycle (and its production of eggs) to be correlated with the endometrial cycle (and that blessing/curse of menstruation).

If menstruation is linked to the moon why don't all women menstruate at once? – to paraphrase James Joyce. It could also be said: why is there such disparity among women? Length, intensity, duration and unevenness all vary, with the normal range being 24–35 days (some say 19 to 37) and normal regularity not altering by more than five days per cycle. Ordinarily the process takes 4–5 days to be completed. Blood loss is about 30 millilitres – a twentieth of a pint. Abnormality is said to exist when the loss is above 80 ml – a seventh of a pint. (As a woman possesses about 1 pint of blood for every 14 lb of weight, a 9 stone/126 lb woman loses about 1/180th of her blood each month through menstruation, but her body also recreates some 9 pints of blood each month.) The duration of her period does not appear to relate to blood loss, but women tend to be consistent as individuals with the loss they each experience. The blood does not coagulate, as does normal blood, because it contains no fibrinogen, the clotting factor. Coupled with blood, which is about 75% of the total, is the sloughed-off lining of the womb, as well as much cervical mucus, degenerated vaginal epithelium and bacteria. The characteristic odour is thought to arise from bacteria acting upon the blood.

That sloughed-off lining is the endometrium. It is – or was before its expulsion – an extremely dynamic tissue. In a short time it prepared itself to meet the unique and most specific functional requirements of a fertilized egg. It created a suitable nest for implantation (whether or not that implantation will occur) but is no more than one cell thick. Each cell is very tall (columnar), varying from 1 to 6 mm according to the state of the menstrual cycle. Then, following construction, comes destruction – should a fertilized egg fail to arrive. The subsequent changes are considerable. Oestrogen and progesterone, the controlling hormones, are withdrawn. There are alterations to the blood supply of the endometrium. Tissue damage results. Special digestive enzymes are liberated. Blood platelets aggregate. Bleeding and sloughing begin. 'Oh, it's my time again,' says the woman, happy or not at the event.

Menstruation stops at menopause or with a pregnancy or (sometimes) with illness or (on occasion) if certain sports are vigorously pursued. 'Athletic amenorrhoea' was first recognized and described in the late 1970s. The causes of oligomenorrhoea (reduced menstruation) or amenorrhoea (cessation of menstruation) are not properly understood, but relate both to training intensity and calorie restriction in the diet. Energy expenditure coupled with reduced intake leads to a loss of body fat, and a lack of menstruation can then follow.

Up to half of top-class ballerinas, cyclists, runners and rowers have been reported as amenorrhoeic, according to Roger Wolman, consultant in sports medicine. He has listed various sports activities and the menstrual lack associated with them.

Sport	Oligomenorrhoea	Amenorrhoea	Total
Badminton	Nil	Nil	Nil
Hockey	19%	Nil	19%
Swimming	32%	Nil	32%
Heavyweight rowing	33%	Nil	33%
Ballet	24%	27%	51%
Running	19%	46%	65%
Lightweight rowing	22%	47%	69%
Cycling	40%	31%	71%
Gymnastics	29%	71%	100%

All athletes are competitive, even if only against themselves, but some are more competitive than others. Psychological stress may therefore be a contributory factor. Some gymnasts and ballet dancers, whose intense training has to begin early in life, may experience delayed menarche and never menstruate until the age of 20.

Failure to menstruate may not be considered a drawback (perhaps re-earning the old name of blessing rather than curse). Unfortunately the low oestrogen levels, with which failure is associated, affect the skeleton, notably reducing bone density. Other effects are stress fractures and delayed union of epiphyses (separate bony structures at the ends of various bones), a joining which normally occurs at puberty. Bone density can be restored naturally, should amenorrhoea only last for six months, but will continue at a sub-normal level if menstruation lack is prolonged. As osteoporosis, or loss of bone density, is often a severe complaint in later life (increased liability to fracture, damage to vertebrae) any extra decrease in bone density is unwelcome. Therefore, states Wolman, the condition of amenorrhoea 'should not be regarded as benign'. Medical opinion should be sought if it lasts longer than six months.

As for menstruation in general, and why it should occur (when it does not in the great majority of species with internal fertilization), there are – inevitably – hypotheses. One, given prominence in the *Quarterly Review of Biology* (in 1993 by Margie Profet), stated that it rids the body of bacteria carried into the uterus during copulation. She considered that any process (such as menstruation) which involves such a waste of biologically useful material must have a positive reason for its evolution, like that elimination of bacteria. Others, such as C. A. Finn, have argued (in an issue of *Human Reproduction*) that menstruation does *not* occur when sperm have (effectively) entered the uterus but does occur when they are absent. 'Menstruation has evolved in response to the absence of spermatozoa, not their presence.'

In evolutionary terms menstruation is a recent occurrence. If bacteria are a problem all the other non-menstruating animals must have found a way round this malevolence. As for the waste of useful material, added Finn, this is a feature of biological life. 'Women probably lose much more valuable biological material in their faeces than in their menstrual fluid.' Cells are lost continually from the skin, and from the gut. Besides, waste in the reproductive system is prodigious. The 7 million foetal cells, each of which might become an egg, should be contrasted with the very few – perhaps ten, or fewer still – which do so. As for the millions of sperm created, liberated and, almost entirely, wasted, there has been plenty of time for evolution to devise a less profligate strategy. With Profet's theory any bacteria taken in during copulation shortly after a menstrual flow would have 3–4 weeks before being expelled, plenty of time (concluded Finn) 'to make their presence felt'. He preferred that menstruation 'should be considered as part of the implantation process which has evolved to provide closer union between embryo and uterus . . . There is no need to envisage a separate function for it.'

MENOPAUSE Menarche is the beginning. Menstruation then follows and Menopause is the end, with climacteric being the time (five years or so) on either side of the menopause. (A word about these words. 'Men' is always related to moon. 'Arche' is from the Greek for beginning. Menopause does not suggest hesitation – 'pause' is from the Greek for cessation. Climacteric is from the Greek for ladder, implying not so much its apex as one more rung.) Change of life is a common term, but hardly specific as so many changes occur during each and every life.

Biologically the menopause is a strange event. Virtually all physical attributes deteriorate with age but they do not cease. Perhaps there are advantages to the menopause. A female no longer capable of reproduction is, in general, more than capable of assisting others, such as her grandchildren bearing her genes. She may even do more for their upbringing than does their actual mother. As Roger Short phrased it (in a recent *Nature* book review): 'Because humans have the longest period of childhood dependency of any animal, it surely makes good sense that the mother should stop producing children well before she is likely to die, so that her lastborn is not seriously disadvantaged by her death.'

Certain other primates also experience the menopause, although relatively later in their total life-span. Average age for menopause in humans is 51 years, with a normal range of 42 to 60 years (and 2–3 years less in the developing world). Conception after the age of 50 is extremely rare in any case. A few years ago the menopause seemed to be occurring later in life (just as menarche was occurring earlier) but contemporary opinion holds that the time of its arrival 'has probably not altered substantially over the centuries', as C. R. Austin and R. V. Short phrased it. For some reason, or set of reasons, about 1% of women experience premature menopause before the age of 40. Unfortunately, for those whose menopause arrives early, there is increased risk of cardiovascular mortality – according to a report from the Netherlands published in the *Lancet* during March 1996. This work supports the belief that oestrogens protect against such disease.

The decline in a woman's ovarian function starts about five years before her last period. The ovaries fail because their supply of primordial follicles has been exhausted. (The million or so such follicles with which she was born do not all mature. Only, as already stated, some 440 will do so during her reproductive span, and fewer if she bears children. Her original supply – of some 7 million when she was a mid-term foetus, to one million at birth, to fewer and fewer as she progressed to puberty, and still fewer every day of her adult life – is eventually drained.) With the increasing failure of ovulation there is a steady decrease in hormone production. This loss of oestrogen results in various symptoms, most of which are unwelcome. There are approximately 500 million women in the world over 50, a number expected to reach 1.1 billion by 2025. Therefore the degree of displeasure caused by the menopause, already considerable, will become greater. The average woman in a developed country lives one-third of her life after the menopause. Globally the proportion is one-fifth.

Any list of symptoms attributed to the 'change' is lengthy. These include 'urinary problems, depression, nervous tension, palpitations, headaches, insomnia, lack of energy, fluid retention, backaches, difficulty in concentrating, and dizzy spells', according to a WHO group looking into the problem. Some relate to ageing in general, and life-stresses in the mid-years, rather than the menopause itself. A shorter and more specific list includes flushes (flashes in the US), sweats, vaginal dryness and depression. Long-term reduction in oestrogen levels is associated with atrophy of the skin, with accelerated bone loss from the skeleton, and an increase in heart disease.

There is nothing new about problems with the menopause but not only have women numbers increased; so too the proportion experiencing it and not wishing to stay quiet. When wealthy women turned from their traditional female healers to doctors in the seventeenth and eighteenth centuries they were generally offered phlebotomy, the letting of blood, as treatment for their unwanted menorrhagia, the bloody effusion of their climacteric. In the nineteenth century sedatives were offered as a safer alternative. Women also, and increasingly, opted for hysterectomy as an aggressive solution to their menstrual problems, despite the initial high mortality of such invasive surgery. With the arrival of endocrinology in the twentieth century a more promising outlook seemed assured, but it all took time. The new experimental hormones of the 1950s and 1960s were very expensive. They did not necessarily banish all the symptoms, and their side effects were poorly understood. Their consequent use was rare, and almost all women still had to confront the menopause as one more cross to bear, a disabling, demeaning, disagreeable unpleasantness added to the rest.

Hot flushes/flashes and night-time sweats are the most particular symptoms, with 80% of women experiencing them at some time, 70% either daily or more frequently, 70% suffering embarrassment because of them, 50% suffering physical distress as a result, and 25% having to live with symptoms for more than five years. Hot flush prevalence varies among different nations/cultures, such as 80% for Dutch women, 12% for Japanese, and 0% for Mayan. The couple of years before the menopause witness such symptoms at their most severe. Vasomotor instability – changing dimensions in blood vessels – is the cause of increased skin temperature, and giddiness, and disturbed nights, and (possibly) the headaches at this time.

Vaginal dryness follows a lack of normal secretions, and leads to another depressing list – discharge, infection, pain, occasional bleeding, loss of libido. Psychologically (according to an excellent summary of menopause by John Studd and Roger Smith, of London's Chelsea and Westminster Hospital) the symptoms can include depression, irritability, loss of confidence, poor memory, poor concentration, agoraphobia, and attacks of panic. On the bright side (which does need emphasizing after such a litany of misfortune), a feeling of general well-being can arise following the menopause, partly because headaches, menstrual pain, bleeding and so forth do cease. There is also no question of becoming pregnant. Unfortunately there is more dark side, like the generalized skin atrophy caused by a loss of collagen, the principal component of connective tissue. Studd and Smith give a further and related list: skin thinning, osteoporosis, aches and pains, loss of hair, brittle nails.

Menopause, whatever its symptoms, is regarded differently in different portions of the globe. The West, with its enthusiasm for youth, connects the change with ageing, with loss of status, loss of sexuality. Conversely life in other areas can improve with the end of fertility. The risks of childbirth have gone; so too cultural restrictions linked with child-rearing. In sub-Saharan Africa a woman can gain respect in her post-menopausal years among her family and community.

HORMONE REPLACEMENT The loss of oestrogen, which lies at the base of so many menopausal changes, has inevitably been countered by administration of oestrogen to negate, or diminish, these effects. Menopause does increase the development of cardiovascular disease and of osteoporosis, both common in ageing Western populations. Therefore anything, such as hormone replacement, which can reduce this development is to be welcomed – or so it would seem. Osteoporosis affects an estimated 75 million people in the combined areas of Europe, Japan and the US, this total including one in three of their post-menopausal women. In such developed regions the risks of osteoporosis and of cardiovascular disease are reduced by the administration of oestrogens, but there is a price to pay.

Oestrogen replacement alone has a most severe effect upon the endometrium, the lining of the uterus. The risk of cancer in this area may increase nearly ten-fold (according to the WHO) after 10–15 years of oestrogen use. The incidence of breast cancer may also increase – some say by 50% for long-term oestrogen users. However, the uterine cancer risk is dramatically reduced if progestin, another hormone, is taken as well. Unfortunately that addition lessens the cardiovascular advantages of oestrogen alone. As for the increased risk of breast cancer for long-term oestrogen consumers that risk may actually increase still further if progestin is added, although hormone therapy for 2–3 years has little or no effect upon breast cancer incidence. You pays your money, as the saying goes, and takes your choice.

The matter of HRT (hormone replacement therapy) is certainly not clear-cut. There are advantages; there are disadvantages. It is difficult comparing the extreme likelihood of osteoporosis (if HRT is not employed) with an additional cancer risk (if used lengthily). The two are hard to equate. As someone said, when also wishing to illustrate a conflicting conundrum, 'Would you rather arrive in California in summer or by car?' British general practitioners have tended to leave choice to the patients, at least since the late 1980s. Rosalind Ladd concluded (at a women's conference in Kentucky in 1989): 'It is imperative that women take an active and determinative role in decision making for themselves.' Such a move requires 'a realignment of the doctor/patient relationship', a shift from the 'priestly model, where the physician speaks from a position of total authority', to a 'more collegiate model, where knowledge and decision making is shared as much as possible'.

Americans have certainly used HRT at some time in their post-menopausal lives more than, for example, their British counterparts, the proportions being 33% and 10%. British *Vogue*'s health editor, Deborah Hutton, attributes this difference to 'gloomy Anglo-Saxon fatalism' or 'unclarified fears' about hormones. There is also the wish – in some – not to interfere with the natural way of things, however uncertain these individuals may be that the menopause ever occurred naturally in the old days (with childbirth such a peril, longevity not long, and women not necessarily welcome in the tribe when rearing days were done).

As for choice in today's different times, the long-term benefits of HRT are still not clear. In *The HRT Handbook*, recently published, by Elizabeth Farrell and Ann Westmore, it is pointed out that 'definite answers are not expected for several years' – if then. The prospect of individual menopause control (to any degree that is possible) is a natural successor to individual birth control, to freedom from excessive pregnancy, and to freedom from steadfast domesticity, but is not entirely a happy outlook. As Hutton explains: it means 'living with uncertainty, keeping pace with conflicting and fast-changing research, and maintaining an open mind ... whether or not to renew one's prescription'. One bright light shines in so much shadow. Unlike the synthetic oestrogens in the contraceptive pill, the natural oestrogens used in HRT do not increase the risk of clotting in the blood, of thromboembolism. (See also pp. 73–4.)

LACTATION 'We do not call ourselves mammals without good reason,' proclaimed Roger Short during a Ciba symposium on this topic. The breast has evolved as the 'umbilical cord of the newborn'. It is not some extra, like fingernails or hair. It is as critical as the uterus, as fertilization, as the event of birth. A mammal that cannot lactate cannot reproduce. Lactation stretches back some 200 million years, and humankind (of a sort) has been no less dependent upon the process, save for very recently.

Only in the past 10,000 years has mankind become a keeper of milk-producing animals, and only in the twentieth century has the widespread use of bottle-feeding come into being. The rise of the female labour force in the early industrial revolution initiated this procedure because women, having to choose between home or work-place for their babies, generally opted for home – and a bottle in someone else's hands. The result was a high mortality rate, caused by diarrhoea and malnutrition, themselves caused by contamination and diluted ingredients. Today the procedure is even regarded as normal in many areas, or more normal than expecting mammary glands to perform adequately their age-old role. D. B. Jelliffe said, at the same Ciba meeting, that this attitude 'perhaps represents the most wide-spread, uncontrolled biological experiment that has ever taken place' and an 'even more profound change than that satirized by Aldous Huxley with his test-tube babies ... Sexual intercourse and breast-feeding were the two fundamental acts without which the human race could not have persisted.' It is therefore extraordinary that one has been so determinedly discarded.

Further alteration concerns the actual donor. Even in ancient human societies the baby's mother may not have been the milk-supplier. In more modern times, as described in *The History of Childhood* (by L. de Manse), wet-nurses have been well employed. 'The ideal image of the Madonna, with a plump happy infant at her breast in so many fifteenth-century Italian paintings, did not reflect the existing situation.' Many middle-class ladies of those days ban-ished their newborn babies for a couple of years to suitably lactating women. Even J.-J. Rousseau, promoter of the noble savage, put his children in a foundling hospital. Queen Victoria, ahead of her times in advocating a more painless childbirth by welcoming chloroform, pronounced that breast-feeding was disgusting. In the earliest years of the twentieth century wet-nurses continued to do well, as a trade, with mothers often sharing beds with a lactating woman rather than their man. Milk can still be sold today, with the quoted price being three times or so the price of beer; but, in general, the wet-nurse has finally yielded to the can, save in regions where such an artificial supplement is unavailable. Unfortunately, stored breast milk suffers and its components change, whether in collection, heating, pasteurization, or even in its storage.

Much further change linked to breasts is irrelevant to lactation. The age-old food source is firmly a stimulant of sexual appetite, an important factor in courtship. Certain women have reached peaks of success solely – well, 99% – on account of their mammary possessions. Such prowess was not diminished even after Erwin O. Strassman, of Houston, following a major survey, concluded that bigger breasts suggested a smaller IQ. In almost all societies the genitals are hidden, or at least partially obscured, but in many the breasts are exposed equally in both sexes. The fashion of toplessness, promoted in spirited fashion during the 1960s, did not last long. Along with total nudity its followers have not swamped the beaches, bars and stages as might have been expected, bearing mankind's wish for novelty and entertainment in mind.

Perhaps the lack of titillation, the whole meal without hors d'oeuvre, is at fault. Some women, as in New York City, have been resentful of men's relative freedom in hot weather to discard T-shirts. They have done likewise, notably in the subway. The intent, said Evelyn McDonnell, was to try to 'free mammary glands and their owners from the binding fetish of the billboard and the tyranny of men's stares'. New York state allows women to go bare-breasted so long as they are 'not lewd'. John Leo, newspaper columnist, wrote after the subway exposure: 'It's hard to argue that women's breasts have no sexual value on a hot day and yet must be protected under sexual assault laws the rest of the year.' In similar vein Karen Lehrman wrote: 'Maybe theirs aren't sexual organs, but mine respond quite nicely to romantic overtures, thank you.'

Breast-feeding (about which more in a moment) is actively promoted in many developed nations, and has encountered problems. Several US states have passed legislation protecting a woman's right to breast-feed in public, but a similar bill in California (killed by one vote in April 1995) experienced adverse comment, like the statement that such feeding was obscene. 'It is a mystery how campaigners for basic family values find the quintessential bond between mother and child "obscene",' wrote Malcolm Potts, considerable activist in matters affecting human reproduction. It could be counter-argued that other forms of bonding, like the procreative act itself, are a matter of privacy rather than display, but the breast – in modern times – has lost its prime role as crucial ingredient of reproduction. The fact is almost forgotten that it was ever crucial. It was not an adjunct to life. It was, as with all mammals, the vital bridge between foetal existence and a later stage in life.

*

An extra virtue of lactation, often critical, is the spacing it affords between successive births. Breast-feeding limits fertility. As Roger Short phrased it: 'In primitive human communities the duration of the birth interval was all-important, and lactational amenorrhoea lasting up to two years was nature's most effective form of contraception.' Nervous impulses from the teat provide the critical input to the hypothalamus (of the brain) for inhibiting further reproductive activity. The exact endocrine mechanisms involved are still unclear but there is undoubted clarity concerning lactation and conception: the former does discourage the latter. As D. B. Jelliffe has added: 'The child-spacing aspect of adaptive suckling is fundamental for the survival of the infant and also allows the mother to recuperate from pregnancy.'

Two surveys from two very different American communities make this point. With poor Guatemalan Indians, all of whom breast-fed their infants and none of whom used contraception, the birth interval if the first child died was 14–15 months, and 28–31 months if it survived. Among Caucasians of Boston, Massachusetts, there was no menstruation – on average – for 55 days if mothers never breast-fed their infants, 85 days if they did so for 60–89 days, 163 days if they did so for 180–209 days, and 183 days if they did for 300–329 days. The US women who breast-fed lengthily therefore began menstruating before they had finished their lactation, but they too were less likely to become pregnant again so soon after the previous child as

were the Guatemalan women. When such figures are extrapolated to
the world stage, and related to population growth, the differences are
tremendous. Even a seven-day postponement of the next birth has a
marked effect if applied to the 280 million births this planet
experiences every year. In straight and simple arithmetic it means
that 5.6 million of those births occur in the following year. And
another 5.6 million are thus postponed for every week, on average,
that pregnancies are delayed.

Human beings, unique in so many matters, are also the only
mammals – or it would seem – in which the mammary gland has
an erotic function, with breast stimulation an overture to intercourse. In no other primate are the breasts fully developed anatomically at puberty, this being two or so years in advance – at least –
of the first pregnancy. (In apes, the closest living relatives, breasts
only develop towards the end of the first pregnancy.) Some
human societies, living traditional lives, do keep breasts covered,
but – usually – only between puberty and the birth of the first
child. There is no universally desired breast shape. In Africa the
long and pendulous are sometimes favoured. Other groups, such
as Masai, prefer upright hemispheres. In much of the world size
is important, the larger the better. In the Western world, where
breasts, display, courtship, and eroticism are all so intermingled, the
concept of breasts as vital baby-feeders can almost seem bizarre.
Besides, the baby's demands may alter breast shape, a shape that
possibly assisted in the creation of that baby by sexual arousal, and
lactation may be undesired – by both partners – for that precise
reason.

As pregnancy can also alter the female figure, perhaps deleteriously,
even babies may be unwelcome for that reason. The sexual act is then
absolutely disconnected from its role in reproduction. Milk substitute
is all too possible, but uterus substitute – as depicted in Huxley's
Brave New World – is currently impossible and will always be
tremendously expensive. As of now each end of a pregnancy can be
manipulated – test-tube conceptions are famously feasible and premature births can be successfully incubated – but the uterus is the
preferred medium for every embryo, foetus and pre-term offspring.

It always will be, in so far as the future can be envisaged, but pregnancy has been labelled as repulsive by certain militants. There may be a desire to cancel it, along with feeding by the breast, but an alternative is horrendously more complex than a bottle full of milk.

Shakespeare, not averse to discussing female merit, never mentions breasts in this context. Charles Darwin, able to write in detail about sexual selection, also omits this topic. It would seem that the twentieth century, more than all its predecessors, has become fixated upon the dugs, the Bristols, the boobs, the paps, the bosoms, the tits, and the briskets without which many a hopeful goddess might as well yield to some alternative ambition.

HUMAN MILK is usually bluish-white, tastes sweet, weighs slightly more than water, and is an emulsion of fat globules in fluid. Its composition varies 'from mother to mother, day to day, and feed to feed', as Ronald Illingworth phrased it, but always has more fat in the early morning. Average composition is 1–2% protein, 3–5% fat, 6.5–8% carbohydrate, and 0.2% salts. All the rest – at least 85% – is water. Only about 6% of the total energy in breast milk comes from protein, with the remaining 94% coming almost equally from fats and carbohydrate.

The greatest difference in milk composition and colour occurs in the first 2–4 days after delivery when colostrum is released (and before milk production has properly begun). This prior substance is a pale yellow, and contains five times as much protein as the milk to follow, but less carbohydrate and less fat. It is more alkaline and slightly heavier. Human milk is markedly different from cows' milk, having less protein, less carbohydrate and about the same amount of fat (although with fewer long chain fatty acids). It has one-quarter of the sodium, and less potassium, calcium, phosphorus, chloride, and vitamins K and D. Despite that relative lack both of calcium and of vitamin D a fully breast-fed baby is unlikely to develop rickets, partly because there is better absorption by human babies of these items in human than in cows' milk. Certainly a baby's stomach empties more quickly with human milk.

'Shall I be mother?'

A malnourished mother's milk contains less of almost everything, certainly liquids, carbohydrates, protein, and several vitamins. Extra protein given to such a mother increases her milk supply but not its protein content. Mothers of premature babies also have different milk from mothers whose babies have proceeded to term. They have more nitrogen (but similar amino acids), more protein (although proportionately less if milk volume is greater), more sodium and chloride but less carbohydrate. When continental Europe was extremely short of food after the Second World War the birth-weight of babies was hardly affected, despite starvation of the mothers, but the quantity of milk produced was less than in normal times. At both times, before and after birth, the infants experienced priority. As Roger Short phrased it: 'The mother may go into severe negative energy balance before milk yield and neonatal growth rates start to suffer.' As farmers express it – the cow 'milks off her back' as she loses substance to cater for the calf. Pre-term human babies fed on breast milk do not do so well. According to Illingworth they 'are said to grow more slowly than those on formula feeds and are said to be smaller in height, weight and head circumference in the second year', but human milk fed to pre-term or at-term babies provides better protection against infection than cows' milk.

Very many drugs taken by the mother find their way into breast milk, such as alcohol, barbiturates, bromides, paracetamol, ergot, oral contraceptives, and tetracycline (which may all harm the feeding offspring). Scores of others, often excreted in milk, are less likely to cause harm, such as aspirin, thyroxin, corticosteroids, senna, cascara. The nicotine of smoking is thought to reduce milk supply (just as it reduces the size of a developing foetus). Nicotine, whether from the mother or others in the vicinity, is broken down into cotinine. This secondary substance is extensively found in mother's milk, as well as in the infant's urine and saliva, if that child has been born into a smoky environment. Alcohol swiftly finds its way into breast milk, and is said to cause delay in the child's muscular (but not nervous) development. Caffeine has been implicated in sleep disturbance, and all addictive drugs can be found in breast milk. So too pesticides, and many other chemicals used in these technological times. You are what you eat, as the old saying goes, and – if you are an infant – you get what your mother consumes, whether good or bad.

The 'breast is best' slogan, used by promoters of maternal lactation, should add that human milk is best for human babies. 'It seems likely,' D. B. Jelliffe has written, 'that the many interacting nutrients are present in human milk with specific purposes.' It has the highest concentration of lactose (milk sugar) of any mammal. It also has a high concentration of cholesterol, a characteristic assortment of amino acids, and a specific pattern of fatty acids. Quite why human milk is so organized, and what are the various benefits, is by no means understood, but Jelliffe (from the Department of Public Health and Pediatrics, University of California) is adamant: 'All these characteristics ensure that human milk is the most appropriate fluid to supply those nutrients needed for the main feature of the newborn human – rapid growth in size and complexity of the brain.'

Lactation makes considerably greater demands on the mother's body than pregnancy in species where the young are helpless at birth, as with human infants (and rodents and carnivores). This extra energy comes both from within (the mother's tissues) and without (extra food). She can lose fat at this time, deposits having been laid down during pregnancy, and perhaps calcium from her own skeleton, particularly if her calcium intake is low. The human mother does not experience as much change as, say, the rat, where the small intestine more than doubles its weight to cope with the additional food. She also does not have so many voracious appetites dependent upon her. A rat's litter multiplies its weight five-fold before the end of full suckling, a mouse's six-fold, a cat's seven-fold, and a woman's twice. The pregnancy of humans, lengthy when compared with cats, mice and rats, takes nine months to create a 7 lb offspring. Weight increase then speeds up, as it also does with cats, mice and rats, but the human still takes another 4–5 months to grow another 7 lb.

The provision of milk for this post-birth increase is a less efficient process than the provision of nutrients during the pre-birth phase. Efficiency estimates (for humans) vary between 60% and 90%, but there is agreement that food via the placenta is less demanding on the mother than the same amount of food via the breast. A baby is (almost certainly) born into an environment cooler than the mother's body temperature. Therefore much of its received energy supply must be consumed in keeping warm. During the first two months, for instance, only 26% is used for growth. A human mother may feel she has been encumbered, both before and after birth, by a considerable burden, but a newborn human is only 6% of her own weight (on average) and only 12% at the end of suckling. Litters are much more demanding, with equivalent figures for a sow at 7% and 40%, a rabbit 15% and 50%, a cat 12% and 100%, a rat 28% and 120%, and a mouse 28% and 160%.

BREAST-FEEDING 'How is it that poor men's wives, who have no cold fowl or port wine on which to be coshered up, nurse their children without difficulty, whereas the wives of rich men, who eat and drink everything that is good, cannot do so?' wrote Anthony Trollope in 1847. Currently the tables are turned. Breast-feeding in developed nations is commoner among middle and upper classes, whether or not supported by fowl and port wine, than the lower, poorer kind. A higher educational level, rather than the coshering, is considered to be responsible. In underdeveloped nations the situation is inevitably more traditional, with the poor having no alternative. Only the wealthy can afford the novelty.

The World Health Organization is stern in its opinions. 'Where hygiene is poor ... artificially-fed babies have more than 20 times the chance of dying ... than babies that are exclusively breast-fed ... In emergency and epidemic situations, the dangers of artificial feeding are even greater.' Approximately 1.5 million infant deaths a year could be averted through improved breast-feeding practices. A recent 'resolution' adopted by member states not only called on all countries to promote such feeding but 'ensure that no donations of free or subsidized supplies of breast-milk substitutes are provided in any part of the health care system'.

Breast-feeding is said to prevent, or significantly reduce, neonatal sepsis, measles, diarrhoea, pneumonia, ear infections, meningitis, and sudden infant death syndrome. With respiratory infection killing

4.3 million under-5s annually, and diarrhoea 3 million, any reduction in these tremendous totals is to be welcomed. Unfortunately the HIV virus can be transmitted through breast-milk, with some 15% of babies breast-fed by HIV-infected mothers becoming infected this way. A greater proportion become infected during pregnancy and delivery, bringing the proportion of HIV-positive babies reared by HIV-positive women to one-third.

As adequate diet is more crucial in infancy than at any other time of life, and as breast milk – says the WHO – is unequalled as a food, the age-old feeding method has to be promoted. It helps to bond mother and child (or should). It protects a mother's health by reducing the risk of after-birth bleeding when suckling starts within an hour. It helps to protect her against ovarian and breast cancer (more on this in a moment). And, as already mentioned, it helps to space births, thereby improving health in general and lowering health costs. This last virtue may be the most important of them all, with the remorseless re-arrival of child after child transformed into a lesser battering. If breast-feeding helps ten babies to become eight, or eight to be six, or six to be four, the family can gain (with fewer mouths to feed), the community can gain (with less pressure on health services), and the world will gain (with a less intensive population growth). Best of all, the child is more likely to survive.

Complete nutritional requirements are obtained from the breast for up to nine months, save perhaps for water, with a post-prandial cry sometimes indicating thirst rather than hunger. Many other mammals are breast-fed longer than humans, such as baboons one year, macaque monkeys eighteen months, chimpanzees up to two years, and orang-utans three to four years. The animal system of breast-feeding is more haphazard than with clock-conscious humans, in that offspring tend to feed when opportunity, impulse and compliance permit. Perhaps humans could change their ways, and even diminish infant crying, the not wholly welcome adjunct to the joys of breeding. Ronald Barr, of McGill University, Montreal, compared breast-feeding mothers of Holland with a Kung San group in Botswana. The African women fed more continually – a minute or two every fifteen minutes. The Dutch gave longer feeds at much longer intervals. There were other differences, such as the San infants being carried in slings, and the Dutch babies crying twice as long as the Africans. Even in Boston, Mass., where Barr investigated both enthusiastic and reluctant breast-feeders, the babies who cried least often were fed most often. 'If parents wish to try feeding more frequently,' concluded Barr, 'it's a perfectly appropriate thing to do.'

Many arguments have been put forward against breast-feeding. It is difficult to know how much the baby is getting. Bottle-fed babies do well and are less likely to be under-fed. Why start if failure may soon follow? A bottle can be handier, partly because others can give it. The desired female form may suffer. ('It would be unfair on my husband.') Nipples may dribble, and become sore. There may even be mastitis and abscesses. There can be pain associated with bulging breasts. It can be a struggle to establish. Worry can halt even a good flow. Some mothers are incapable of producing enough. ('I may be one of them.') The loose stools associated with breast-fed babies make more work. There can also be embarrassment.

As for advantages, the reduction of gastro-enteritis linked with breast-feeding is the most important diminution of infection. Infant death-rate (in a study of Chilean offspring) was three times greater during the first three months for bottle-fed babies. Babies in that country were dying more frequently among 'upper class mothers' because more were receiving formula feeds. The milk-substitute itself is not necessarily causing or encouraging gastro-enteritis. Contamination of bottles, teats and milk is more likely to be responsible. Quite why sudden infant death is reduced among the breast-fed is a mystery, along with most reasons for that distressing event. Allergy to cows' milk is the most common food allergen to affect infants. Other lactation blessings are less likelihood of overweight among babies, some long-term protection against obesity, less soreness around the anus, some protection against caries, less tetanus, and possibly less diabetes mellitus.

Many studies have shown that breast-fed children do better in intelligence tests than the bottle-fed variety. This assertion was examined by a team from Southampton University (whose results were published in April 1996). A total of 994 men and women, all born between 1920 and 1930 in Hertfordshire (whose infant-feeding procedures had been recorded by health visitors), confirmed the general consensus that breast-feeding was linked with a (slightly) higher IQ. This fact, concluded the researchers, 'had more to do with the child's social environment than with the nutritional qualities of the milk'.

As dominant feature, save for that spacing of births, breast-feeding costs less – for the family and the nation. Ronald Illingworth, in *The Normal Child* (10th Edition, 1993), quotes some figures. A 10% decline of breast-feeding in Tanzania cost £4 million in foreign exchange. If 20% of urban mothers in developing countries do not breast-feed, the cost of replacement food is £160 million a year. Some 200 million children suffer from malnutrition as a result of the decline in breast-feeding. Chile needed another 32,000 cows to make up the loss as mothers switched to the bottle. Mothers short of funds tend to dilute the bottle-milk when need be, whereas the breast keeps up the quality and, if possible, the quantity to the detriment of the mother, with her body becoming malnourished rather than the infant's.

Human babies can appear most incompetent (certainly if compared to animals) when groping, mouth agape, for a nipple, and often sucking on any projection to come their way, but it seems that smell rather than sight is acting as guide. Some Swedish researchers washed one breast of each participating mother immediately after delivery. Each newborn baby was then placed prone precisely between its mother's breasts. Of the thirty infants, the thirty mothers and sixty breasts involved, twenty-two of the infants chose from the thirty unwashed breasts.

POLYTHELIA Between 1% and 5% of men and women have additional nipples. Less commonly these are associated with extra breasts. Polythelia (as the condition is called) usually occurs along the mammary line. This streak is clearly visible in a seven-week embryo/foetus and then disappears leaving only a small spot of tissue on each side in the chest area, a spot to become the nipples of both sexes. In an adult the line (if it were still manifest) would run from the centre of each armpit down through the corresponding nipple to a point on either side of the groin.

Asymmetry in breasts is also quite common, with the left usually larger than the right. Bearing in mind the fluctuating number of nipples among the mammalian community, varying from two (but never one) to twenty-two and varying from the pectoral (chest) end to the inguinal (groin) end, or sometimes (as with pigs) stretching from one to the other, it is not totally surprising that humans are sometimes abnormal in this regard by having more than two. Bats also have pectoral teats but, to accommodate a life so frequently upside-down, have two false teats at the inguinal end from which the young can hang.

Males can wonder why they possess nipples when these will never mature into life-giving glands. As with bodily and facial hairs, which are just as numerous in both sexes (but grow more conspicuously in men), males and females are often more similar than our gender-sensitive society appreciates. At birth the breast tissue of both sexes appears histologically identical. During childhood this substance remains quiescent in both boys and girls. Only during puberty is there differentiation, and a change happens in both sexes. In most males there is a sudden proliferation of ducts and of alterations to the surrounding tissue, but this is followed by an atrophy of the new growth. Only in females is this proliferation succeeded by a general enlargement of the ducts and the surrounding tissue. Both oestrogen and progesterone are necessary for this process to occur.

With trans-sexuals wishing to become females the simplest part of the alteration from one sex to another is the development of breasts by the administration of hormones. Gynecomastia, the benign enlargement of male breasts, can also occur naturally. Males do produce oestrogen, the female hormone, but usually its effects are swamped by androgens, the male hormones. At puberty a male's production of oestrogen goes up three-fold, but there is a thirty-fold increase in the concentration of testosterone. Gynecomastia results when there is an imbalance of male to female hormones. It can occur at birth (in 60–90% of all newborns because maternal hormones have crossed the placenta), at puberty (when so much is being re-arranged) and also in old age – 50–80 years (when hormone supply is altering yet again) – or when some disorder is having this extra effect.

BREAST IMPLANTS result from humanity's enthusiasm for alteration (if possible) and for breasts in particular. It is thought that 1–2.2 million women 'may' have received silicone breast implants within North America since 1962. Other countries than the US and Canada are even hazier about numbers, often producing no estimates whatsoever. A figure of 'about 100,000' for the UK has been quoted.

A 1994 article in the *British Medical Journal* summed up the situation, partly because of allegations (and impending litigation) concerning implants and connective tissue diseases. Back in 1964 the disease of hypergammaglobulinaemia was reported in two patients who had received silicone and paraffin injections. In 1982 the first three patients with both silicone implants and connective tissue diseases were reported. In the subsequent twelve years a further 290 individuals were reported (in English medical journals) who not only had such diseases but also an implant. The US Food and Drug Administration, diligent watchdog of American intake, placed a moratorium in 1992 on the use of implants, save for the purpose of research. France took the strongest line among European nations by establishing a complete moratorium on such implants in 1992. In May 1996 this ban was officially extended 'until market approval conditions are reinforced'. The French Ministry of Health did not recommend the 'explantation' of existing implants but women possessing them should be informed of the risks.

In 1994 some manufacturers of implants and their suppliers set aside $4.225 billion to deal with potential legal suits. Women were given until June of that year to decide whether to join 'a class action suit', allegedly guaranteeing $200,000 to $2 million (for each successful claimant), or to litigate separately, or to do neither.

More than 90,000 women joined the class action, ready to accept any settlement to come their way, and 15,000 opted out (with half of those living outside the US). In September 1994 a judge approved the

$4.25 billion compensation deal on behalf of the claimants, who all had to show that they had suffered from one of a range of illnesses. These included 'serious systemic diseases, such as lupus; neurological syndromes; and so called silicone disease', according to the *BMJ*. Women claiming in connection with this third category had to show 'at least five' of an assortment of symptoms, including rashes, chronic fatigue, muscle weakness, and memory loss. As the deal eventually averaged out at $47,000 per claimant, and as the payments varied from $105,000 to $1.4 million, it must be assumed that many women did not qualify. The thirteen-fold difference between high and low payments reflected the severity of the complaint and the age of the victim.

Despite the massive pay-out, the biggest product liability settlement in American legal history, the story has not yet ended. Women free of symptoms at the time of the settlement could still register before the end of the year, with claims permissible for the next thirty years. Only 5% of the $4.25 billion was set aside for non-American claimants. British women have had little option save to accept US assessments of compensation, partly for the difficulties in getting US courts to accept foreign claims. There is also the problem that the (British) Department of Health maintains that there is no scientific evidence to justify a ban on the use of silicone implants. The *BMJ*, in its report, summed up the problem of association between implants and disease: 'No precise data exist on how many women have received silicone breast implants; no systematic follow-up data have been collected ... Virtually nothing is known about how many women have had repeat implant procedures or how many ... have died ... Despite the increased number of cases reported in the literature no association has been convincingly established.' The FDA (of the United States), notwithstanding its ban on implants 'for purely cosmetic purposes', permits their use for 'reconstruction' as part of clinical trials.

The *New England Journal of Medicine*, not always so strait-laced as its British counterparts, has described a hitherto unreported disadvantage of breast implants. A 51-year-old woman, after arriving in Frisco, Colorado (9,300 ft) from her sea-level home, complained of a 'swishing' in her breasts. Boyle's Law was promptly implicated as the cause. As with toothpaste tubes and crisp packets, their confined gases expand with decreased pressure – and with altitude. The patient's breasts stopped swishing, one assumes, when she went back home again – or leakage into the Frisco air had evened matters out.

Exciting travel has also stimulated the National Hospital for Aesthetic Plastic Surgery, Bromsgrove. In 1995 American women were being offered – for £3,898 – several nights in Cotswold inns, trips to York and Stratford-upon-Avon, *and* breast implants in a novel form of package tour, now that Britain permits the implantation and America does not. Shakespeare would be amazed. Macbeth, so troubled about his wife, requested the doctor to 'Cleanse the stuff'd bosom of that perilous stuff which weighs upon the heart'. Today's stuffing is of quite a different kind. As for the peril no one is quite sure.

CANCER There is an extra cruelty that the most frequent sites for cancer in women should be the most female of locations, namely the cervix and the breast, being 15% and 18% of the total respectively. There are 570,000 new cases worldwide of breast cancer every year, and it is the single commonest cause of death among women aged 40–50. It is the second most common cancer among all women in the world. Rates vary five-fold in different countries, suggesting environmental influence. Hong Kong, Mauritius and Japan are low incidence countries (in that order). Argentina, Cuba, Chile and Panama are higher, with Panama as high as Hong Kong and Argentina twice as high. In eastern Europe the order of incidence is Hungary, Czechoslovakia (when united), East Germany (when not united), Poland, Bulgaria and the Soviet Union (when a union). In northern Europe the high to low incidence is Denmark, Iceland, Norway, Sweden and Finland. In the rest of continental Europe, in Australasia and the United Kingdom the incidence is similar, with the rate varying between 24 deaths (per 100,000 people a year) for France to 37 for England and Wales. The three British regions are not entirely equal, being Northern Ireland 32, Scotland 35, and England and

Wales 37. Plainly all sorts of influences are at work to cause these different sets of figures.

The lifetime incidence of breast cancer in the developed world is approximately 10%. If a woman in North America or Europe lives to the age of 80 she stands one chance in nine of developing the disease. Within the UK there are 30,000 new cases every year and 15,000 deaths. In Western Europe and the United States approximately one in twelve women develop breast cancer. Survival figures for England and Wales show that, on average, 62% of all women diagnosed are alive five years later. Expectation depends upon the cancer's stage. Five-year survival is only 18% for stage IV but 84% for stage I. Figures for Holland are comparable. In that country the breast is the site for about 30% of all primary cancers and leads to 22% of all female cancer deaths. In general about 40% of all breast cancers occur in women over 70, this cancer quantity being likely to rise with an increasingly aged population. (Breast cancer also occurs in men, causing about 1% of all male cancers.) In the United States breast cancer kills 46,000 women a year, as against 56,000 for female lung cancer. If the global epidemic, as it has been called, of breast cancer continues in similar fashion it will be claiming 1 million lives annually by 2000.

For Britain, at least, there is some good news. The death rate in England and Wales stopped increasing in the late 1980s and even started to fall. In the three decades from 1960 to 1989 it rose remorselessly, making the subsequent drop yet more welcome. There were 9,517 deaths at ages 20–79 in 1993 and 10,538 four years earlier. The decline has been greater in younger than in older women. One recent survey reported a drop in breast cancer mortality by 14% in women aged 20–49, 11% in women aged 50–69, and 5% in women aged 70–79. It would be helpful to know the cause.

Earlier diagnosis and better treatment are undoubtedly involved, but so too a change in childbearing patterns. Routine screening was only offered (freely by the National Health Service) to women aged 50–64 during those years. Other factors must therefore be relevant. There is also that matter of regional differences within the UK, with some districts allegedly offering excellent diagnosis and management, and others not (said Roger Blamey, of the British Association of Surgical Oncology). 'At best a woman who finds a breast lump may be seen in a specialist clinic ... receive all the diagnostic tests ... and be assured she is not suffering from breast cancer at the one visit ... At worst she may be referred to a non-specialist unit, wait three months for an appointment, be referred for a mammogram weeks later, be seen several times by junior staff ... and finally receive an inappropriate operation for the removal of her lump.' Only one in ten women referred to specialists is found to have cancer.

A major trouble for cancer epidemiologists is the length of time before an aggravating cause can have its carcinomic effect. About 40% of all breast cancers occur in women over 70. Yet having babies, especially at a young age, is known to protect – to some degree – against breast cancer. The die is therefore cast long before its effects show themselves in full. A kind of confirmation for breast-feeding's association with a lower incidence of cancer has arisen from Hong Kong. Certain villagers there only feed their babies from their right breast, and carcinoma more frequently occurs on the left side. Contrarily a colossal study (on 89,887 American women), reported in the *Lancet* of February 1996, concluded that 'there is no important overall association between breast-feeding and the occurrence of breast cancer'. An Italian report, published in May 1996 (and after studying the diets of 2,569 women), concluded that the risk of breast cancer decreased with increasing fat intake but increased with increasing carbohydrate consumption.

The surge in births after the Second World War, the famous baby boom, reduced the proportion of childless women from 20% to 10%. This sudden deluge of births among the young is thought to be partly responsible for the drop in cancer deaths among women in their fifties and sixties three to four decades later. Similarly there was both a drop and a postponement of births during the earlier depression years. Depression women, either postponing or denying motherhood, were in their seventies and eighties five to six decades later – and their age-group showed a rise in cancer deaths. Cause and effect are not precisely clear-cut, just as they are not with smoking, where lifetime smokers may fail to develop cancer and life-time abstainers may get it.

The debate over a possible relationship between the use of oral contraceptives and breast cancer has been fierce for a long time. The current and general opinion is that no association exists but some Dutch work (published towards the end of 1994) reported that there were 'higher risks for some subgroups of users'. This study group concluded 'that four or more years of OC use, especially if partly before age 20, is associated with an increased risk of breast cancer developing at an early age'. It added: 'There is limited evidence that the excess risk disappears as the cohort of young OC users ages, but this issue needs confirmation.'

As for the association between hormone replacement therapy (HRT) and breast cancer that too is fraught with conflicting evidence. There is some agreement that early menarche, late first pregnancy and late menopause are each linked, however loosely, with increased breast cancer, indicating that ovarian hormones, notably oestrogen, are involved. HRT is the replacement of hormones, and therefore ought to be influential, but this does not seem to be the case. One summary of the situation, published in 1994, reported that there had been twenty-eight studies since 1972. Of this number nineteen had failed to be significant, six had shown an increased risk of breast cancer with oestrogen, and three a reduced risk. Despite this disparity the report concluded: 'It is prudent to conclude that postmenopausal HRT is associated with a slight excess risk of breast cancer of the order of 10–30%.' This does not mean that almost a third of HRT users will get the disease. It merely suggests that the developed world's expectation of breast cancer will go up by 10–30%, namely from 10% to 11% or 13%.

Such an increase should therefore be set against the positive advantages of HRT. These include not only the relief of unpleasant symptoms, improved quality of life, and protection from osteoporosis, but a reduction in the risk of heart disease and stroke. Such ischaemic (insufficient blood) disease is a much more important cause of death in later years than breast cancer. In this respect HRT confers a big advantage whereas its relationship with breast cancer only creates (if this case is proven) a small disadvantage. Hence a conclusion from the Los Angeles Leisure World Study that women taking long-term HRT live for 2–3 more years than women who have never taken it.

SCREENING is another world of confusion, with choice blurred by conflicting statements. Joan Austoker, Director of Oxford's Cancer Research Campaign Primary Care Education Research Group, wrote in the *British Medical Journal* (of July 1994) that 'Breast screening by mammography has been shown in randomized trials to reduce mortality from breast cancer in women aged 50 and over.' In the *Lancet* (of July 1995) Charles J. Wright and C. Barber Mueller, of British Columbia University's Department of Health Care and Epidemiology, wrote that 'the early trials of screening mammography, reporting 30% relative reduction in mortality from breast cancer in women over 50 years of age' led to strong demand for such programmes. But 'there has been little publicity about the subsequent trials showing no significant benefit in any age group, or about the harm and costs associated with screening mammography'.

The Canadians elaborated on the harm and costs. 'About 5% of screening mammograms are positive or suspicious, and of these 80–93% are false positives that cause much unnecessary anxiety and further procedures including surgery. False reassurance by negative mammography occurs in 10–15% of women with breast cancer that will manifest clinically within a year. Our calculations confirm others that the mean annual cost per life "saved" is around $1.2 million.' Their conclusion is adamant: 'Since the benefit achieved is marginal, the harm caused is substantial, and the costs incurred are enormous, we suggest that public funding for breast cancer screening in any age group is not justifiable.'

In the *BMJ*, also in 1994, R. W. Blamey, consultant radiologist at Nottingham City Hospital, and others, steered a path between the opinions already quoted. After stating that 'screening tests should be simple to apply, cheap, easy to perform, easy and unambiguous to interpret, and able to identify women with disease and exclude those without disease' they added: 'Mammography is expensive; it requires high technology machinery, special film and dedicated processing, and highly trained radiologists to interpret the films; and it detects only 95% of all breast cancers.' Nevertheless 'it is the best screening tool available for detecting breast cancer and is the only screening method for any malignancy which has been shown to be of value in randomized trials'. It would seem that such screening is costly, is difficult, is not as good as might be hoped, but is better than nothing at all.

Blamey and his colleagues concluded that ignorance 'of the pathogenesis of breast cancer' means that 'primary prevention is currently a distant prospect'. Consequently screening 'presents an alternative approach to try to reduce mortality from breast cancer'. In Britain 1.6 million women a year undergo breast cancer screening. During a recent year 6,500 cancers were detected, a ratio of 1:246 among those tested. In March 1995 the age range of women to be screened was expanded from 50–65 to 50–69 and there was specu-lation that the lower age of 50 might be lowered further still. The expense to the nation will therefore increase proportionately, and the number of radiologists able to spot tumours on the mammograms will also have to rise. Or perhaps the detection can become comput-erized? It seems that various skills acquired for spotting distant stars and galaxies might be transferable for spotting minute and suspicious objects on mammograms (and astronomical techniques are being adapted). Any automated system will have to match – and better – the success rate currently achieved by radiologists, say those involved, and 1997 might witness the arrival of such a system.

In 1995 the screening programme again became a major topic in the *Lancet*, and several correspondents made several points. Yes, there has been a fall in breast cancer rates (in England and Wales) between 1985 and 1993, and the 'NHS breast screening programme is delighted to have played a part in improving breast care'. But the first NHS screening units 'were not operational until 1988', and the country as a whole was not covered 'until at least 1990'. To call any part of the 11% fall 'attributable to the screening programme is intellectually dishonest'. 'My view,' wrote M. Baum, of London's

Royal Marsden Hospital, 'is that much of this fall was a result of the widespread adoption of adjuvant tamoxifen in the management of postmenopausal women with early breast cancer.'

A major trial of tamoxifen is now being run in the US, despite the possibility that takers of this drug are at greater risk of developing uterine cancer. (Tamoxifen, widely prescribed for women who have experienced surgery for breast cancer, is a chemical that opposes, but sometimes mimics, the effects of oestrogen. It is often favoured over the 'slash, burn and poison' – so-called – approach of many therapies.) Swapping breast cancer for uterine cancer might not have immediate appeal, but if this means a major risk is being replaced by a minor one the appeal becomes more convincing. At all events the US Food and Drug Administration did not ban the clinical trial (which is intended to involve 16,000 women, half to receive the drug and half a placebo).

Baum is well aware (as were Wright and Mueller) of the screening costs. The UK Health of the Nation, a survey of the country's medical status, 'hopes to reduce the rate of breast cancer deaths among women invited for screening by at least 25% by the year 2000'. If, states Baum, the NHS programme achieves half of its intended target, 'this would suggest a price tag of £1 million for each woman who benefits ... It is invidious to put a price on a woman's life ... [but] could these large sums of money be redirected ... in such a way that more women would benefit and fewer well women would be harmed?' If younger women were included in the screening programme 'we can expect one woman out of 10,000 screened to benefit ... at an estimated cost of £2 million'. Others in the correspondence columns put the cost of each life saved nearer £100,000 in the older age group (and therefore better value than lives saved via haemodialysis).

Cost is also related to breast cancer quite differently. The *New England Journal of Medicine* stated that a failure to diagnose the disease is 'among the most expensive errors a physician can make'. In 1988 average payment for such failure was $115,000 for family doctors and $207,000 for gynaecologists. If the undiagnosed victim was aged less than 40 the average payment went up to $330,000. The United States may lead the world in litigation, with everyone suing everyone at the drop of a hat (so it is alleged, particularly if that hat lands heavily/painfully/disturbingly upon another's foot), but Britain and others can rival the pay-outs. A British 38-year-old mother of two, after referral to a consultant surgeon, learned her lump was benign. Nine months later, when it had begun to give great pain, the same man agreed to operate and discovered the lump to be cancerous. Her £215,000 was awarded for the 'loss of an 80% chance of survival'.

*

With one in nine (or so) women developing breast cancer during life it was inevitable that a search would be made for a controlling gene (or genes). Unfortunately the disease does not run (much) in families. It is not like haemophilia or Huntington's chorea, for example. Most breast cancers arise without obvious cause, and there is no simple environmental villain like the nicotine of smoke, but at least a proportion of breast cancer cases can be explained by inherited mutations. Estimates of their frequency vary, but they confer on female carriers a lifetime probability of breast cancer of over 90%, as opposed to less than 10% in non-carriers. The search was therefore valid.

Rumours of a genetic finding began in 1993. Then, in September 1994, the breast cancer gene – as it was named, or BRCA1, was discovered by a group working at the University of Utah. It was located on the long arm of chromosome 17, and found in about one in every 200 women. Unfortunately it was no straightforward gene, either causing or not causing the disease. *New Scientist* labelled it 'distinctly messy', with its role in non-inherited breast cancer 'not at all clear'. *Nature* reported that BRCA1 'is probably responsible for about one-third of families with multiple cases of breast cancer only, but more than 80% of families in which there is both breast cancer and epithelial ovarian cancer'.

For a while BRCA1 reigned in solitary (and messy) fashion; but, fifteen months after its discovery, BRCA2 was found (in 1995) on chromosome 13. The Institute of Cancer Research, in Sutton, Surrey, which reported this finding, thought that No.2 is responsible for nearly as many cancer cases as No.1. Unfortunately – for simplicity –

most cases of breast cancer do not run in families, and therefore genetics is not immediately implicated.

The advance in knowledge has not led to an advance in management of cancer cases. The work seems to have disclosed more problems than solutions. What should a woman do if tests are positive? Would such tests be reliable with such a 'messy' gene when it comes in the form of dozens of different mutations? What of the patent application that was filed immediately the gene was discovered? Like some traveller suddenly observing a longed for destination, but realizing simultaneously the crevasses in between, the isolation of BRCA1 has not provided much relief. *Breakthrough*, a book published in 1995, excitedly describes the discovery but admits 'the stark truth is that little will change in terms of treatment in the near future'. It is fascinating that a gene can be located, but frustrating that such advance gains so little benefit – for the time being.

In mid-1995 the first UN World Conference on Women was held in China. This prompted many reports on the status of women, their life expectancy, their level of income, their literacy and school enrolment. The United Nations Development Programme compiled an index which combined all such factors. It then concluded that no country exists where women fare better than men. They make up 70% of the world's poor, and 66% of its illiterates. Globally, two-thirds of women's work is unpaid (compared with one-third of men's work). 'If women's work were accurately reflected in national statistics,' stated the UNDP report's principal author, 'it would shatter the myth that men are the main breadwinners of the world.' In many countries with high levels of education and health the improvements have not filtered through to women. Britain provides near equality in education and health but is ranked thirty-ninth in the world for disparity in income (lower than, for example, Turkey, Tanzania, Brazil and Zambia). In every nation women work longer hours.

As counterweight to such inequality the report ended optimistically. Save for parts of Eastern Europe and of the former Soviet Union the work of men and women is becoming more equal. The gender index (adding together those several aspects of female status) has improved during the past twenty-five years within the seventy-nine nations which were examined. With complete equality registering 1.0 (in the index) Sweden was top country with 0.919 (having been 0.764 in 1970). Then came, in descending order, Finland, Norway, Denmark, US, Australia, France, Japan, Canada and Austria. At rock bottom was Afghanistan with a score of 0.169.

Complete equality everywhere may never come to pass – there will always be the awkward differential of women having babies; but, as was mentioned so vociferously at the China conference, there is current room for improvement right around the globe. There is also the other face to this coin: women are biologically superior in that they live longer (unlike a century ago when fewer than one in three women survived repeated childbirth to reach the menopause even in the healthier communities). Their current longevity is not immediately comprehensible. They do suffer less than men from all major diseases (save for breast cancer) but they visit doctors more often, consume more prescribed drugs, are more often sick from work, spend more days in bed and also in hospital, have more operations, more mental disorder, more headaches, more stomach aches, more constipation, more varicose veins, more back pain, more joint pain, more bladder disorders, more foot problems and then, on top of all this malfunction, have more days in which to suffer. 'They are', as was said with such touching clarity by the two male heroes in *Some Like It Hot*, 'a completely different sex.'

SEX

WHY SEX?   SEX RATIO   INFANTICIDE   GENDER CHOICE

SEX APPEAL   SEX EDUCATION   SEXUAL DISEASE

AIDS   TRANS-SEXUALITY   INTERSEXES

If the mother is definitely female then a girl is born.
A school-child's essay

Too much of a good thing is wonderful.
Mae West

The pleasure is momentary, the position ridiculous, the
expense damnable.
(allegedly) The Earl of Chesterfield

WHY SEX? A personal anecdote can start this most complex of
subjects. Three of us once landed on a small island in the north of
Hudson Bay. All looked well until, frighteningly close, we spotted a
polar bear. We knew such a bear could outrun us on land, outswim
us in water, and certainly outscramble us should we aim for nearby
rocks. I suddenly regarded my two companions in quite a different
guise. Forget their amiability, their cleverness, their general expertise.
Could I run faster than at least one of them, and thus escape the bear?

Two issues were therefore involved. There was undeniable interspe-
cific competition – man versus bear. There was also intraspecific
competition – my two friends versus me. Both forms of competition,
relevant to species and to individuals, are enhanced by the sexual
system of reproduction. Sexuality serves the species, as innovation
achieved by genetic mixing from two parents provides for variety,
and variability is crucial in a changing world. New forms can have
unforeseen advantages, and that variability also serves the individual.
His or her assortment of genes is not eaten by the bear, and it is good
for the species if some individuals manage to escape, but the seeming
simplicity of such statements must not imply that the origin of sex is
immediately comprehensible. It is not.

Evolutionary biologists debate this topic eternally, and well over a century since Charles Darwin first set the scene when he wrote: 'Sexual Selection depends on the success of certain individuals over others of the same sex in relation to the propagation of species; while Natural Selection depends on the success of both sexes, at all ages, in relation to the general conditions of life.' Such straightforwardness can vanish when the different sexual strategies of nature are encountered. Immature female (but sexless) honey-bees die readily in defending the hive. Alligator eggs vary their sex ratio (male to female) according to temperature. Sex ratio is often 50:50 even in polygynous or polyandrous species. Greenfly females breed asexually for several generations and then, at each summer's end, mate with the males. Males (when 50% of any population) consume half (or more if they are bigger) of the available food supply, and yet fewer males could service all the females. Many plants, such as dandelions, make replicas of themselves. Many others, if hermaphrodite, are extraordinarily complex in their procedures for ensuring that fertilization only occurs with another plant's pollen. Some animals change their sex during their lifetime. Many males (as with bachelor seals) live and die without procreating. Certain other males, as in harem species, achieve all the offspring.

For humans, and for (almost) every vertebrate, there is no choice in the matter of sexual reproduction. 'Male and female created he them,' states Genesis, and offspring can only be created by a mixing of gametes from both kinds. Sex *is* here to stay. It has been here for hundreds of millions of years, and cannot be readily amended, any more than four-footed animals can suddenly grow another pair of limbs. Human parthenogenesis ought to be feasible, in that females possess the blueprint necessary for creating other human beings, but the sexual system is too deeply entrenched for such asexuality to be likely. For us sex is a basic fact of life. Only for biologists does it present the extra dimension of its evolutionary value as a means of procreation.

One major trouble is the role of altruism. In running from the bear one of my companions might (had ill fortune and bear both come our way) have sacrificed his individual life for the sake of the species

– in that two of us would have escaped; but genuine altruism is rare. Jackal offspring often stay with their parents, acting as helpers to rear the next generation rather than rearing a family themselves, this form of assistance built into their behaviour just as much as the negative form (employed by various eagles) when the strongest sibling jettisons the rest. Any single animal, in fighting for itself, in acquiring a mate, in caring for offspring, in destroying offspring sired by others (as with lions), is obeying instincts. It certainly is not caring whether its survival (or sacrifice) is benefiting the species of which it is a member.

For most plants and animals the system of reproduction is entirely pre-ordained, along with its programmed style of life. The merits – of sexual reproduction versus asexuality – have to be discussed either theoretically or with reference to life forms that do have choice. One early theory stated that sex existed to rid the species of damaging mutations. Asexuality is much like a photocopier: any blemish is replicated, and any new blemish will similarly be repeated. The sexual system, by this theory, is better because it is constantly making an original rather than a copy. Such an argument was popular (as from 1964 when first properly proposed) until a counter-argument (originating in 1982) undermined it. The new proposition stated that, if a mutation caused a death or an inability to breed, that mutation died with the animal. Most mutations are harmful, and it is therefore good to lose them (with the most famous analogy being of a finger jabbing at the mechanism of a watch. Practically every jab will do damage but, just occasionally, one can do good.) Therefore any system, such as sexual reproduction, for getting rid of mutations will be beneficial even though the good mutations enable change to occur.

To this counter-argument there then came counter-counter-arguments, principally that mutations are insufficiently common for sex to be necessary in getting rid of them. An asexual breeder, producing offspring more liberally, would have benefits above the slower sexual breeder. Disease then entered the discussion (in the 1980s) as the most potent force. Of course life is a struggle for food, for territory, for survival; but disease, whether from viruses, bacteria, protozoa or multi-celled organisms, is always a formidable foe. Malaria (which still kills millions of humans every year) is highly relevant to the sexual discussion. To be inbred is to lose both heterozygosity (mixed gametes from both parents) and polymorphism (a range of characters within the population as a whole).

Where malaria is common the error known as sickle-cell anaemia comes into its own. There is advantage in this malformation (of red blood cells) in that individuals who have inherited the trait from only one parent are less likely to be attacked by malaria. Those who have inherited the trait from both parents suffer from anaemia. Those who have not received it from either parent are more likely to be assaulted by malaria. Sexual reproduction, as against the asexual kind, steps up the likelihood of heterozygotes, the more-likely-to-be-successful intermediates.

For human beings asexual reproduction is not an option. The sickle-cell story illustrates one advantage of sexual reproduction, a form of advantage which has parallels among life-forms where the option still exists. Food may be scarce, conditions may be difficult, finding a mate can be troublesome, but disease can be a killer, of individuals, of populations, of species. It is therefore reasonable that a system as fundamental as sexual reproduction should be linked with something as lethal as disease. Without the trait of sickle-cell the disease of malaria would be even more hazardous to humankind. The same goes for thalassaemia, another hereditary anomaly of haemoglobin. Known after the Greek for sea, because the feature is so prevalent among Mediterranean islanders, it too is beneficial among human heterozygotes against malaria. (The chapter on genetics gives more information about both these diseases.)

Computers have recently come to the aid of sexual theoreticians. Bill Hamilton, of Oxford, a pioneer thinker in this field, has also been a pioneer in computer simulation. He has constructed model populations, possessing both sexual and asexual individuals, into which he introduces random death. Time after time the asexuals then win. Next he introduces parasites into the equation, a few more virulent than others, to infect their hosts, some of whom are more susceptible to parasitism than others. Death is then no longer random, it affecting more often the least virulent parasites and the less resistant hosts. No longer does asexual reproduction win the game, and it wins least often when most genes are involved, both for parasitic virulence and for host resistance.

Sex is so widespread – 'Birds do it, bees do it, even educated fleas do it,' as Noël Coward wrote and Eartha Kitt later sang so beguilingly – that it is hard to envisage a world without boys and girls, men and women, male and female. Human existence without the convolutions linked to mating is also hard to visualize, being as much part of us as legs, arms and brain. The reason for sexual reproduction can often seem of minor consequence, relative to the fascinating manifestations, all of which relate to the parity, more or less, between the sexes.

SEX RATIO In Britain about 105 boys are born for every 100 girls, but the proportion oscillates. It varies as the years pass – the gap was widening before the 1960s but has since been narrowing – and according to the seasons of each year. The figures are nearer 104:100 in November and 106:100 in May. American figures match those of Europe for ex-Europeans, but are nearer 102:100 for those ex-Africa. Some countries have a customarily high ratio, such as Greece and Korea which can be 113:100. Others are usually low, like Cuba at 101:100.

As males die more speedily all along the line, whether as miscarriages, still births, infants or children, it is easy to presume that the ratio has somehow been engineered to create parity when it is critical – at mating time. In Britain such evenness is currently reached by the age of 30. In the US it is nearer 50. In both countries it then alters dramatically, with men suddenly dying much faster than women. By the age of 95 every man is outnumbered by four women. From conception time, when the ratio is thought to be 150:100, to extreme old age when it is 20:100, males steadfastly lose their numerical superiority.

There are strangenesses among these general rules. In France and Britain, following the slaughter of both world wars, there was an increase in boy births, notably when conceived by soldiers. Divers, test pilots, clergymen, anaesthetists, older fathers, unskilled and manual worker fathers, women with schizophrenia or hepatitis A, and parents born either illegitimate or with older brothers and sisters are slightly more likely to have girls. ('It takes a man to make a girl,' say Cockneys.) The offspring of black/white matings, of AB blood group mothers, of irradiated fathers (from Hiroshima or Nagasaki), of mothers aged 50 or more, of transport aircraft pilots, of first-born individuals, and of O blood group parents are slightly more likely to be boys. Among the Havasupai Indians (of Arizona) the first two children in each family are boys more often than not (130:100). Fifth or later children are more likely to be girls (100:130). There is no evidence of infanticide among these people (whose births have been noted since 1868). Therefore something is happening to affect the boy–girl ratio.

People of high social status create more sons than do their opposites. American presidents have produced 90 sons as against 61 daughters (Chelsea Clinton being the 61st). National *Who's Who*s also emphasize this trend, with listed Britons having 1,789 sons and 1,522 daughters, listed Germans 1,473 sons and 1,294 daughters, and equivalent Americans 1,180 sons and 1,064 daughters. There is argument that daughters are under-reported, but Ulrich Mueller (of Mannheim, Germany) who compiled the figures considers this improbable. (He made comparisons with families possessing famous daughters to reinforce his convictions.) Mueller also discovered that 59% of children born to 'eminent British industrialists' between 1789 and 1925 – self-made men in general – were boys. The children of such an élite, who also appeared in *Who's Who* (after it first appeared in 1848) and who were therefore worthy in their way (possibly thanks to inherited wealth or position), were much less capable of siring boys, their fathers' 59% dropping to 48%.

As to the mechanisms involved in such discrepancies, there is less certainty. Robert Trivers and Dan Willard, of Harvard University, believe sex ratios have evolutionary significance. Parents in poor condition (unlike US presidents, for example) are more likely to have females because females can always find a mate (or be mated) and thus strengthen the population. Males in good condition (like those industrialists) can have many strong babies from many mates (although not within the family strictures and monogamy of industrial Britain). Well-fed opossums, given extra sardines by Stephen Austad and Mel Sunquist (also of Harvard), produced 1.4 males for each female, a ratio which dropped to 1:1 when the rich fish diet was stopped. To equate well-fed opossums with US presidents is possibly stretching the point, but there does seem to be a relationship between prosperity and males. Even if that turns out to be false, there is growing conviction that sex-ratio is not random, whether for animals or humans. It is not the spinning of an evenly balanced coin which can, if spun 100 times, give heads and tails differently from the 50:50 of expectation, such as 47:53. Something is happening other than the vagaries of chance.

The certain fact at the basis of any controlling mechanism for humans is that sperm are either Y-bearing or X-bearing, causing respectively either boys or girls to be conceived. This X or Y system is the same for all mammals. Strangely, and in birds, the situation is reversed, with Y causing females and X causing males. Y-sperm are said to swim more vigorously (being likened to hares rather than the X-sperm tortoises). If the egg is ready to be fertilized the hares will reach it first. If not ready the hares may burn themselves out, and the more laggardly tortoises can triumph by their longer survival time.

If sex is frequent, as with those Arizona Indians (one can presume) at the start of their married lives, an egg is more likely to encounter fresh sperm (and the Y-bearers are more likely to succeed). When the marriage is more casual, and later children are being conceived, the more relaxed X-bearers may have their turn. (Quite what this says about test pilots as against transport pilots, and clergymen as against irradiated Japanese, can be put – as the phrase goes – on hold. Quite what it also says about the British is confusing, as the British sex ratio does not change, either with birth order or with age. Contraception may play its part, often being determinedly used by newly-weds whose lives would be disrupted by a pregnancy. When contraception is less strict, and couples are thinking that a child might be welcome, the frequency of intercourse may also have subsided.)

Another theory states that hormones are the controlling influence. William James, of University College London, believes that both testosterone (the male hormone) and oestrogen (the female equivalent) encourage the creation of boys. It is the gonadotropins, formed in the pituitary gland and acting upon both kinds of gonad, that encourage girl conceptions, perhaps by affecting the cervical mucus, or perhaps by restraining the Y-sperm. Some firm evidence has surfaced following the injection of gonadotropins into previously infertile women: they produced more daughters. Conversely testosterone injected into previously infertile men produced more sons.

An extra fact, also supporting the hormonal theory, is that dominant women (as defined by personality tests) give birth to more boys, and the less dominant to more daughters. Plainly such assertiveness, or lack of it, needs to be correlated with the hormone levels of these women. In any case the old belief that nature relies upon something as arbitrary as chance for something as critical as male and female numbers is less popular these days, even if there is still uncertainty as to the method used to ensure an appropriate balance between the sexes.

Children of either sex, as is well known, insist on saying they are 4 years old, or 8, or 12 right up to the very day when they switch to 5, 9 or 13. Schools do likewise, taking the date of birth as sacrosanct. If two children – infant buddies one day apart in age – were born on 31 August and 1 September they may be separated – in Britain – into two distinct 'years' at school. This is allegedly the reason why top British footballers are more likely to have been born in the last four months of the year and least likely to have been born in the four principal summer months. In a recent survey over 260 of these sporting heroes had been born in September or October but less than 120 in May, June and July. The difference, first demonstrated by a Yorkshire insurance executive named Les Gudgeon, is most marked.

There is no neat explanation, save that boys born in August will be playing at school with boys older by 1–11 months. Those born in September will be older (and therefore probably bigger, stronger, faster) than their immediate competitors by that same margin of 1–11 months. Children tend to be most enthusiastic about activities at which they are best, and least likely to enjoy what they perform least well. Hence – so this argument runs – early disillusionment with football can lead to withdrawal, and early success may lead to greater enthusiasm.

That enthusiasm will be maintained as, year by year, the autumnal-born boys shine over the summer-births. When adulthood arrives, and birthdays are forgotten, enthusiasm will have been well engrained (or not) and Gudgeon will find his figures reinforced by a further crop of (autumnal-born) youngsters joining the ranks of the élite.

Publication of these football figures prompted some Portuguese, from the University of Porto, to examine the records of medical school entrants, always selected from 'the most successful high school students'. The asymmetry was most conspicuous when student birth-dates for each quarter of the year were totalled, being 53, 94, 58 and 58 for the four quarters. No reason was given for the disparity (with April, May and June births being superior), save that seasons are not uniform – in temperate Portugal – and neuro-biological development may gain or lose accordingly.

The Portuguese figures, in their turn, prompted the Faculty of Medicine in Florence, Italy, to examine its own records. Sure enough these mirrored those from Porto, with successful (in that they had gained admission) students for the four yearly quarters being 221, 296, 222 and 218. These birth-date differences are most marked, and continue to be so even when the slightly different birth-rates per quarter for all Italians are taken into consideration. December tends to have fewest births, and the Florentines recommend that some similar school in the southern hemisphere examines its own records to see whether there are parallels, perhaps six months out of phase with their northern colleagues.

Michael Holmes (in *Nature* in 1995) drew attention to some further strangenesses. Of the nineteen physicists who took an early position on relativity 80% were born in the winter months (of October to April). Similarly, of the twenty-eight biologists who took an early position on evolution, 92% were born in those months. As for US presidents, 100% of those who held office in the eighteenth century were born in wintertime, 66% in the nineteenth century, and 44% in the twentieth century. This parallels, according to Cesare

SEX

Emiliani, of the University of Miami, the change of the United States from a revolutionary society to a bastion of traditional values. Most appropriately he then concluded: 'All of the above is of course a paean to the glory of small samples, which are wonderfully user-friendly in offering strong support to untenable theories.'

INFANTICIDE This procedure inevitably swamps any relatively minor differences there may be in the human sex ratio. In some areas of China, India and Pakistan the variance in adult population numbers may be five males for every female, according to Olivia P. Judson of Oxford University. Not only are girls killed at birth but, these days, prenatal sex determination and sex-selective abortion can increase boy births. Infanticide is assumed to have been practised in many cultures over many centuries (and even certain animal parents favour one sex over the other) but today's human opportunities for sex selection are greater. There is no longer such need for neglect, abandonment or even murder of the less desirable sex. For wealthier citizens the modern diagnostic techniques may be frowned on in many communities as a means of sexual selection, but not where the birth of a girl is traditionally accompanied by tears.

In former times girls may have been more favoured, but boys are now preferred, outstandingly in China with its policy of one child per family. In Ethiopia, during the great hunger of the early 1980s, feeding agencies assisted only those families where at least one member was desperately in need. Parents would often arrive at the food centres, looking thin themselves but with at least one miserably emaciated girl among their children who served as ticket for extra rations. A report in *Nature*, based on figures supplied by a Bombay social worker, told of a sample of 8,000 abortions at Indian clinics of which 7,997 involved female foetuses. (The use of sex determination techniques, such as amniocentesis, to identity female foetuses for abortion became illegal in India as from 1 January 1996. Punishment for transgressors was set at 10,000 rupees and up to three years in prison.) A psychologist from the University of Colorado estimated that sex-biased infanticide 'characterized some 9 per cent of the world's cultures', and 'more often than not' the unwanted sex has been female.

GENDER CHOICE Infanticide has not been practised solely on wrong-sex offspring. It was also used in various cultures following unusual births, such as breech deliveries, twins or even of babies born with much hair. Few steps could be taken to prevent such variations, but many suggestions were made, and believed, concerning gender. Timing, location (of intercourse) and position used were all considered relevant. Even Marie Stopes, correctly hailed as pioneer in birth control, recommended that marital beds should be aligned with the Earth's magnetic field if boy births were required. Long before pre-birth sex determination became feasible or easy (as today) there were arguments not only against such a practice but in favour. H. G. Wells, the forecaster and novelist, half-welcomed choice as an effective counter-threat to militant feminism: 'The modern world has no room for sexual heroines,' he wrote in 1914. 'If woman is too much for us we will reduce her to a minority.'

One decade later gender treatment centres were starting to flourish. Norman Haire operated a nursing home in East Sussex. This provided a wholesome diet for its patients as well as placebo pills coloured pink or blue. A money-back guarantee was provided for those couples where the undesired sex was born. If the desired sex arrived the fee was doubled, and the happy parents trumpeted the clinic's success to all who would hear. (Simple arithmetic shows the financial wizardry behind this deal.) Later, and in the 1930s, a professor of obstetrics (from one of London's teaching hospitals) was a touch more cunning. He listened to the foetal heart-beat, pronounced upon its owner's sex, and then wrote down the opposite sex in his diary. If his pronouncement proved to be correct everyone was happy. If not, and if angry parents arrived at his surgery, he would look up his diary, point to his alleged diagnosis, and chastise the parents for their faulty memory.

When the 1950s arrived, along with human artificial insemination, there also arrived the first theoretically plausible means of scientific sex selection. Female X-carrying spermatozoa are heavier than the Y-carriers. Therefore attempts were made to centrifuge and separate the two kinds of sperm, a system well tried with the semen of cattle (providing an 80% success rate). Unfortunately, for those desiring such an effective means of sex determination, human semen was not so obliging. Not until the late 1960s, when more liberal abortion laws came into force (as in the UK), was a truly practical method of gender choice available via amniocentesis. The new legislation allowed for a broad interpretation of 'social' factors acceptable for abortion, but a preferred sex was not one of them.

Nevertheless mothers (and fathers) can be a determined lot. They can learn of their foetus's sex at one clinic, nod approvingly at the staff's refusal to permit abortion, and then hurry off to another clinic having thought up the necessary social grounds for a termination. For a time within the United States, notably in 1987, there were 'Gender Choice Child Selection Kits' retailing at a mere $49.95. These contained instructions, thermometers, and equipment for measuring vaginal mucus, all to determine the best moment for conceiving a male or a female. Packets were appropriately coloured in pink and blue. The Food and Drug Administration did not much like the implied claims, and the kits speedily left the shelves. (The FDA is a colossal organization, regulating about 25 cents of every dollar spent by the American consumer, or $1 trillion worth of goods and services annually.)

In 1994, founded by the biochemist Peter Liu and with Alan Rose as medical adviser, the London Gender Clinic Ltd opened for business at Hendon. Considerable publicity ensued when Gillian and Neil Clark, having paid £650 for treatment in the hope of acquiring a sister for their two sons, produced Sophie May Clark apparently to order. The technique used for gender selection had been pioneered and licensed by Ronald Ericsson, American scientist and entrepreneur. London's clinic was then the only one of its kind in Britain, but there were fifty-five operating under licence within the United States. As reported in *Nature* in 1973, Ericsson is thought to use a system which enriches semen with the desired kind of sperm, Y sperm for boys and X for girls. Male sperm, being slightly lighter, are believed to be able to swim through a solution of the protein albumin more proficiently than heavier female sperm. The clinic claims it has had a 70–80% success rate, higher than the 50% rate (almost) it might expect if it did nothing whatsoever.

Australian researchers, led by Sean Flaherty and Huai-Xiu Wang, reported in *Fertility and Sterility* (in June 1994) that they had tried to repeat Ericsson's experience but had failed even after using two different albumin methods. Indeed, although trying for male sperm, they achieved slightly more female sperm but not at a level likely to be of practical significance to prospective parents (or to science). London's clinic reported that, of the first 200 couples to be treated, most had been Asians wanting male children whereas most of the British couples (like the Clarks) had wanted girls. Not everyone is in favour of such clinics and of the service they aim to provide. The UK's Royal College of Nursing voted (in 1994) by almost two to one that the government should curb 'commercial gender selection clinics', part of the reason being that 'often gullible couples are being persuaded to part with very large sums of money for marginally increased chances'.

This Royal College was not in existence when couples, perhaps even more gullible, were instructed to achieve a boy by having 'marital relations' with the pillows at the foot of the bed and the husband keeping his cap on, or by conceiving under a tree, or after the woman had anointed her 'privities' with the 'juice and seed of male mercury [which has] round seeds hanging in pairs', all such advice having been tendered in the past. Boys do seem to have been the favoured sex in most societies for several centuries, whether or not the sex ratio could be influenced. Without doubt that influence will increase, almost certainly beyond the claimed 70–80% rate. Scientific advance will inevitably make gender choice quite feasible, and the days of becapped gentlemen enjoying well anointed privities at the foot of the bed will finally be concluded. Or, if such traditions are preferred, they will not be primarily intended for the acquisition of a male.

*

Should the disproportion of five boys for every girl, as in China and Pakistan, become more widespread, should the business of dowries continue to be such a penalty upon those with daughters, and should boys remain favoured, as providers for the elderly, or as status symbols, or as traditionally welcomed, there will have to come a time when the pendulum swings the other way. Girls do provide the only mechanism for the birth of boys. They are the only means whereby all those boys – when turned into men – can acquire male offspring. The dowry offerings may then be reversed, with young men providing bounty to gain their share of the limited supply. A computer could provide some of the answers concerning reversal, but a machine can never predict when the wish for a boy, so deeply rooted, will be supplanted by the contrary wish for a girl.

SEX APPEAL Interest may be intense but solid information on this subject is in short supply. More scientific investigation has probably been dedicated to the pancreas or spleen than to the sexual attraction of one human being for another. 'Our eyes met across a crowded room,' write novelists (when not averse to clichés), but what does the remark entail? Why those two pairs of eyes? And how can it be that one mutual glance will lead to a bonding for life? Everyone else in the room may be unaware that something akin to a couple of assegais has just hurtled past them, and will then be astounded that she (of all people) is suddenly going out with him (of all people). We each know the scene, in its many manifestations, and can still find it both fascinating and inexplicable.

Martin Johnson, expert on human sexual habits, wrote that 'few studies on human sexuality would qualify as good science, especially when compared with contributions from sociology, history, psycho-analysis, anthropology, literature study, ethics and even theology'. He agreed there were reasons for such lack. There was 'the private and personal nature of sexual expression'. Moreover 'the complexity and diversity of sexuality, which can vary with age, situation and experience, is daunting for science'. Many stimuli are involved – gender clothing, other inanimate objects, particular body parts, specific situations, power, danger, appearance, behaviour. As Johnson added, providing further reason for science's reticence, 'our responses to these stimuli are also highly individual'. Science prefers simpler situations than those in which humans meet, are attracted and then form lasting partnerships. (Or maybe not so lasting. Animals which pair, or bond, for life tend to retain their unity, with only death intervening to make the survivor seek another mate.)

The Economist frequently surprises its readers (and perhaps itself) by reporting on topics not expected to be within its pages, such as sex. In October 1994, under the innocuous heading 'Not much of it about', it discussed sexual proclivity. 'The America we thought we knew from previous sex-surveys – a heaving, panting maelstrom of adultery, experimentation, pregnant cheerleaders and rampant bedroom politics – turns out to be a quiet, almost tired place.' It was quoting from *Sex in America: A Definitive Survey*, recently published, allegedly the first sexual survey based on a truly random sample of people, rather than on boastful volunteers. 'About once a night,' revealed *The Economist*, 'all over the country, as night falls, faithfully monogamous couples – same race, different gender – indulge in the most ordinary form of sex imaginable, and then fall asleep.' The survey had also revealed that 20% of men (and 30% of women) had had only one partner since the age of 18, 21% of men (and 32% of women) had had 2–4 partners, 22% of men (and 20% of women) had had 5–10 partners, 17% of men (and 5% of women) had had 10–20 partners, and 17% of men (and 3% of women) had had 21 or more. Such percentages do not add up to 100 because a number of individuals had had no partners since the age of 18. In short, not much of a maelstrom, not much rampancy; in fact, not much of it about.

There is a nature versus nurture divide in sex appeal (as in so many aspects of human development). To what extent is attraction genetic? And to how great a degree are we attracted because our particular society has infused in us a distinct form of desirable partner? On this aspect Martin Johnson had further points to make. 'With so much variety (of attraction, stimuli and response) is a large genetic component likely? Is it not more likely that we construct our sexuality socially?' (The matter of homosexuality is much involved in such questions. It is confusing that attraction for the same gender might be inherited, but the possibility cannot be denied.)

Modern societies, in all the different lands, have strong notions concerning appeal, even if hard to define. Youth, health, maturity are all important, but older men – via power/wealth – can also be attractive. Extreme youth, as at puberty, is not necessarily appealing. In areas other than sexuality, according to various studies, the good-looking are judged to be nicer, their work is more appreciated, they are more likely to be offered a job and, even if they commit a felony, are more likely to receive a lighter sentence. Such judgement begins early. Even 3-year-olds seem to prefer pretty children (as adults judge prettiness) and the better-looking are less likely to be blamed for misbehaviour. Research from the University of Connecticut dis-covered that women 'hospitalized for psychiatric disorders' were rated by men as less attractive than a random selection of other women. One easy explanation is that women with psychiatric prob-lems often look as if they have problems and take less care of their appearance, but school photographs can help to contradict this assumption. Children who subsequently become psychiatric patients are thought to be less attractive. Perhaps, as further explanation, the very fact of their unattractiveness is part of their undoing.

When confronted by such suppositions, and by their counter-suppositions, it becomes easier to understand why science, which likes to compare one single variable with another single variable, has been so poorly applied to the merits of appearance, whether or not in a sexual context. Somerset Maugham, the writer (and homosexual), is worth quoting on many a subject: 'Human beauty is determined by sexual attractiveness. It is an intensification of traits common to a certain people at a certain time, but a slight one, for too great a departure from the normal excites aversion rather than admiration. Sexually the aim of both men and women is to distinguish themselves from others and thus call attention to themselves. This they do by accentuating the characteristics of their race. So the Chinese compress their naturally small feet and the Europeans constrict their naturally slim waists. And when the characteristics of a people change, their ideal of beauty changes too. English women have added to their stature during the last hundred years; the heroines of the older novels were far from tall, and literature had to wait for Tennyson to learn that inches add to beauty.'

Without doubt society does influence preference, particularly when alleged gods and goddesses are portrayed so relentlessly. Even in the ancient past similar role-models probably existed, perhaps as partners of a chief. As for national, or geographical, preferences, these certainly differ. I was personally struck, a long time ago, when having to choose cover-girls in Africa. Of the several international offerings my opinion was formed very speedily. (To say that my mind was made up is possibly erroneous, as the decision-making process is scarcely intellectual.) An African friend would regularly disagree with my choice. We tested this disagreement by co-opting other Africans from the street and they, without exception, agreed always with my friend. Skin colour was not the issue, there being many shades among the photographs. Something else (or many somethings else) were serving as guide for choice.

Both Africans and Europeans would probably disagree with choices made in New Guinea, such as those of the Foré tribe (examined by anthropologist Jared Diamond). Its members explained to him that:

The most beautiful women in the world are Foré women; they have gorgeous black skin, thick dark frizzy hair, full lips, broad noses, small eyes, a nice smell, and perfectly shaped breasts and nipples. Women of other New Guinea tribes are less attractive, and white women are unspeakably hideous. Just compare your women with our women to see why – white skin like sick albinos, straight hair like strings, sometimes even hair coloured yellow like dead grass or red like a poisonous snail, thick lips and narrow noses like axe blades, big eyes like cows, a repulsive smell when they sweat, and breasts and nipples of the wrong shape. If you want to buy a wife, find a Foré if you want someone beautiful . . .

Facial appeal is important to humans and, so it is thought, has been for millions of years. A study of the herpes virus, with two forms in humans but only one in monkeys, led Glenn Gentry and others of the University of Mississippi to believe that face-to-face copulation began 'at least' 10 million years ago. The big apes mainly copulate front to back and the hominid alteration certainly brought the face into sharper focus. Nevertheless the beauty of a face is extremely indefinable, certainly by scientists who long for definition. *New Scientist* once published a lengthy correspondence which concluded with a yardstick (but no means of measurement). If Helen of Troy's beauty could launch one thousand ships a millihelen had sufficient excellence to launch just one. As for a microhelen, accredited to those singularly bereft, it was sufficient to launch a plank.

Such flippancy underlies a real problem. Beauty, they say (with justification), lies in the eye of the beholder. There may be paragons in the vicinity but these may be socially out of reach. An attainable goal may therefore seem more desirable, notably to those equally short of the correct attributes. (A look at the photographic section portraying recent marriages in the local newspaper will show the truth of this assertion. Those who score −2 on a scale of 1 to 10 beam happily beside consorts similarly endowed.)

There does not seem to be a relationship between attractiveness and progeny. Numbers of children per woman vary but not in proportion to those women's mental and physical excellence. Such a situation may also have existed in the cave or wherever our habits were first engrained. There must have been paragons, better than the great majority, but no reason exists to believe their pregnancies were commoner than with other sections of the community. It is easiest to believe that all women, save for restraints imposed by lactation, ovulation or old age, were pregnant for all of their adult lives. Helens and microhelens were equally enshrined.

Males, as is known, have a tendency towards polygyny. Females, as is also known, favour males who will provide the necessary support during both pregnancy and motherhood. It is highly probable that choice was always more in their feminine domain. Strong, youthful and powerful males could provide security for them and their dependants. If women were indeed the arbiters of choice their preferences would have become paramount. Their notion of a good-looking male, embodying desirable features, would then have become rooted – over time – in their offspring. If such females wished for relatively hairless males the children of such unions would also, whether male or female, become less hairy. If they wanted men of gentler disposition, or greater belligerence, or taller or shorter, those qualities would also become more prominent in their offspring. Most importantly, however much all females might be pregnant and fulfilling their reproductive capability, there was no such certainty of sexual fulfilment among the males. Many a woman might favour one compliant man, and many another man might therefore fail to find a partner throughout his adult life. Charles Darwin (and many others) saw men as the sexual selectors. By no means is that earlier opinion so warmly held today.

Sex appeal is, as Johnson said, both private and personal. Therefore private and personal reflections may be (briefly) permissible. I do find it strange that Girl A, well made, a credit to her parents, and with eyes, lashes, lips and ears as well formed as the rest of her, is most uninteresting whereas Girl B, also well made, equally credit-worthy and with eyes, lashes and all else wonderfully made, doubles my heart-beat upon the instant. Having read treatises by Desmond Morris I can understand that breasts, lips, smooth skin, sleek hair and clear eyes are telling me of fitness as a sexual partner for reproduction, but Girl A can appear no less fit than Girl B. And what of ankles? They too can show fitness, and therefore desirability, but one pair of ankles striding ahead of me might as well be logs for all I care. Her friend's ankles, walking with an equal pace, are exquisite by compari-son. The pairs are chalk and cheese, one as exciting as – well – chalk but the other whetting appetite most enticingly.

Before leaving this reflection, with its admissions mounting apace, I find that trees, for example, are more like ankles than does seem reasonable. One tree can have most excellent form, being beautifully proportioned. Its neighbour, also well made and without deficiency of any kind, is much less attractive or is even ugly. So where does aesthetics spring from and why is assessment, so easily made, almost instantaneous? In regarding lips, eyes, ankles or branches of a tree I do not consider the act an intellectual process. My conclusions surface without comprehension of any kind. I just know at once which tree I prefer. So too, most emphatically, with Girl A and Girl B. One is paragon; the other is nothing of the kind.

Greto Garbo, considered excellent by many, had big feet, a down-turned mouth and little laughter. Her expression, described in *Nova*, was of 'an aching sadness' and 'her looks were not pretty'. To be unconventional in appearance does seem advantageous. Francis Bacon considered, four centuries ago, that 'there is no excellent beauty which hath not some strangeness in the proportion'. Perhaps exceptional beauty is irrelevant to most pairings. We are, most of us, more pragmatic in our approach and prepared to adapt. 'Gentlemen prefer blondes,' as the saying goes, 'but marry brunettes' – if only because there are more of them. 'You're not my type,' said the girl sitting solo at a bar to a similarly solitary male in a memorable cartoon. 'But why not come back in an hour's time?'

Most of either sex do not understand their compulsions. They may use words like generosity, sincerity, intelligence or thoughtfulness in describing a consort's appeal, but are attempting to rationalize a form of preconviction. The same individuals may also like beef or sunsets, the chill of winter or dappled light reflecting on a boat, but feel no need to explain or describe these other attractions. To suggest that Cupid is to blame for sex appeal, a winged archer hitting targets as he chooses, is perhaps not wide of the mark, alternative reasons being so inadequate as to why A should suddenly love B, and B love A, or (much to every novelist's delight) some variant on this general theme, such as the role of C.

H. G. Wells, dazzling in many areas, was also famously successful in the sexual arena. Photographs of the man can make us wonder at his achievements, in that his body and general demeanour were as unlike Apollo as an ungainly man in a crumpled suit could possibly contrive. I once asked a lady, whose own accomplishments had been considerable, what was so appealing about the great H. G. A distant look entered her eyes, a soft look, a memory of days long passed. After sighing, she said he smelt of honey. The cause of this aroma, to be less ethereal about the matter, may have been his diabetes (with sugar excreted on the skin) but the role of smell should never be casually disregarded. A group of women (in Switzerland) were once

asked which male-worn T-shirts most appealed. They preferred those whose owners had immune system genes which did not match their own. Curiously, as an addendum to this experiment, the women taking oral contraceptives were most attracted to the shirts whose owners had similar genes.

Assortative mating definitely exists, the tendency for husband and wife to be alike for certain characteristics. In one Spanish study 'spousal similarity' was found for occupation, educational level and, rather more surprisingly, number of siblings in the families each partner came from. There was also a correlation between total height, ileospinal (backbone) height and total arm length. Mates also converged in fatness during marriage, which was either a 'cohabitational effect on spouse resemblance' or another form of similarity, a proneness to fatness. Other investigations have shown a preference by the blue-eyed to partner with blue-eyeds, and for red-heads to behave likewise. It has been suggested that such choosing reflects a wish for males to marry their mothers – or even (let it be said) for them to marry themselves.

One major matter, in this section on Appeal, needs to be emphasized. Unlike food, most of which is agreeable, there is no such unanimity of enthusiasm for most of the opposite sex: there is frequently revulsion. Authorities of zoos often wonder, having put one male and one female within a cage, why 'nothing happened'. Have they never considered that plight for themselves? And have they never thought, when contemplating desire in a broader context, how odd it is that sexual attraction is so specific? Eyes *do* meet across a crowded room, just two pairs of eyes, while everyone else is unaware that a bond has been initiated, possibly for life. And a good many of those others, if not all, might be appalled by the notion of such a bond for themselves, either with her (perish the thought) or with him (heavens above!).

A final point, raised in *Nature* by two zoologists from the University of Stockholm, brought nature, art and appeal most conveniently together. 'It is striking that many signals used for communication by organisms are judged to be beautiful by humans. Examples include the colours and symmetries of flowers, the patterns on butterflies' wings and coral reef fish, and the elaborate courtship displays of birds. The almost universal appeal of such signals to humans is surprising ... This raises the possibility that human aesthetic sense is based on general principles of perception that have been important during the evolution of biological signals.'

In short, sexual appeal is as old as sex, and the rules for humans have been handed down through many million years. Small wonder, therefore, they are so incomprehensible when, out of the blue, they suddenly hit home.

SEX EDUCATION is both more necessary these days, and also less so. The arrival of AIDS has boosted the need for greater protection and awareness. So too the lack of inhibition relative to old days concerning sexual experiment and experience among the young. It is harder in modern times to be as unaware of sex as people used to be. Films and television are becoming more explicit every year. Schools teach reproduction under the general heading of biology. (My 9-year-old, in filling in a form, put 'Not yet' in the heading marked Sex.) Books are plentiful, and there is no vacuum like that entered by Marie Stopes when her *Married Love* hit the stands in 1918. A recent list of titles, all aimed at children, include: *Sex: How? Why? What?*; *The Family Guide to Sex and Intimacy*; *Understanding the Facts of Life*; *What's Happening to My Body?*; *Where Do Babies Come From?*; *We're Going to Have a Baby*; *How Sex Works*; and *Mummy Laid an Egg*. There are advertisements for condoms, in the cinema, on TV, in newspapers. The old stories of wedding night confusion, with neither partner knowledgeable or competent (let alone happy), are undoubtedly less probable these days. According to a British survey 'fewer than 1% of men and women then aged 16–24 were married at the time of their first sexual intercourse'. The average age of that first intercourse has plummeted in the past forty years, from 21 to 17 for women/girls and from 20 to 17 for men/boys.

'Should one try to impose a measure of sexual restraint on exuberant youth?' asked the *Lancet*'s leading editorial (in October

1994): 'The answer seems to be "Yes".' More than half the females who had had first intercourse before 16 thought it had been 'too soon'; a quarter of the males thought so. The median age proposed as 'about right' was 19 and 17 respectively. Despite parents being more willing to broach the subject than ever before, and despite schools providing more sex education than ever before ('Biology' can seem to concentrate almost exclusively on this topic), most people learn facts of life from their peers rather than parents or teachers, according to the *Lancet*. Its editorial stressed that 'girls in particular need to appreciate that they can catch a sexually transmitted disease 365 days a year but can only become pregnant 60 days a year'.

The double threat of disease and pregnancy has still not penetrated sufficiently into the minds of those most at risk, according to the organizers of the largest survey ever undertaken in Europe into the sex lives of 15 to 18-year-olds. CNRS, France's national research agency, documented the lives of 6,182 French teenagers to discover how often youngsters indulged in high-risk sexual activity. The interviewees were either attending state high schools, or private schools, or had left school to join state-run job training programmes.

Those with school behind them were more ignorant about HIV than those at school, with twice as many in this group believing the virus could be transmitted by mosquitoes or from toilet seats. More than half the total were sexually active, namely 75% of the 18-year-olds and 30% of 15-year-olds. A tenth took no precautions the first time they had sex, but 75% used a condom. The girls moved on to oral contraceptives when sex became more regular, and half said their partner had not used a condom on the last occasion they had had sex (with pregnancy apparently more worrying, or simpler to circumvent, than the possibility of infection). About 40% of the males and 27% of the females had had sex with more than one partner in the year preceding the survey, with 12% of males and 5% of females having had five partners or more. (An AIDS survey came through a friend's letter-box one day, asking him to state – anonymously – how many partners he had had in the previous 5 years. Was he in the category 1–10, 10–100, 100–1,000, 1,000–10,000, or over 10,000? Having considered himself quite a Lothario he was appalled by apparent negligence, and threw the form away.)

Sex education is almost universally condoned, but a sex manual published by Britain's state-sponsored Health Education Authority in the spring of 1994 was almost universally condemned. Following its publication, at £3.99, the product was withdrawn from shops on the orders of Brian Mawhinney, then Health Minister. The *Daily Mail*, among other newspapers, was adamant in its criticism: 'In the 140 pages of *Your Pocket Guide to Sex* there is more crudity, foul language and obscenity than most parents would wish their children to learn in a lifetime – let alone at the age of 16.' The book was reputedly aimed at 16- to 25-year-olds but there was no official age restriction on its purchase. Among those on the HEA's board was the Bishop of Peterborough, a man particularly incensed by the book's publication: 'If a fellow wants to write that, and someone's prepared to publish it ... well, it's a free society. But when it goes out over the name of people like ourselves, we feel very betrayed.'

Plainly, as all agreed, there is sex education and sex education. The *Pocket Guide*, with its cover picture of an angel/cherub riding on a condom, had affronted many, particularly in its chosen quotes, as from a 21-year-old male: 'If you've never licked chocolate mousse off your girl-friend's nipples, you don't know what "sexy" means.' Many a 16-year-old, if not most, did not even know what sex meant not so long ago. Everything has since changed, not necessarily for the better in every aspect, or so it is easy to believe.

SEXUAL DISEASE In these days of initials the individuals who know of AIDS and HIV are much more numerous than those acquainted with STD. Yet Sexually Transmitted Diseases are more plentiful than the Acquired Immuno-Deficiency Syndrome, so linked with the Human Immuno-deficiency Virus. There are more than 250 million new STD cases every year as against 1 million new HIV infections. In parts of the developing world 10–20% of women are infected with an STD. The most widely known, apart from AIDS, are gonorrhoea and syphilis but there are twenty others. According to the WHO about 685,000 people are infected every day with some form of STD. This incidence therefore almost rivals the prevalence of malaria.

When Anne, Britain's Princess Royal, declared that the arrival of AIDS was an 'own goal' she was soundly chastised for the remark, particularly as the disease can assault unborn children. Nevertheless her intent was clear. AIDS would not be such an epidemic if people were less casual in their sexual relationships, if most individuals restricted themselves to a single partner. The same is true for all STDs. They are, as their name indicates, transmitted during sexual intercourse. If people were less profligate there would be less disease. There is, as the Princess Royal might have added, a rough justice to such licentiousness. A major pleasure is partnered by an extremely unpleasant crop of diseases. To abstain is to be free of such infection. To participate is to be at risk.

STDs are a mixed group, involving protozoa (single-celled animals), bacteria and viruses. Those caused by bacteria include gonorrhoea, chlamydia, syphilis (allegedly from the New World), and chancroid. Trichomoniasis, the commonest STD by far, is caused by a protozoan. HIV, as its name states, is caused by a virus; so too are HSV (herpes simplex virus), HPV (human papillomavirus), HBV (hepatitis B virus), and CMV (cytomegalovirus). The bacterial and protozoal STDs are usually curable with antibiotics; viral STDs are not. The viral diseases are now more feared in developed nations, having overtaken the anxiety previously held for gonorrhoea and syphilis. (Films shown to novice servicemen during the Second World War, apprising them of sexual hazards, were sufficiently vivid and distressing almost to render such recruits impotent for the remainder of their days. The ET room, for early treatment with various unguents, was well used by those returning to camp after forays beyond its wire.)

Worldwide the number of new cases of trichomoniasis is 120 million per year. For genital chlamydia it is 50 million, for genital papillomavirus 30 million, for gonorrhoea 25 million, for genital herpes 20 million, for syphilis 3 million, for chancroid about 2 million, and for HIV about 1 million. As with virtually every public health problem the situation is worse in the developing countries. In some the prevalence of gonorrhoea is as high as 18% (of women attending ante-natal, family planning, or gynaecological clinics), of syphilis up to 17%, and of trichomoniasis up to 30%. The differential between developed and undeveloped nations may be as much as 100 times for syphilis, 10–15 times for gonorrhoea, and 2–3 times for chlamydia. In large African cities (with Africa as most affected continent for STDs) the acquisition rate for gonorrhoea per year may be 3,000–10,000 for every 100,000 inhabitants. For the United States the rate (in 1991) was 233 per 100,000, and in Sweden about 30 per 100,000. Youth is always vulnerable, whether in developed or undeveloped worlds.

Women bear a greater burden from STDs than do men. Chlamydia, as already mentioned, is less likely to show symptoms in women, and most of the common bacterial STDs are asymptomatic with them. Gonorrhoea, for example, shows no symptoms in 1–5% of men but in 30–80% of women (according to different surveys). Consequently the presence of the disease will only be discovered when women present themselves for specific screening. As further complication STDs are harder to diagnose in women and, yet more unfair, the risk of transmission of several STDs to women is higher. Many STDs frequently result in awesome reproductive outcomes, such as (according to one American report) 'spontaneous abortion, premature labor, fetal infection, or neonatal infection'.

What is not known is the real extent of STDs, and for several reasons. The bearer of such a disease may not possess symptoms. With chlamydia about two-thirds of women and one-third of men are asymptomatic, but such people can still infect others. The suspicion of an STD does not necessarily send the victim hurrying to a doctor, as shame or embarrassment can act as hindrance. Screening for STDs, particularly in developing countries, is not always part of the routine when people attend a clinic for other reasons. The stigma of an STD can have patients seeking some alternative form of help, thus preventing them from appearing in official statistics. The stigma also affects countries, in that governments may be reluctant to admit the prevalence of such disease (as happened in the early years of

AIDS). What is known is that the situation is not stable. AIDS is increasing everywhere, but gonorrhoea has been decreasing, for instance in the UK, the US, Canada and Sweden. There is even talk in Britain of the eradication of endemic gonorrhoea. Its incidence has certainly been falling – by 38% in 1993 over 1992 – but that still left 11,800 cases. As counterweight the incidence of herpes between those years rose by 5% to 25,500. The incidence of chlamydia has been rising in both North America and several European nations. Chancroid and primary syphilis have also been gaining in the United States.

Without doubt AIDS has brought the risk of sexual disease into the open. Condoms have been widely publicized as protectors against STDs in general, AIDS in particular. The threat of disease is generally considered a greater threat than the chance of pregnancy (abortions are always possible), but HIV has tended to eclipse other STDs. Of course it is serious, particularly when leading to what is called full-blown AIDS, but the number of cases is few when compared with STDs in total, being 1/250th of that massive quantity.

*

These other STDs, even if only infrequently lethal, can have severe consequences. For women they can cause inflammatory disease which, in its turn, can lead to life-long pain, infertility, and ectopic pregnancies. For men they can also lead to infertility. Infants may suffer, acquiring blindness or pneumonia along with the infection. Syphilis can certainly kill either infants or adults, and may do so long after the original infection. The presence of an STD can certainly be disruptive to families. If an asymptomatic male (who acquired the disease from elsewhere) discovers this disease's symptoms in his partner (she having acquired the disease from him) she may be abandoned by him. The various possibilities on this theme are endless, and always unsatisfactory.

Particularly distressing is the fact that common STDs contribute to the spread of AIDS. A World Health 'Population Report' stated that 'Infection with chancroid, chlamydia, gonorrhoea, herpes, syphilis, trichomoniasis, or the less common STDs donovanosis and lymphogranuloma venereum makes a person more likely to become infected with HIV', a likelihood that may be increased nine-fold. The link between HIV and other STDs helps to explain why HIV in heterosexual populations is so prevalent in Africa. In Europe and the US all other STDs are more often treated and cured; not so in Africa. Genital ulcers, common with various STDs, make it easier for the virus of HIV to leave a person's fluids.

As if this form of extra punishment is insufficient, those individuals already infected with HIV can be less easily cured of other STDs. 'Thus,' as the WHO phrased it, 'HIV enhances its own transmission.' The longer that STD symptoms last the more likely are their possessors to acquire HIV. The longer these individuals have HIV the more will their other STD symptoms prove resistant to cure. It is not an own goal, as Princess Anne proclaimed. It is a circle more vicious than most, an ever-widening circle entrapping more and more (and see in a couple of pages a fuller account of HIV, this latest sexual epidemic).

In initiating a major conference on STDs the president of the Kaiser Family Foundation (in the US) first looked backwards. 'When school principals were surveyed twenty years ago (in the 1970s) concerning the biggest challenges facing their schools, the top-ranked answers included running in the halls, talking in class, tardiness and chewing gum ... The same principals will tell you today that they worry about violence, gangs, drugs, and teen pregnancy,' with 'the most heated battle of all' (between parents, children, educators, public health experts and various activists) being adolescent sexuality. The thought of it, he added, may cause adults to squirm but 'the facts are incontrovertible. Each year between 2.5 million and 3 million US teenagers become infected with a sexually transmitted disease'. So what to do?

The major suggestion, debated actively, involves the provision of condoms for schools. Some 68% (of a sample of 1,316 adults) believed that state schools should distribute them. Of this number almost half wanted the contraceptives to be freely available, and a quarter wanted parents first to give permission. As for the opponents of such hand-outs, whether paid for, free or requiring parental consent, they were generally in favour of sex education. This did not mean they were in favour of helping children practise what had been preached. As a minister from Washington DC, said: 'When they teach you drivers ed, they teach you how to drive. They don't give you a car.' The provision of condoms is more akin to giving seat-belts, helping those in the driving seat to achieve experience without unnecessary risk to each participant.

As with STDs promoting the disease of HIV, so do other major problems create additional distress. Crack cocaine, the addictive smokable form which gained such widespread use in the 1980s, also promotes STDs. A 'Special Article' in the *New England Journal of Medicine* (of November 1994) on 'Intersecting Epidemics' reported that the association between crack and STDs existed 'because of high-risk sexual practices among crack users'. A total of 2,323 young adults, aged between 18 and 29, from inner cities (of New York, Miami, San Francisco), were investigated. Of those who smoked crack 15.7% were HIV positive as against 5.2% of the non-smokers. The prevalence of HIV was worse among women. Of those who had had sex in exchange for money or drugs 30.4% were HIV infected as against 9.1% who did not take the drugs. In short, crack promotes the heterosexual transmission of HIV. An earlier household survey discovered that 1 million Americans were using the new drug, including 1% of all those aged between 18 and 25.

Drew E. Altman, the Kaiser Family Foundation president who said that 'violence, gangs, guns, drugs, and teen pregnancy' were the present concern of school principals, could have added that these intersecting epidemics all encourage teenage pregnancies and there-fore STDs to follow in their wake. 'Ill fortune seldom comes alone,' wrote John Dryden almost four centuries ago. There is no reason for him to mend his lines if he were here today.

AIDS Not until 1981 was AIDS first reported (in the United States). Within fourteen years the WHO was stating that the causative virus existed in 'virtually all countries', having affected 18 million adults and 1.5 million children. It estimated that 'between 30 to 40 million men, women and children' would be infected by the year 2000, that '4.5 million people' have already developed AIDS, and developing countries will soon account for 'over 90% of all people with HIV infection'. Males are still the principal victims, but 15 million women will be infected by the end of the century, and 5–10 million children may have lost either their mother or both parents to the disease. The illness mainly strikes adolescents, young adults and people in early middle age. Therefore the economic loss due to this single disease is considerable. Thailand expects that this one epidemic will cost its country 'close to $11 billion' by 2000.

'AIDS, without a doubt, has caused the biggest upset so far in the history of public health,' wrote the *Lancet* in June 1995. 'Nowhere is this defeat more obvious than in sub-Saharan Africa.' The US Bureau of the Census considers that AIDS will double the death-rate of thirteen African countries by the year 2010. In Kenya 15% of tested ante-natal patients are positive for HIV. In Francistown, Botswana, the proportion quadrupled between 1991 and 1993, rising to 34%. In sub-Saharan Africa as a whole 3% are thought to be HIV infected, a total of 10 million individuals. Of this number 2.5 million have already developed AIDS. The disease has been particularly severe in Uganda, where four-fifths of young adult death in rural areas is linked to the virus. Worse still, the speed of progression from infection to death in Africa is twice as rapid as in developed nations. One-quarter of individuals found to be infected (at their first interview) were dead within two years.

For a time Asia seemed to be escaping the onslaught, partly for its reluctance to gather and publish unwelcome news. The official figures for India (in mid-1994) were 15,000 HIV positive and 559 with AIDS. The Indian Health Organization considers the true figures were nearer 2 million and 100,000 respectively. The WHO reckons 1.75 million is the figure for HIV-infected adults and has predicted that, by the turn of the century, India will have more cases of AIDS than any other country in the world. (Commercial sex in India plays the dominant role in transmission, but the second most important method is through infected blood and blood products.) In parts of Asia HIV incidence is now rising as fast as in Africa ten years ago.

Current estimates state that Asia will overtake Africa by the late 1990s in the number of newly infected people per year. Between 1993 and 1994 there was an eight-fold increase in AIDS within south and south-east Asia. In a single year the proportion of Asian AIDS cases rose from 1% to 6% of the world total.

The figures, percentages and increments are all formidable, but should be set beside those relating to some other gross maladies. Pneumonia, diarrhoea, malaria, measles and malnutrition kill in total 25,000 a day. Diarrhoea alone kills 3.5 million children every year. (The AIDS total for everyone since 1981 has not yet equalled that annual infant quantity of death.) Some 20 million people died from the Spanish influenza epidemic of 1918–19 (more than were killed in the preceding war). As for plague, which so assaulted Europe in various earlier centuries, it cut swathes through populations, taking a third or more at a time. Perhaps it is the novelty of AIDS that is so disarming, or the current inability to produce a cure in an age when more cures are known for more diseases than ever before. Or perhaps, by striking at well-known people, it immediately gained renown, particularly when other well-known people promoted the need for money and research. The US March of Dimes undoubtedly assisted in the conquest of poliomyelitis. Something of the same sort, it was widely believed, could therefore do likewise half a century later. One man returning home from an AIDS conference sat next to a couple of elderly Californian women. 'Were you at that AIDS thing?' they asked. The man assented, and one then took him to task: 'I don't agree with all those demonstrations. Three of my family died of cancer and I never went demonstrating on the street.'

Cancer is old, AIDS is new, and AIDS is increasing apace. Between 1981 and 1992 $6 billion was spent on research intended to end the AIDS epidemic (or confine it to the kind of trough into which polio has been placed). Even Australia (with only 2,630 deaths) committed $322 million (Australian) to the fight against AIDS for 1994–7. Overall there were 36,000 AIDS-related scientific papers published between 1981 and 1992. Strangely, and somewhat disturbingly, 11% of original AIDS research papers were never mentioned in the reference list of any other scientific paper. (Were they so irrelevant, or unimportant, or merely neglected in the deluge?) The Tenth International Conference on AIDS, held in Japan in August 1994, was attended by over 11,000 delegates from 128 countries. The first to be held in Asia, it was welcomed by the WHO's director-general: 'The world has sufficient knowledge about HIV/AIDS to contain the spread of the pandemic through effective prevention. Fear, indifference and denial, along with poverty, have been our main enemy.'

Containment can be successful in reducing the spread. Thailand had 'negligible' HIV numbers in 1988. The incidence then spread rapidly and, by mid-1993, 4% of men drafted into the army were HIV-positive. As prostitution in Thailand is widespread, and 'almost all' men attending government STD clinics reported that 'commercial sex' had led to their infection, the authorities reacted with a programme to enforce condom use. Between 1989 and 1993 the use of such contraceptives, according to surveys of prostitutes, increased from 14% to 94%. Simultaneously cases of the five major sexually transmitted diseases fell by 79%. A joint American/Thai survey estimated that the risk of HIV transmission from prostitutes within those four years declined from 2.6% to 1.6%. Unfortunately the number of infected prostitutes had risen, and there are not enforceable laws about condom use at home.

News from Uganda has also improved of late. Some figures published in May 1996 reported a drop in the number of young adults with HIV. The percentages are still considerable, falling from 20% or so to 10% or even from 30% to 20%, and an increased awareness of the perils of 'slim' (as it is known) is thought responsible for the decline. As 11 million people in Africa are now believed to have been infected, there has to be much more awareness if that colossal figure is not to become even more horrendous.

European nations vary widely in number of AIDS cases per million of population (as reported in 1994). In Spain it was 619, in Switzerland 538, in France 525, in the UK 156, and in Germany 139. In the United States the comparable figure was 1,644, but that considerable number is eclipsed by sub-Saharan Africa. Malawi has 3,076 per million, and Uganda 3,442. There is no need for extreme sexual waywardness, and numerous partners, for HIV to travel speedily through a population. An editorial in the *Washington Times* outlined what it called 'a little mathematical scenario'. If 15-year-olds have their first sexual experience, and enjoy the company of two partners per year until reaching the age of 24 (a not impossible situation in many communities), they will each have been exposed to over 500 partners during that time. Washington DC has a 2% HIV-prevalence among 15 to 17-year-olds. This means that all its young-sters will be exposed to the probability, rather than the possibility, of an HIV-infected partner. One survey of sophomores in US colleges asked, in essence: Have you ever been dishonest with a partner to get them into bed? Men and women 'in alarming numbers' said Yes. When also asked if they would withhold HIV information in order to have a sexual relationship 'many of them said "Yes"'. The spread of HIV is comfortably assured if deceit assists in its transmission.

Within the United States the increase in AIDS is slowing, but that shift is not universal among the various categories of victims. Between 1989 and 1992 the cases rose, in yearly increments, from 35,000 to 43,000 to 45,000 to 47,000, a decline in the increase, but men and women were not equally involved. Male cases rose from 31,000 to 40,000 (29%) but female cases went from 3,900 to 6,500 (69%). Exposure to the virus from homosexual or bisexual contact increased from 19,600 to 23,900, a rise of 22%, but exposure from heterosexual contact increased by 108% for women and 114% for men. The hetero-numbers were much smaller than for homosexual exposure, being about one-fifth, but the increase was considerable (and alarm-ing for individuals who had comforted themselves that AIDS was a problem only for homosexuals, drug-injectors and those who received wrongful blood transfusions). The United States witnessed the initial rise of AIDS, and has since focused most of its attention on the homosexual and drug-injecting communities, but the country is not typical. In the developing world, where most AIDS victims are now found, the disease is mainly spread by heterosexual intimacy. This explains the wildfire manner in which this newest of STDs became so rife so speedily.

Haemophiliacs and others with a coagulation disorder who have received HIV along with blood need special sympathy. For one ailment to be compounded by another seems particularly severe, and approximately 300 Americans a year during those same four years from 1989–92 were exposed to HIV in this fashion. During 1977–91 a total of 6,278 males diagnosed with haemophilia were living in the United Kingdom, and 1,227 of them were infected with HIV between 1979 and 1986 as a result of transfusion therapy. In Japan 200 haemophiliacs infected with HIV were each promised $450,000 in compensation for receiving contaminated blood, with 40% to be paid by the government and the rest by the relevant pharmaceutical companies.

This particular story was concluded in March 1996 when a one-off payment was agreed of 45 million yen to each plaintiff, or to the plaintiff's family if that person had died. An additional 150,000 yen a month would be given to individuals who develop, or have developed, AIDS. Of an estimated 4–5,000 haemophiliacs some 1,800 have contracted HIV, 580 have developed AIDS and 400 have died. Certain families have filed murder charges against Takeshi Abe, the doctor in charge of the governmental research team set up in 1983. Abe's report – in March 1984 – had recommended that non-heat-treated blood products should continue to be sold in Japan. He has also been charged with accepting 43 million yen in donations from the five drug companies involved.

Japan amended its earlier ruling in 1985. Coagulants (treated with heat to kill HIV) were then introduced, but more than two years after they had been approved for use in the US. In France, government officials were jailed for delaying the introduction of such coagulants. A major difference for the US is that American blood donors do it for money whereas the British, for example, receive no more than a cup of tea. Consequently there is little reason for Britons to hide information about lifestyle, hepatitis, or HIV. Needy Americans are not quite so forthcoming.

As for children, who either have mothers with AIDS or who acquire HIV during their mother's pregnancy, they have a poor start to life. There are 3,000 in the UK whose mothers are infected, and their childhood is likely to be blighted by a sick parent who may die before they reach their teens. Every year in the US about 7,000

HIV-positive women give birth, and 15–30% of their babies are HIV-positive. What is not known is why, after such a close relationship as exists between foetus and mother, the remaining 70–85% escape infection. They do so despite the placental intimacy, the upheaval of birth, and the lactation that may follow. Once again there is disparity between nations. In Kenya 45% of babies born to HIV-positive mothers were (in a recent year) infected, as against 14% for all of Europe. The escapers are fortunate, but globally, and in another depressing WHO statistic, a million babies are born each year who have not escaped, and who therefore, in all probability, will die before many years have passed.

In May 1996, and after ten years of debate, the US Food and Drug Administration finally approved a home-use-kit for checking on antibodies to HIV. Known as Confide, it will initially only be available from pharmacies and clinics in Texas and by mail order in Florida and Texas. If this pilot scheme works well it will be launched nationally in 1997. Included with the kit is an information brochure and a lancet with which to leave a drop of blood on a sample card already encoded with a PIN number (as used with credit cards). One week after mailing the card to a laboratory the patient/inquirer can phone to get the result, and will receive it after identification via the PIN number. Those with negative results will merely hear a recording. Those who are seropositive will be 'routed' to appropriate counsellors. Inevitably there has been opposition to the scheme, principally from resentment that telephone counselling is to be given for a fatal disease. There is also unhappiness that the counselling is only given during normal (Eastern Stardard Time) working hours.

The AIDS pandemic 'finds its ugliest manifestation', as *Scientific American* phrased it, 'in the proposition that AIDS has arrived in time to stop the population explosion'. Opinion has been expressed that the 'problem' of AIDS will act as counterweight to the 'problem' of Africa, where the rate of population growth is highest and where poverty is deepest. At its present rate of transmission, according to Gerard Piel, chairman emeritus of *Scientific American*, HIV will infect some 200 million people by 2010. The African share of the casualties might then approach 100 million. 'That,' he added, 'as a disciple of

Thomas Malthus observed of the million Irish who perished in the 1845–50 potato famine, would scarcely be enough.' As for the 'paroxysm of violence' that seized the industrial world between 1914 and 1945, the absence of the many tens of millions killed 'was not remarked in 1970, when the rate of world population growth reached its all-time peak, at around 2 per cent'.

England and Wales are not Africa, but most of their inhabitants are able to name individuals who have died from AIDS. The death total from this single cause was 1,680 in 1993 and is expected to rise to 2,375 in 1997. The annual number of deaths in England and Wales from all causes is in excess of 500,000, whether for 1993, 1997 or any recent or foreseeable year. Therefore the AIDS proportion is approximately one in 200. In Uganda, so sorely hit by AIDS, it is officially expected that 1 million of its citizens will have died of the disease by 1998. This is a huge figure but the human reproductive system is more than capable of replacement. Uganda's population will continue to rise unless other factors, such as a wish for fewer children, become paramount.

*

Some anomalies can end this dismal section on the rise of HIV. In March 1996 the California Medical Association overthrew its policy demanding that cases of HIV should always be reported to the authorities (thus joining thirteen other states where such reporting is not mandatory). All US states require the reporting of AIDS cases. Several recent studies show that a proportion of infected people remain not only free of AIDS but also free of symptoms for ten years or more. Studies of homosexual and bisexual men (largely in San Francisco) indicate that 'about half' remain free of AIDS ten years after becoming infected. So too a proportion of those individuals who receive the virus either from their mothers or from blood transfusions. The reception of contaminated blood carries a 95% risk of infection but the risk with sexual intercourse is between 0.1% and 1%. Positive infection is followed by the development of antibodies (as is typical with all viral infections) and these are clinically detectable within three months, within a year, or sometimes longer.

No one knows why there should be this variation. Nor is it known why the process of developing antibody, known as seroconversion, leads to the patient feeling ill or completely well. The majority of HIV-positives feel entirely fit, either all or most of the time. Certain prostitutes have apparently developed an immunity to the virus – and are therefore being actively studied. Other individuals are far less fortunate, such as the British nurse in Zimbabwe who 'had never had a boyfriend' but once stubbed her toe on entering the operating theatre. That small wound led to her acquiring HIV, and then to her death from AIDS.

New laws have been initiated in response to AIDS, such as that Thai compulsion for prostitutes to use the freely donated condoms. In Denmark anyone carrying the virus is a criminal if they do not use condoms or inform their partners that they are HIV infected. The Australian Football League can now demand that any player suspected of carrying the virus should have a blood test or be fined £1,000. There are also rules that referees send off bloodied players until their bleeding stops, the blood has been cleaned, and the player's clothing has been changed. In India, and in April 1996, AIDS was positively deleted from the list of contagious diseases whose carriers cannot officially travel on trains. (The suggestion – via a parliamentary bill – had first been made in 1989, but seven years had to pass before, following intense lobbying and public exposure in the media, the government finally realized this proposal should never become a law.)

In Kenya the testing of pregnant women for HIV has caused the testers to think again. Most of the women did not 'actively request' the results, fewer than one-third informed their partners, and those that did do so often experienced violence if the tests had proved positive. The men, it can be presumed, were often the cause of their wives' infection. When concluding their report on Kenyan HIV the medical team – from Nairobi and Belgium – stated that any similar investigators 'should weigh the benefits of the study for women involved against possible risks such as increased violence and loss of security'. It is bad enough acquiring the virus. It is even worse if family breakdown is to follow.

Finally, with all ill winds able to blow most contrarily, three Italians have adopted a novel life-style. Sergio Magnis, Antonio Lamarra and Ferdinando Attanasio rob banks as a blatant threesome. They do not bother to wear masks against the security of cameras. In one eight-month period between 1994 and 1995 they collected 250 million lire, a tidy sum in anyone's currency. The police know all about them, and wish for some reprieve as frequent arrests of these three unhappy thieves lead to equally frequent dismissals. A 1993 law prohibits the imprisonment of people with AIDS, and all three have the disease. They need the money, they say, for food and medicines, and wish to draw attention to the plight of AIDS victims abandoned by the state.

The three are not alone. Over 3,000 prisoners have been released since the law became effective, and more will encounter this form of instant parole because 5.47% of Italy's prison population of 50,000 have tested positive for HIV. One policeman, tired of the re-arresting, has warned that he may shoot on the next occasion. 'I told him I was glad,' said Magnis. 'A quick bullet is better than counting the days.' Meanwhile the Justice Ministry is deliberating on whether to amend the law of 1993.

TRANS-SEXUALITY The notion of changing one's sex is, for most people, about as bizarre as returning to childhood – or even more so; but for a few the wish is paramount. The incidence of 'gender dysphoria' is increasing, having been estimated (for the UK) as one in 100,000 in 1980 and one in 18,000 ten years later. As for the nature of the surgery, should a wish to become female be contemplated, many a male can be appalled even by its essentials. The penis and testes have to be removed and, if possible, a form of vagina has to be fashioned using skin, as from the scrotum, which is both sensitive and available. The operation takes four and a half hours, and ten days must then be spent in hospital before a man can emerge, so far as modern techniques permit, as a woman. (It is believed that about 8,000 operations have so far been carried out within the United Kingdom, most of them in London.)

'You see, I'm having a sex change,' said a 25-year-old man to the gently astonished sales assistant as she and he ruffled through some pretty dresses hanging in the store. A television documentary was following one individual who told of his wish for change, of his childhood preference for reading a sister's magazines, and of his hatred for his father. He described homosexual encounters, his initial interviews with extremely antagonistic psychiatrists, and his evening role as female impersonator. Daytime work was in a catering concern 'where nobody minded' what sex he was. At first he dressed in male clothes but later swapped to skirts and dresses. He did not like 'the gay scene' and married a girl after making her pregnant. Later the two of them produced another child, but always the father felt 'as if he was a woman in a man's body'.

He then met more sympathetic psychiatrists, but was informed bluntly of some harsh facts. He would have to live as a woman for at least a year. Only then might an operation be considered. He would not be able to change the gender on his birth certificate. Therefore it would be a criminal offence to marry a man. If sent to gaol for any reason he would be imprisoned in a male establishment. Did he therefore wish efforts to be made that might cure him of his dysphoria? 'No,' he replied, most incisively. And 'No' is what has been said, equally incisively, by scores of other trans-sexuals.

Almost always these individuals have been men wishing to become women. The alternative operation, of attaching a prosthesis to serve as penis, is far less satisfactory. First operations of either kind were performed in Berlin in 1931. Harry Benjamin, of the US, was also a considerable pioneer. The action later moved to Casablanca, Morocco, where a surgeon named Biru achieved considerable success. Currently London has acquired eminence, but operations are also possible in France, Germany, Italy and the US, for example. In Britain there has been opposition, mainly from those who feel such transformations should not be permitted under the National Health Service. The waiting list for other operations cannot be lengthened, it is argued, by any form of cosmetic procedure which causes, for example, cancer cases to be further delayed. According to one surgeon, who has performed over 1,000 trans-sexual operations, his patients do not tolerate the term 'cosmetic'. Their very identity is being jeopardized, with suicide or extreme misery as alternatives.

The same surgeon described the intense joy among his patients as they surfaced from the anaesthetic following their sex-change. The intensity is allegedly even greater than for individuals learning that a cancerous growth has been eradicated. No longer are the trans-sexuals' lives to be, as one man-woman said, 'a living lie'. This joy is not necessarily maintained at a similar level. However much the new woman, fully equipped with breasts (having been given suitable hormone treatment for at least a year), with the female form, and with no awkward bulge around the groin, may feel more female than ever before, there are still difficulties, often in the permanent acquisition of a male companion. The created vagina, cleverly and sensitively contrived by the surgeon, may suffer a form of prolapse. Sexual relations may therefore be endangered and relationships may not endure. They can often fail to last with ordinary male–female partnerships, but the trans-sexual individuals can feel doubly wounded by unhappy terminations.

As for the surgeons they can feel ostracized, aware their work is 'a bit *infra dig* even though, or because, it can make a lot of money', as one reported. They find it easy to detect the new women (in a crowd), partly because the allegedly female hands and feet are too big. In Britain surgeons and patients are annoyed by what they consider backward legislation – the governmental insistence that born males must bureaucratically remain males. They point to the United States, Germany and Italy where marriage is permitted between the new women and ordinary males. 'Gender dysphoria' is generally preferred as a term over trans-sexual, and the medical teams call their work gender reassignment surgery. Patients are informed of self-help groups, such as (for the UK) 'Change', BM Box 3440, London WC1N 3XX and (for the US) the Harry Benjamin International Association for Gender Dysphoria, 1515 El Camino Real, Palo Alto, California 94306. The patients are almost always standard XYs, as with conventional males, rather than any form of chromosomal abnormality. Dysphoria is opposite to euphoria, a state of feeling amiss as against feeling well. Gender is a state of mind however much, as with the great majority, it is based upon the physical distinction of possessing either two X chromosomes as the twenty-third pair or one X and one Y.

Or perhaps that state of mind is actually determined by the brain's structure. Trans-sexuals, always intriguing, have been thoroughly investigated for any physical differences between them and run-of-the-mill males. Their genetics, genitalia, and hormone levels have all been studied, but to no avail. So too their brains at post-mortem, and the result of one such Dutch study was published in November 1995. The hypothalamus of six male-to-female trans-sexuals had been compared with the same organ of thirty-six ordinary men, this 'material' having been collected over eleven years. A difference had been found, namely in the 'central sub-division of the bed nucleus of the stria terminalis' (an area of the hypothalamus of little concern to most of us, save that we possess it and may possess it differently if males wishing to be female).

Although this difference exists, and is some sort of answer to the male to female longing, the dissimilarity may itself have been formed by foetal hormone levels. And the foetal hormone levels may themselves have been triggered by some other variable. The cause of trans-sexuality has therefore a long way to go before being finally unravelled. As *Nature* phrased it, in commenting on the Dutch success, 'the difficulties inherent in studying the diverse sexual behaviour of humans ensure that this will be far from the final word on this subject'.

INTERSEXES Chromosome errors almost always cause infertility, and therefore relate to human reproduction. In this current age, when the entire human gene arrangement is being mapped, it is intriguing to remember that human chromosome numbers were only accurately counted in 1956. Before then many a textbook had listed the chromosome total as forty-eight rather than the forty-six it should have been. Only three years after that somewhat embarrassing readjustment it was realized that various forms of physical sexual abnormality were based on wrongful chromosomes. Individuals who had been labelled as pseudo-males, pseudo-hermaphrodites or other variations on this theme could suddenly be comprehended by looking, not so much at their aberrant sexual organs but down a microscope. Klinefelter's syndrome, for example, already known for its intersexuality, was discovered to be based on an extra X chromosome. Affected individuals, instead of being either XY (as for males) or XX (as for females), were XXY. It was therefore apparent why their masculinity had been trampled upon and accompanied by female characteristics. Similarly individuals with Turner's syndrome, also well known beforehand, were found to have forty-five chromosomes, possessing one X but no X or Y to partner it. It was small wonder, therefore, that XO people (as they were named) were immature females, possessing a diminutive uterus, small genitalia in general, and sometimes no ovaries at all.

The floodgates then opened. Further peerings down the microscope revealed XXX females, XXXX females, XXXY males, XXYY males, and XXXXY males, these last possessing forty-nine chromosomes instead of the standard forty-six. Included among this assortment were XYY males, among the last to be discovered. It was inevitably presumed that such individuals would be doubly male, perhaps possessing some of the less welcome masculine attributes, such as criminality, a compulsion for sexual assault or wild behaviour. A search was therefore made for XYYs in establishments harbouring 'difficult' criminals. It appeared initially as if the presumption was correct, with a higher proportion than might otherwise have been expected of XYYs, but further work eroded this finding. At Nottingham's more conventional prison one fifth of the men over 6 ft (tallness being a male attribute) were found to be XYY. As for their crimes, these were modest, being of the petty pilfering variety. To be incarcerated does seem to be a prerogative of males. Their Y chromosomes can, in part, be blamed but a double ration does not appear to be associated with the severer forms of criminality which were, at first, suspected.

PREGNANCY

PREGNANCY RATE   FERTILIZATION   EMBRYOLOGY

THE TIMETABLE   FERTILITY   ECLAMPSIA   FOETAL RIGHTS

MISCARRIAGE   ECTOPICS   SCANNING   RUBELLA

LITIGATION

Let me then say it bluntly: pregnancy is barbaric.

Shulamith Firestone, US feminist

The history of man for the nine months preceding his birth
would, probably, be far more interesting and contain events
of greater moment than all the three-score and ten years
that follow it.

Samuel Taylor Coleridge

PREGNANCY RATE Within the generally severe pages of the
British Medical Journal a space is allocated to personal anecdotes.
These can be refreshingly human and different from the formal
articles. They can also be revealing.

> She was a thin, 11-year-old child. The only complaint was vomiting
> ... unassociated with any pain ... She had not reached the menarche
> ... no signs of secondary sexual characteristics. While in hospital her
> vomiting continued for a week or so, eventually settling ... It was
> thought appropriate to allow her home to be seen again in outpatients
> if necessary. She never returned.
>
> Almost six months later a letter was received from an obstetric
> colleague some miles away. 'You will be pleased to know I recently
> delivered, vaginally, a normal 2.8 kg baby boy from this 12-year-old
> young lady. I believe she was admitted to (your) hospital with early
> morning sickness ...

The Cornish doctor chided himself, somewhat harshly bearing in
mind the youth and appearance of his patient, but sex and pregnancy
can both happen long before physical maturity is reached. In *The
Health of the Nation* (published for the UK in 1992) the conception
rates for 13- to 15-year-olds were recorded as 9.5 per 1,000. These

rates have been fluctuating in recent years. The 1991 figure of 9.3 per 1,000 was the first reduction for a decade but was still higher than the 7.2 per 1,000 recorded in 1980. As for the teenage rate as a whole, the under twenties, this rose during the 1980s to 65.1 per 1,000 in 1991, a proportion still lower than the 82.4 per 1,000 in 1970 (when contraception was not generally available for young people). As pregnancies are actually occurring for the youngest teenagers at around 1% of the total it is entirely feasible, biologically and socially, for earlier pregnancies to occur. The London Brook Advisory Centres reported in 1994 that 8% of boys and 4% of girls started sexual intercourse before age 14.

The United States has a greater pregnancy rate among the young, this being 'three to ten times higher than among industrial nations of Western Europe', according to a social ecologist at the University of California. These conceptions are less frequently between similarly aged teenagers; in most cases an older partner was involved both for the teenage girls and the teenage boys. Almost a quarter of births fathered by school-age boys involved post-school mothers. Similarly 71% 'of the 46,511 marital and unwed births in California among school-age girls (in 1993)' were fathered by adults. The average paternal age was 22.6 years, making them some five years older than the mothers. For the 'junior high school mothers' (aged 15 or less) most of their births were fathered by males six to seven years their senior. (As it is unlawful in California for any male who is not the husband to have sexual intercourse with any girl under 18, the breaking of this law must be one of the commoner pastimes. Also the very many births to young girls must be overshadowed as a statistic by the numbers of abortions in that age-group and, to an even greater extent, by the quantity of intercourse which did not – by luck or management – end in conception.)

As with politics the selection of facts can give misleading impressions. It is true that the US pregnancy rate among schoolgirls has been rising recently, but it is also true that teenage pregnancies as a whole had been falling – from 68.3 per 1,000 in 1971 to 50.9 per 1,000 in 1987. As for those older fathers, at least one-third of them become married to the respective mothers by the end of the pregnancy and, so stated a report from the New York Medical College, 'perhaps half that many will marry her later'. It is also intriguing, according to the same source, that 'mothers who never marry do not appear to be substantially worse off than those who divorce soon after marriage'.

Poverty is related to American teenage births. The 40% of girls said to live 'near or below poverty income levels' account for six out of seven teenage births. In areas where the US poverty rate is no lower than in other Western nations the teenage pregnancy rate is also low. Despite considerable prevention programmes aimed at reducing juvenile US pregnancies the numbers have lately been growing. California University reported that 'in every US state and major city for which figures are available, birth rates among school-age girls increased from 1986 to 1991'. Blame is put upon 'the excessive economic attrition inflicted on US youth' and the 'thoroughly integrated nature of sexual behaviour among adolescents and adults'.

*

Conversely, at the other end of the age range, women are postponing motherhood. In 1992, and in the UK, more babies were born for the first time to women in their early thirties than in their early twenties. The figures were 166,800 for women aged 30–34 and 163,300 for those aged 20–24. One decade earlier the comparable figures had been 126,600 and 194,500. Some celebrities helped to publicize this trend. Britt Ekland was 45 when her son was born, Patricia Hodge was 42 and 45 when her two were born, and the singer Joni Mitchell was 52. The preference for putting a 'family on hold' (as one newspaper phrased it) is particularly strong among well-educated, middle-class women whose partners/husbands are either in the professions or doing well in business. 'Let's not interfere with our flamboyant, go-getting, upwardly-mobile, exciting, travelling life-style just yet' is the kind of sentiment brought to mind when contemplating these delaying figures. So too, a few years later in each couple's advance: 'I don't like to feel that we're getting old, but they do say, don't they, that a woman's reproductive capabilities begin to acceler-ate downhill after the age of 30, and certainly by 35; so oughtn't we to think about a baby now?' In 1980 in Britain 171,000 babies were born to women in their thirties and forties. By 1990 the equivalent figure had risen to 218,000, a gain of 30%.

FERTILIZATION The egg is thought to be viable for eight to twenty-four hours. During this time it moves about one-quarter of the way down the uterine tube. Spermatozoa suffer extreme attrition on their journey to reach the egg. Each ejaculation may contain 200 million sperm, but only a few hundred thousand reach the uterus, only a few thousand enter the tube, only about a hundred reach the egg, and only one is the actual agent of conception. The minimum time for sperm to meet egg is thought to be forty minutes, at which time the mother may be stepping on to a bus, or watching a movie, unaware of the dramatic event occurring within her: there is a common belief that rest or the prone position may assist conception and, as corollary, that violent movement acts as contraception, but there is no evidence supporting either claim. Syngamy is the name for two gametes coming together. Without this event the egg begins to degenerate within twenty-four hours (having perhaps waited forty years before its expulsion from the ovary).

Although only one spermatozoon penetrates the vitelline membrane to enter the egg, and therefore supplies the male's genetic material of twenty-three chromosomes to match the female's twenty-three, a number of other sperm usually penetrate the egg's outer layers, the corona radiata and the zona pellucida. These assist in the process of fertilization. Only the single invader which reaches the innermost sanctum causes the egg to undergo changes that will result in the first cell division and, if all goes well, in the birth of a human being thirty-eight weeks later.

In fact the egg on its own possesses all the material necessary for such creation. Many an invertebrate achieves natural parthenogenesis, with an offspring arising solely from an egg. Artificial parthenogenesis has been achieved not only with invertebrates but in various verte-brate animals, including mammals. The resulting embryos tend to die when still embryonic, but living young have occasionally been produced, with rabbits for example. With humans the most famous parthenogenetic birth of all, allegedly occurring some 2,000 years ago, is a matter more of faith than strict observation. (If the rules of biology are considered in the least relevant, Mary's offspring should have been a girl, her sex chromosomes both being Xs and her lack of a Y quite forbidding a masculine child.)

On the other hand nature does break its own rules, in all sorts of areas, from time to time. In 1955 the *Sunday Pictorial*, urged by J. B. S. Haldane (its scientific columnist), sought to discover – via its tremendous readership – if any woman believed she had given birth parthenogenetically, and without the involvement of any man. Nineteen responded, with most immediately rejected for their belief that an intact hymen inevitably indicated a virgin birth. After further sifting – the child had to bear a striking (and uncanny) resemblance to her mother – the nineteen originals became one. Mrs E. Jones and 11-year-old Monica were then subjected to further tests, involving blood, saliva, tasting powers and skin transplants. The first three showed uniformity but the grafts did not take (as they should have done had Mrs Jones and her daughter been of identical stock, as with one-egg twins). J. B. S. Haldane was forced to conclude that, despite the extraordinary female-to-female similarities, Monica must have had a father. Nevertheless it was an interesting attempt to use a massive readership for the advancement of science. Human parthenogenesis is, for the time being, unproven, save where faith and conjecture are the rule.

EMBRYOLOGY The first division of the fertilized cell takes place some thirty hours after sperm and egg have coalesced. (The mother is still unaware that such a momentous event is happening, save for those who say they 'knew right from the start' of an impending pregnancy.) Further divisions occur every half-day or so, with the developing bundle progressing along its uterine tube. On reaching the uterus, after perhaps four days, it may be thirty-two cells in size. The cell cluster then starts absorbing fluid, this eventually causing a cavity to open up. Soon after arriving in the uterus this expanding body, now known as the blastocyst, becomes tightly linked to the uterine wall. Between six and seven days post-conception the developing form is not so much attached to the uterus as solidly embedded in it, a process known as implantation.

Yet another cavity, distinct from the blastocyst's fluid-filled centre, is formed on Day 8 within the growing mass of cells. This, the amniotic cavity, will contain the embryo and then the foetus on their nine-month voyage to become a baby. A yolk sac forms at about this time. So too a system, the uteroplacental circulation, for acquiring nutrients from the mother and donating waste products in return. In fact, much like a bird creating a nest, virtually everything occurring at this time is a form of preparation for the task ahead. Two weeks after conception, with the mother still (probably) unaware of her condition and expecting the next menstruation to arrive on cue, less than 0.1% of the very great quantity of cells now in existence at the implantation site will divide and become human being. All the rest will become everything else, the placenta, umbilical cord, amniotic sac, and so forth.

This fact is pertinent to the ethical discussions concerning embryonic research. There *is* an embryo within all the development, but it is only a minute portion, being less than one-thousandth of the whole. Hence the term pre-embryo, now used in such discussion, to describe the quantity of material gathering within the uterus. The actual embryo is such a modest fraction at that time that its name does not accurately portray the considerable development occurring all around it. Nevertheless it is there, it is growing – rapidly, and at two weeks post-conception it abruptly earns the name of foetus. To the mother, whenever she learns of the situation, or suspects it even earlier, the development is referred to as a baby from start to finish.

*

The embryology of any individual, human or animal, may appear simple in the earliest stages of development, but the complexity becomes more daunting by the hour, let alone the days and weeks of pregnancy. A newborn baby is indeed at the beginning of its life but is also a finished product. During the previous nine months of existence it has expanded from one cell to 20 billion cells, or maybe 100 billion cells or more. (Who, as they say, is counting?) Similarly, although the newborn offspring may seem inept as it lies immobile in its cot – urinating, defecating, drooling, crying – it is a functionally independent organism, able to breathe, to eat, to digest, to use its senses and to learn. It needs only food, air and care to achieve maturity. Its growth in weight, from birth to adulthood, is not several billion-fold – as the conceptus has already experienced – but a mere twenty-fold. Its length will only increase less than four times. Nothing truly startling will happen during the eighteen leisurely years of growth, as it learns to walk, to speak, to think, and achieve its true potential. It already possesses all its necessary parts – the germ cells, brain cells (these will not increase in number), and every other form of cell for every kind of organ.

In fact such items were already in existence six months or so before birth, with the final two-thirds of each pregnancy a matter of growth more than of differentiation. The three-month foetus weighs less than a gram (and will be 120 times heavier at birth) but that early modestly sized portion of living matter – less than 3 inches long – *is* a miniature human being, replete with nerves, blood vessels, and tissues of all kinds. It can swallow. It will move its foot away from tickling. Its heart pumps 6 gallons of blood a day as the diminutive quantity of blood is steadily recirculated through arteries and veins. All this development is achieved in the first trimester of pregnancy, the earliest twelve weeks.

*

There are several conundrums involved in pregnancy, quite distinct from the amazements as a single, scarcely visible egg fulfils its programme to become an independent offspring. First is the ability of the human body (or any other body) to contain such explosive growth within one portion of it without this growth urge disastrously affecting the mother's growth as a whole. Growth stimulants pour from her pituitary glands, and yet her structure does not change. The only parts to enlarge are those needing to enlarge. A second strangeness concerns her own weight gain. Typically a woman who gives birth to a 7.3 lb baby also expels a 1.5 lb placenta and 1.85 lb of amniotic fluid. Her uterus then weighs 2 lb, and her breasts have increased almost by 1 lb. She has 3 lb more blood, and slightly less extra-cellular fluid. This all adds up to 19 lb and yet, again on average, her total weight gain is 27 lb. Plainly the extra weight must be somewhere, but there is little certainty about its location(s).

The third conundrum, yet more bewildering, concerns the production of a foreign organism (for that is what a foetus is) within the mother. A skin graft, for instance, between mother and newborn baby, or vice versa, will not be accepted. Mother and child are two distinct beings, much like siblings, and yet one is made within the other. Peter Medawar, when director of London's National Institute for Medical Research, suggested that the foetus's privileged position was due to one (or all) of three things: the placenta's 'lack of antigenicity' indicating that it could not reject foreign tissue; an immunological inertia (or tolerance to foreign tissue) of the mother; or an anatomical isolation of the foetus from the mother. Having outlined these three possibilities, Medawar then added that: the placenta *is* antigenic; there is *ample* evidence that pregnant mammals are not immunologically inert; and there is *no* question of placenta and foetus being anatomically isolated. In other words, game, set and match to the notion that pregnancies are impossible – save, of course, for the overriding fact that they do occur, again and again and again.

*

Embryologically the human embryo starts in similar fashion to every other mammal, and indeed to creatures far lower down the animal kingdom. Three layers begin to develop. Somewhat like the layers of a cake they are initially flat, and have been named ectoderm, mesoderm and endoderm (or entoderm in the US). The flat cake then becomes a circular cake, with all three layers moving to link with their other ends so that endoderm forms the central tube, ectoderm the outer tube and mesoderm the middle tube. One extremity of the three-layered tube is the head end, with the other being the tail end. So far, in a sense, so good and reasonably straightforward. The outer layer, the ectoderm, then forms an extra tube entirely from its own material, this further tube also running from head to tail. It forms all nervous tissue – the brain, the spinal cord, and every nerve leading from and to them.

As for the three basic layers, they each have quite separate roles to play, all becoming manifest as the embryo swells into a foetus. From the ectoderm will arise: outer skin, nails, hair, sweat glands, eye lens, mouth lining, tooth enamel, nose lining, salivary glands, and all nervous tissue. From the mesoderm will develop: muscle, bone, cartilage, inner skin, blood vessels, kidneys, and connective tissue. From the endo(ento)derm will arise the lining of the gut (from oesophagus to large intestine), liver, pancreas, bladder, lining of lungs, and glands, such as thymus and thyroid. In essence, with all higher forms of life having evolved from simpler ancestors, the original worm design has been maintained. Such a lowly animal has an outer skin, an inner gut running from front to rear, and tissue in between. The human is basically no different, with the outer layer being skin, the inner layer being gut, and the median layer being muscle, blood vessel, and supportive skeleton.

So much for simplicity and the fundamental framework. What follows is also straightforward in its fashion, as limb-buds protrude to grow into limbs, as blood-vessels grow to every region, as nerves do likewise, and organs develop to grow in their appointed places; but this degree of organization is hard to absorb, particularly when one remembers that the blueprint for all this creation was contained in such a minute fragment of tissue. The human egg is not visible to an unaided eye, but most of its form – as with each far larger bird's egg which is also only a single cell – is not part of the inheritance mechanism. Only the genes, and only the DNA (about which much more later) organize development, and nothing of their structure can be seen. Microdots, allegedly used by spies to contain a page or two of information, are gargantuan by comparison.

*

It is necessary to appreciate, before complete incomprehension takes over, that every aspect of each developing foetus has to be arranged, whether the many hundreds of connections linking each of the many thousands of millions of nerve cells, whether the total ramification of the circulatory system so that a pin-prick anywhere will encounter blood, or whether the chemical laboratory of the digestive system so that (almost) anything eaten is consumed and then transferred into something else, such as kidney tubule, fingernail, eyelash, ear-wax, spleen cell, tendon, muscle, tooth or any other part. Virtually all of this is made, and in place and functioning to some degree, within twelve weeks of that union between sperm and egg. 'I will give thanks,' wrote the psalmist (of No. 139), '. . . for I am fearfully and wonderfully made.' And so speedily, he might have added. And so astoundingly.

THE TIMETABLE Physicians tend to time a pregnancy from the start of the previous menstruation, this being the only known fact. The relative uncertainty of conception's date is no reason for doubting that its occurrence is a most positive event (and far more relevant than the earlier menstruation). For a true timetable of pregnancy the fact of fertilization is the only starting point, it being the beginning of everything that follows. The calendar (beginning on the next page) therefore assumes that the previous menstruation happened in mid-December, with conception occurring (very conveniently) on the opening day of a brand new year. The first of January is appropriate for the start of anything, not least the initiation of a human being who will be born thrity-eight weeks later, give or take the days on either side that each foetus unhelpfully adds to or subtracts from the average span.

An analogy, or comparison, can be made here with the creation of an assembly-line car. Anyone witnessing such manufacture, and amazed to observe the transformation from initial portion to final product, will repeatedly fail to notice when such-and-such an item – windscreen wiper, reversing light, horn – was actually added. So too, but differently, with foetal growth, the difference being that nothing is added as a complete entity. Every piece of the offspring has to start invisibly, with one cell, and then a group of cells developing in distinct fashion from its neighbours. By the time that limb-buds, for example, are visible as minute protrusions they are long past the moment when that budding actually began. Similarly, as mere buds, they are in advance of the time when arms and legs become noticeably different, and when hands and feet are also quite dissimilar. The nine-month development is, in one sense, an assembly line, but far more intricate, far more complex and infinitely more bewildering than any man-made production process. The making of every human is in quite a different league.

First week: 1–7 days (1–7 January)
Fertilization. First cell division (after thirty hours). Whether dissimilar twins or not now established. Cell mass moves along the 4 inches of oviduct, and becomes a hollow blastocyst en route. Enters uterus (third to fifth day). Implantation occurs (sixth to seventh day). Mother probably unaware of the situation. Customary menstruation not due for another week. Uterus still weighs about 2 oz.

Second week: 8–14 days (8–14 January)
Bleeding may very rarely occur (on the thirteenth day) as a result of increased blood flow to implantation site. May be confused with menstruation which is due at the end of this week. Pre-embryo is now plate-shaped with hundreds of cells, a fragment of which will form the true embryo. The rest is creating structures like the yolk sac, umbilical cord, placenta, and amnion, with the yolk sac and amnion being the first to be formed.
   Mother still expecting menstruation at its usual time.

Third week: 15–21 days (15–21 January)
Embryo $\frac{1}{10}$ inch long, being a flat, egg-shaped (some say pear-shaped), three-layered disc. The primitive streak is formed on Day 15. All tissues to the right of it will give rise to the right side of the body, and on the left to the left. Menstruation now overdue. Morning sickness and nausea can start. So can breast tenderness. Primordial germ cells, which will eventually produce the cells that produce the sperm and eggs, already present but the embryo's sex cannot as yet be determined. Both eyes and ears start developing by the eighteenth day. On Day 16 the primitive streak spans about half the embryo's length (and it disappears by Day 26). On Day 21 a set of constrictions appear in the heart tube (which will form the several heart chambers during the

next five weeks). The first sign of blood vessel formation can be
detected on Day 17.

Fourth Week: 22–28 days (22–28 January)
Embryo $\frac{1}{4}$ inch long. Heart ($\frac{1}{10}$ inch – although more of a tube than a
heart) starts beating on twenty-second day, and a circulatory system
of a sort exists. (The heart is an organ that has to start work almost
the moment it is formed, and then continue to work for the rest of its
days.) The head end of the embryo begins to bend sharply, this bend
marking the site of the future mid-brain. Tongue has started to form.
Upper limb-buds appear on the twenty-fourth day.

The lung divides on Day 27, forming the rudiments of the two
lungs. (Between Weeks 5 and 28 these will branch sixteen more times
to form each lung's bronchial 'tree'.) The yolk sac, often said to be
useless as its function is not known, is now as big as the embryo (but
will grow little more, although it will never quite disappear). Umbilical
cord still very rudimentary. The neural tube folds over, closing at the
head end on Day 24 and at the tail end on Day 26. (Failure of the tube
to fold over properly can lead to errors, such as anencephaly (head
end) and spina bifida (tail end).)

Mother can now often be given medical confirmation of her state.
Given a date thirty-four weeks ahead. Relative size increase is never
again so great as in this first month. The embryo is now 10,000 times
larger than the egg. The extent of physical change will never again be
equalled.

Fifth week: 29–35 days (29 January–4 February)
Heart now pumping frequently. External ears start taking shape. Arm
buds differentiate into hand, arm and shoulders (on the thirty-first
day). Finger outlines appear (on the thirty-third day). Foot still a flat
and budding protuberance. Nose and upper jaw start to form. So do
stomach, and oesophagus, and duodenum. Embryo $\frac{1}{2}$ inch long.

Sixth week: 36–42 days (5–11 February)
Tip of nose visible (thirty-seventh day). Eyelids begin to form. Five
separate fingers. Toe outlines appear. Marked skeletal growth, but still
of cartilage and not bone. Stomach, intestines, reproductive organs,
kidneys, bladder, liver, lungs, brain, nerves, circulatory system are all
being actively developed. The gut tube is fully formed by the end of
this sixth week. Sexual differentiation now begins. (The rate at which

embryo and gonads develop is said to be the first detectable difference between future males and females. It is argued that embryos destined to be males – because of their Y chromosome – do so to avoid being swamped by the rising levels of the mother's oestrogen hormones. Allegedly males develop faster right from the two-cell stage – which may help to explain why they also beat females to the finishing line, all the way to death.)

Embryo now ¾ inch long.

Seventh week: 43–49 days (12–18 February)
1 inch long. Stomach already producing some digestive juices, and it rotates so that the greater curvature is directed to the left side. The liver and kidney have started functioning. Muscular reflexes can work (as can be shown experimentally, but nothing yet felt by the mother). Outer ear and hearing part of the ear almost complete. Upper and lower jaw very clear. Mouth has lips, something of a tongue, and first teeth showing up as buds. Thumbs markedly different from fingers. Heels and toes definite. First true bone cells appear. Circulatory system now effectively operational.

Eighth week: 50–56 days (19–25 February)
A sort of neck now visible. The head of the embryo is very large compared to the rest of the body. The maternal uterus is now about 4 inches long, and wide, and deep. The amniotic cavity has greatly expanded since the fourth week. By the eighth week it properly surrounds the developing individual (and has obliterated the chorionic cavity which, separately, was performing a similar function). The expanded amnion, as one investigator phrased it, 'creates a roomy, weightless chamber in which the fetus can grow and develop freely'.

The corpus luteum which produced the egg is now about the size of the original ovary. The ovary itself has not grown. The corpus luteum will continue to grow until it becomes almost hen's-egg size. It will then regress, having fulfilled its hormone-producing role.

Ninth week: 57–63 days (26 February–4 March)
Embryo often called foetus after this time. (Some call it foetus (fetus in US) after fifth week, with embryo only existing from the third to the fifth week, with previous state being called pre-embryo. Some don't care, and just talk about 'product of conception'.) Sex of individual can be detected externally. (Until now organs have appeared

similar even to expert scrutiny, although chromosomal assessment of sex possible under the microscope). Head is $\frac{3}{4}$ inch long and is half crown to rump length (i.e. total length of foetus minus the legs).

Footprints and palmprints now indelibly (and uniquely) engraved for life. Spontaneous movements occur. Also eyelids and palms sensitive to touch (as shown by reflex squinting and gripping attempts). Nails start to grow. Dissimilar twins start looking dissimilar. Eyelids close over the eyes for the first time, and eyes still look outwards rather than forwards. Amount of hormone HCG (used in pregnancy tests) now reaches its maximum level. Eighth and ninth weeks generally considered best time for abortions, for 'terminating pregnancies vaginally'. (By no means is the foetus a 'protoplasmic blob' at this time, as some allege particularly when wishing to be rid of it.)

Tenth week: 64–70 days (5–11 March)
Quarter stage reached on sixty-sixth day (but foetus will have to multiply its weight over 600 times in remaining three-quarters of pregnancy). Mother's average weight gain is now 1 lb. (But this average figure can be misleading, as many women can lose weight in first three months and many start gaining weight from the start.) Weight of placenta is less than an ounce but may be four times heavier than the foetus. Uterus weighs 7 oz and contains 1 oz of amniotic fluid, perhaps 2 oz, perhaps 3 oz. Breasts have increased in size, some say by about 2 oz. ('No one', said a gynaecologist, 'has been enthusiastic enough to cut a breast off and weigh it. What is certain is that women have either let their brassières out by now, or bought themselves a larger size.')

Commonest time for miscarriages is at tenth week, i.e. time of third missed period. The second most common time is at the sixth week, i.e. time of the second missed period. It is much rarer for miscarriages to occur at time of fourth missed period.

Eleventh to fourteenth weeks: 71–98 days (12 March–8 April)
Many more reflexes possible, such as frowning. Thumb can be moved to the fingers. Swallowing starts. If foot is tickled (experimentally) the whole leg will be withdrawn. Two halves of palate fuse together. Vocal cords completed. Urination has begun (and urine is removed with regular renewal of the amniotic fluid). Swallowed amniotic fluid can be digested. Sperm cells or egg cells exist. Uterus moves up out of pelvis, and can be felt, especially in thin people. At twelfth week

foetus's crown to rump length is $2\frac{3}{4}$ inches, and crown to heel is $2\frac{3}{4}$ inches. Its weight is $\frac{3}{4}$ oz.

End of Development Period, even though the developed form weighs so little. Start of Growth Period. (Growth has of course been considerable in first three months, but in subsequent six months foetal growth is from less than 1 oz to $7\frac{1}{2}$ lb.)

Foetus and placenta roughly equal in size. Heart pumps 50 pints a day. Sex of individual now physically definite, as external genitalia begin to differentiate. Uterus half way between pubic bone and navel. By this month the placenta is producing the hormone progesterone in sufficient quantity to maintain pregnancy. This job was formerly done by the ovary's corpus luteum.

Fifteenth to eighteenth weeks: 99–126 days (9 April–6 May)
Growth of head hair starts. Eyelashes and eyebrows also begin. Nipples appear. Nails become hard. Heart-beat can be heard externally by listening to the mother's abdomen. The separate heart-beat of twins should be detectable with a stethoscope. Mother starts feeling foetal movements (the quickening) although certain movements started less obtrusively six weeks before. Thin women may notice movements earlier than normal or fat women. So do women not having their first child. Foetal hiccups also occur. By now amniotic fluid (which has been steadily increasing) is over $\frac{1}{2}$ pint. Placenta weighs 6 oz (at eighteen weeks), amniotic fluid 9 oz, and uterus 20 oz. Half-way time, with mother well aware of nineteen more weeks to go.

Foetus's crown to rump length is 6 inches, and crown to heel length is 9 inches (approximately half that of the newborn baby). Its head is now one-third of total body length. The foetus's weight is 11 oz (approximately one-tenth that of a newborn baby). Mother's weight gain (at eighteen weeks) is 9 lb (or twelve times that of foetus). Her breasts may have gained 8 oz in weight by now, and colostrum may be expressed from them. Breasts do not usually increase much in size between the twelfth and twentieth weeks. The customary picture is of breast growth at the beginning and end of pregnancy, rather than the middle. The mother is now putting on 1 lb in weight a week, and will do so for the next two months; thereafter the weekly gain will fall slightly. If foetus is aborted at this time it may possibly survive but, if it does, probably only for a few minutes.

By Week 16 the respiratory 'tree' starts producing very many

'Congratulations, it's a 650-pound boy!'

branches, as terminal bronchioles become respiratory bronchioles which, in their turn and starting from the head end, become terminal sacs. Some 20–70 million of these sacs are present in each lung before birth (and there may be 300–400 million in each adult lung). If there is a dominant weakness (akin to a design fault in structural engineering) in the development and birth of a human embryo it lies with the lungs. Even with a full-term baby these can serve poorly for the immediate task of receiving oxygen from air (rather than the uterine arrangement). With premature babies their lungs' inability to cope can be the cause of their demise.

The six pre-birth weeks (assuming birth at Week 38) are the period of accelerated lung development, with each day improving the chances of survival. Babies born before twenty-eight weeks, and when lungs are still poorly developed, are likely to suffer for that reason. There is reason for this deficiency. All the other foetal systems – blood circulation, kidney excretion, nervous control, even the digestive system (which has been passing substance through its gut) – can carry on more or less as before when converted from an internal to an external state; but the lungs, which have been absorbing oxygen from the amniotic fluid, suddenly have to absorb their oxygen from air, an entirely different medium.

Nineteenth to twenty-second weeks: 127–154 days (7 May–3 June)
Eyelids can and do open. Premature life now possible, although infant only size of a large man's palm. It can now grip firmly with its hands (although some degree of the grip reflex was seen twelve weeks earlier). Lanugo, the hairy growth on arms, legs, and back appears. This is generally seen on premature babies, but has normally gone by birth. The foetus's face is red and wrinkled, although fat is now being deposited. Uterus at end of month is up to the navel, but height varies according to the mother's posture. Foetus crown to rump length is 8 inches, and crown to heel length is 12 inches. (Leg length is gradually becoming relatively more important.) Weight is 1 lb 6 oz.

Twenty-third to twenty-sixth weeks: 155–182 days (4 June–1 July)
Many prematures (strictly called immatures until twenty-eighth week) of this age able to live. More than 50% usually survive after twenty-four weeks, and perhaps 80% at twenty-five weeks. Volume of amniotic fluid perhaps 1½ pints, but after thirtieth week may either

not increase or even decrease to allow for foetal growth. Head hair may grow long (although many babies are born bald). Most of lanugo goes. Thumb-sucking a frequent habit. Umbilical cord has reached a maximum length. At end of month uterus 2–3 inches above the navel. Foetus's crown to rump length is 9 inches, crown to heel length is 14 inches. Weight 2 lb 11 oz.

Twenty-seventh to thirtieth weeks: 183–210 days (2–29 July)
Three-quarter stage reached during the twenty-ninth week (and existing foetal weight needs to be multiplied two and a half times in remaining quarter of pregnancy). Chance of survival if born now good. Fingernails reach fingertips (and may actually need cutting at birth). Foetus (generally) settles into head-down position. Earlier acrobatics cease. Fat is being deposited, and smooths out the skin.

Mother's average weight gain is (at twenty-eight weeks) about 19 lb (or nearly six times that of foetus). Her breasts have increased by about 14 oz. Placenta weighs 15 oz (at twenty-eight weeks). Foetus now four times heavier than placenta. Uterus is half-way between navel and lower end of breast bone and weighs 2 lb. Its height above the navel may fluctuate as the baby's head does or does not fit into the mother's pelvis. This fitting in (or engaging) usually comes later. Efforts to rearrange breech presentations often made now, sometimes as late as thirty-second week. Foetus's crown to rump length is 10½ inches, crown to heel length is 16 inches. Weight at 3 lb 12 oz is half final weight (this 3 lb 12 oz being very much an average figure. The biggest babies at this time may be twice the weight of the smallest).

Thirty-first to thirty-fourth weeks: 211–238 days (30 July–26 August)
Premature babies of this age reasonably handsome (less like pink old men) as body is rounded and skin is smooth. Crown to rump length is 12 inches, crown to heel length is 17½ inches. Weight 5 lb 1 oz. Most airlines do not like women to fly after the thirty-third week (with five weeks to go) for long-distance flights, or after thirty-fourth week (four weeks to go) for short trips.

Thirty-fifth to thirty-eighth weeks: 239–266 days (27 August–23 September)
Heart pumping 600 pints a day (although total blood content only slightly over half a pint). Growth usually stops for each individual shortly before its birth, but by then the weight of that original fertilized

egg has been increased 5,000,000,000 times. Now, and in the next twenty years, weight only has to be increased another twenty times. Mother's average weight gain is now 27½ lb (but may be nil or 60 lb). Breasts have gained about 1 lb. Maternal skin surface area has increased by 1½ sq. ft.

Uterus, now 14 inches long, reaches highest point a couple of weeks before birth. It weighs 2¼ lb (or some twenty times original weight). Placenta weighs 1½ lb at term, and is 7–9 inches in diameter. Baby's crown to rump length is 13 inches, crown to heel length is 20 inches. Head is now a quarter of total body length. Weight at thirty-eight weeks is highly variable, but an average for European communities is 7 lb 4 oz, and slightly lower among Asian communities.

FERTILITY About half of all embryos suffer spontaneous abortion, often without the mothers knowing that anything untoward has happened. There is also loss when two eggs have been successfully fertilized. Richard W. I. Cooke, of the Royal Liverpool Children's Hospital, wrote (in April 1996) that 'the early embryonic loss of a twin is extremely common, so common in fact, that as many as one in eight of all single live births could result from an original twin conception.' The cause, or causes, of this considerable rejection are poorly understood, but some defect (or defects) in the developing embryo is thought to be responsible, quite apart from the major defect of failure to proceed to term. Nevertheless human reproductive efficiency is high. According to William J. Larsen, in *Human Embryology*, 'an average couple who do not practice contraception and have intercourse twice a week (timed randomly with respect to ovulation) have a better than 50 per cent chance of fertilizing any given oocyte'. Bearing the 50% rejection rate in mind this suggests that the chance of one month's intercourse producing an effective pregnancy is better than 25%. There is also a 25% chance that a coin will produce heads twice in a row when being tossed two times; but, as all coin-tossers know, there is no certainty that this will occur.

Similarly many couples can fail for a long time before achieving success. In normal circumstances (and without contraception) a couple can expect to produce at least ten infants in their reproductive lifetime (and many a nineteenth-century family did so). Exceptions, as published in the *Guinness Book of Records*, include a nineteenth-century Russian woman who, in twenty-seven confinements, gave birth to sixty-nine offspring (sixteen sets of twins, seven of triplets, and four of quads, but no singletons). Without the aid of multiple births a Brazilian woman who married at 15 produced fourteen sons and twenty-four daughters. If ten is considered normal, and if all ten survive to adulthood, the next generation (assuming incest and gender parity for arithmetic convenience) will be fifty, and the generation after that 250. Biologically that is a suitable increment. Even if two parents produce only two offspring they maintain the status quo.

Globally the business of pregnancy is awesome. Nine conceptions every second (an already quoted figure) means that 777,000 women become pregnant every day. (Miscarriages of all kinds are omitted from this total.) With each conceptus taking thirty-eight weeks before it is delivered, such a pregnancy rate indicates that over 200 million women are pregnant at any one time. With half of the world's population of 5 billion being under reproductive age, and with half of the remainder being masculine, this means that about 1.25 billion women are available for reproduction, save those too old to bear offspring. If this elderly quantity is assumed to be 250 million that leaves a neat 1 billion who might have become pregnant, with the actual number being one-fifth of that total.

The human species does not immediately become pregnant after delivery (as with many animals). Moreover, human breast-feeding delays pregnancy. So, of course, does contraception. So does infertility. And so does spinsterhood, and unwillingness (for whatever reason). The net result, assuming all those round figures approach correctness, is that 20 per cent of the available stock is actually pregnant at any moment. A further 20 per cent has a child less than 9 months old, and yet another 20 per cent has a child between 9 and 18 months. In short, whatever women of developed nations may think as they create their 2.2 children (or the average number in their area), a very large number of women are in the business of creating children for the major part of their reproductive span.

The other side of this particular coin is disturbing. Reproductive health problems 'account for over one-third of the total burden of disease in women', according to the WHO, 'compared with around 10% in men'. 'Death in childbirth is almost always preventable,' reported Tomris Türmen, director of its family health division. Of the 500,000 women who die during pregnancy or childbirth 99% live in developing countries 'where maternal mortality rates are often 100 times higher than in industrialized countries'.

ECLAMPSIA, the convulsions sometimes occurring at pregnancy's end, is thought responsible for 10% of maternity deaths, but infection and haemorrhage are the main causes. Eclampsia (from the Greek to explode) occurs in nearly one in 2,000 maternities in the UK, but can be as high as one in 100 in some developing countries. There is also pre-eclampsia – high blood pressure linked to pregnancy, said to affect 10% of the 3.5 million US pregnancies every year. One definition of eclampsia is the occurrence of convulsions in association with the signs and symptoms of pre-eclampsia, these being oedema – excess of fluid, proteinuria – protein in urine, and new hypertension. The cause, or causes, of both eclampsia and pre-eclampsia, are – as the saying goes (so frequently) – poorly understood.

The WHO's health figures are always disturbing, and certainly so in the matter of child-rearing – 23 million women a year experience 'serious pregnancy-related complications' (with 12 million 'long-term disabilities' as a result), 20 million women a year seek to end pregnancy 'by risking an unsafe abortion' (and 70,000 of these desperate women die). As for the 2.8 million babies who die during the first week of life these deaths 'are largely a consequence of inadequate or inappropriate care during pregnancy'.

FOETAL RIGHTS One major change, not so widespread in developing countries, is that the foetus itself is being regarded as a patient more than ever before. In Victorian times, if the mother was in trouble, the offspring received little attention. Even newborn babies were placed on one side and half-forgotten if the mother was ailing (as she often was from puerperal fever). Advances in ultrasound, in Doppler, and various invasive procedures mean that much foetal disease can now be diagnosed long before birth, and can also – on rare occasion – be treated. An immediate problem, running counter to the notion of foetus-as-patient, is that many an investigation reveals the presence of a congenital malformation. This discovery is then followed by a recommendation, or demand, for abortion.

A problem of timing therefore arises: when does the foetus acquire rights of its own? One answer, favoured by the US and UK, states that viability provides a satisfactory turning point. These two countries (and others) have selected twenty-four weeks post-conception as that time. Most babies born before then die, and most born after it manage to survive (albeit with considerable assistance). However, the UK and France, for example, still permit termination of pregnancy for serious foetal abnormality long after viability. The rights of the foetus have then plainly been brushed aside, with the compliance (one assumes) of the mother. A newborn baby undoubtedly has rights, as with any human being, but these can also be subjected to the wishes of its parent. Every minor is a form of second-class citizen, short on rights as well as responsibilities, but no one considered a foetus to be more than some form of maternal appendage – until very recently.

That situation altered, so far as the UK is concerned, in November 1995. The Court of Appeal Criminal Division decided that a charge of murder or manslaughter could be brought if someone had deliberately injured a pregnant woman and/or her foetus and if that foetus was born alive and then died from that injury within a year and a day. Liability was the same whether the child's death resulted from injury to the mother or a direct injury to the child when a foetus. This decision resulted from an incident when a man stabbed his twenty-six-week-pregnant girlfriend in the abdomen. Three weeks later she gave birth, after apparently making a good recovery, but the baby was found to have been injured by the stabbing. Unfortunately the mother died a few weeks later and the baby, having been given intensive care and surgery, also succumbed at 4 months. The man admitted he had killed the mother and was also charged with the murder of the child.

When this case was brought to trial the judge decided the man could not be doubly charged because the child had not been – at the time of the killing – an independent being. It was this judgement that was overturned by the Court of Appeal. The ruling is therefore important medically. If a pre-birth injury is caused by a doctor's gross negligence there could be a charge of manslaughter in respect of the dead infant even if the mother survived. In short, the second-class citizenship of pre-birth human beings has advanced in the direction of equality with post-birth individuals.

MISCARRIAGE Whether half of all embryos suffer spontaneous abortion, as already stated, or one-third (another frequently quoted estimate), and whether one-quarter of these expulsions occur before pregnancy is detected clinically (one further estimate), there is no doubt that successful fertilization does not always lead to a successful pregnancy and birth. Miscarriages do occur, either early or late (with some so early that no one appreciates what has taken place). Some are associated with considerable pain; others produce hardly any discomfort. Vaginal bleeding is the fundamental sign, but only about half the women who bleed during a pregnancy will actually miscarry. Small spots of blood ('spotting') are usually the first sign. These may cease (and be of little significance) or may lead to heavier bleeding. Similarly, a heavy flow (rarely necessitating a blood transfusion) may not be partnered by the loss of the foetus, and a slight spotting may be followed by such loss. There may also be cramping, either slight or very severe. In short there are no formulas. Each woman's experience – blood loss, time, cramps, pain – tends to be unique. The only similarity, for the 50% whose blood loss leads on to miscarriage, is that a developing human being has been cancelled.

Unfortunately, for those for whom miscarriages are most unwelcome, one such event is slightly more likely to be followed by another. In a major survey, published in the *Lancet* in October 1994 (and with information gathered from South America, Italy, and Washington

DC) the rate of loss was 8.6% for women who had had no previous pregnancy. This rose to 9.6% if there had been a previous pregnancy (or pregnancies) but not previous loss, to 13% if there had been one previous loss, to 21.4% if there had been two previous losses, and to 25% following three previous losses, but even this final figure means that the odds of a successful subsequent pregnancy are still 3 to 1 in favour.

ECTOPICS (Greek: out of place.) The manner of implantation, after a bunch of cells has travelled from the oviduct, reached the uterus and become attached to its wall, does seem a touch haphazard. On occasion the bunch of cells wanders right out of the uterus, with no one much the wiser. On other occasions, and far more seriously, the developing bunch attaches itself in quite the wrong place. Ectopic pregnancies, as they are called, occur far too frequently, bearing in mind (as a review article in the *New England Journal of Medicine* affirmed) that 'the gravity of ectopic pregnancy cannot be overstated'. In 1989, for example, there were (an estimated) 88,400 such pregnancies in the US, a frequency of about one in forty to normal pregnancies. The wrong attachments occurred within the uterine tube (commonest), on the ovary, at the bottom of the uterus, or even outside the reproductive system and on some other organ, such as the large intestine. They – wherever they are – are always unwelcome, and all pregnant women are at risk. (One ectopic study published in the *Lancet*, 6 May 1995, put the proportion differently but stated that 'ectopic pregnancy is a potential surgical emergency that occurs in 1–2% of reported pregnancies'.) The *New England Journal* article added that an ectopic 'may be the only life-threatening disease in which prevalence has increased as mortality has declined'. (As with crime figures extra police vigilance may cause them actually to rise. Increased medical awareness – about everything – and technical expertise are thought to be mainly responsible for the relative rise in ectopics.)

Modern methods enable these wrongly placed pregnancies to be diagnosed at eight weeks post-conception, or perhaps even earlier. A trouble with ectopic symptoms – pelvic pain, vaginal bleeding – is that they mimic those linked to spontaneous abortion. If the misplaced foetus is left where it has chosen to implant itself there can, eventually, be 'catastrophic intra-abdominal hemorrhage' (as the same article described it). Some women are more at risk, such as those with earlier problems linked to their uterine tubes, or previous ectopics, or pelvic inflammatory disease. The standard procedure for investigating ectopics has been to take a look – via the technique of laparoscopy. Another method involves assessing the level of progesterone, the hormone being produced by the ovary's corpus luteum. Only pregnancies correctly situated lead to the correct quantities of progesterone. It is very important, as the review article stressed, for doctors to be aware of ectopic possibility, and diagnose this problem, 'before the onset of hypotension, bleeding, pain, and overt rupture'.

The standard method of removing ectopics used to involve laparotomy, a cutting open of the abdomen. Then came operative laparoscopy, a form of 'key-hole' surgery becoming commoner with surgery in general. The wrongly placed foetus can also be removed by non-invasive means. The modern procedure costs less, speeds recovery, and is less likely to lead to subsequent infertility. On occasion, and following investigation, the surgeon may encounter a well-developed (albeit wrongly located) foetus capable of survival. Even normal weight can be achieved outside the uterus.

Finally, and most bizarre wrongness of all, one foetus may even grow within another. An infant has (very rarely) been opened, following its distress, to reveal a dead and poorly formed foetus somewhere within its body. A 3-month-old boy in Hong Kong had three foetuses removed from his abdominal cavity after his mother had grown suspicious at his swelling condition. A photograph was published in the *British Medical Journal* (of June 1994) which showed a 35-centimetre foetus found 'with identifiable head, trunk and limbs' within a Nepali boy aged 4 'of small stature' who had complained of an expanding mass in his upper abdomen. Following an operation, he made a full recovery.

Lithopaedion – stone child – is the name given to a foetus which dies within its mother's body and becomes calcified. Such an incident was commoner in earlier days, before surgery for ectopics was even contemplated, and the earliest documented case was in 1200 BC. This misfortune can still occur today. A 92-year-old Austrian woman, admitted to hospital when gravely ill (before dying one week later), was found to have within her a stone child estimated to have died thirty-one weeks post-conception. Her son reported that she had given birth to three children before 'thinking' she was pregnant again when aged 32. She then became very ill, with considerable abdominal pain, but recovered to start menstruating once again. Her lithopaedion was therefore (probably) 60 years old when eventually discovered.

Obstetricians (and mothers) were much encouraged in August 1994 to learn of the first successful transplant of an ectopic pregnancy. Minerva, the *BMJ*'s (otherwise anonymous) columnist, reported that a consultant at St George's Hospital, London, had performed a salpingotomy (cutting of the uterine tube), had removed the embryo developing there, had re-implanted it within the uterus, and had been gratified to record the birth thirty-three weeks later of a 6 lb baby. This exciting revelation was then published widely. Eventually came news that the case report, first detailed in the *British Journal of Obstetrics and Gynaecology*, was untrue. Compounding this error, Sir Geoffrey Chamberlain, president of the Royal College of Obstetrics and Gynaecology and also editor-in-chief of its journal, had his name at the head of the erroneous article, along with Malcolm Pearce, who had falsified medical records in the claim of an ectopic transplant. Doubts arose mainly because nobody at the hospital could recall having assisted in such an operation.

From an outsider's point of view the truth was likely to break out, sooner or later. It was such a happy story – a wrongly placed but developing foetus becoming rightly placed – that it was bound to arouse exceptional interest, being no run-of-the-mill medical research. It was a first, and a most welcome one. Following the inevitable inquiry Pearce's name was struck from the Medical Register and Chamberlain resigned his presidency. (The principal curiosity for the rest of us, apart from disappointment that an ectopic had not been successfully transplanted, concerns the degree of scientific fraud.

One estimate, published in *Science* in 1987, stated that only one in a million scientific papers was anything other than accurate and truthful. The *Lancet* later reported that 'a compilation of 46 documented instances of scientific misconduct in biomedical research since 1974' implied that fraudulent misbehaviour is much commoner. Presumably some instances never come to light, and the malignancy is therefore yet more frequent.)

For the time being, while the medical profession berates itself for blatant misdemeanour, it is sad that ectopics cannot – yet – be removed from unsatisfactory locations and shifted, like seedlings from the potting-shed, to the environment that serves them best.

What can be done, and what has been done, is foetal surgery. Scott Adzick, and others at the University of California in San Francisco, performed eighteen such operations even by 1990. Earlier, and in work on animals, they had learned that foetal wounds heal without any scarring. (It had long been known that wounds on newborn infants heal better than on older children, and children's wounds heal better than an adult's wounds, but foetal wounds heal best of all. This manifestation of an ageing skin – and of ageing itself – therefore occurs even before birth.) Foetal surgery can correct conditions such as diaphragmatic hernia. This error provides no hazard for a foetus, but can kill a newborn via suffocation.

SCANNING Checking on a developing foetus would have been developed whether or not ectopic pregnancies ever occurred, and the checking has increased immeasurably in recent years. Originally the world of the foetus was a closed book. No one knew what might emerge on the birth day until, suddenly, one infant was there to see, to weigh, to examine, to check for errors. Now there is a battery of techniques permitting the mother (and father) to know much about their infant before its actual arrival – a bit like skilful handling of still-wrapped presents before their opening.

Ultrasound imaging can detect structure, malformations and growth (to within a millimetre). Doppler ultrasonography can provide information about blood flow (and therefore about the organs being supplied). Cordocentesis – the aspiration of umbilical blood – can confirm or eliminate chromosomal abnormality, and also be helpful about placental and foetal health. Amniocentesis, the extraction of some amniotic fluid, enables cells discarded by the developing foetus to be examined, and thus demonstrate some of the signs discovered by cordoentesis. (Amniocentesis is traditionally performed during the second trimester – at sixteen weeks' gestation – and has been associated with a 1% risk of subsequent and spontaneous abortion. There is a similar slight risk with cordocentesis, but there is chance of greater gain, with its blood samples being more informative than amniotic fluid.)

So much for the good news. On the other side is contrary opinion about the actual benefits. A major article in the *New England Journal of Medicine* recently stated that the detection of various conditions – congenital anomalies, multiple pregnancies, foetal growth disorders, foetal age, placental abnormalities – is undoubtedly enhanced by ultrasonography but 'a beneficial effect on perinatal outcome has not been substantiated'. There is 'concern about unnecessary testing, overtreatment, and cost' and this concern is growing. The new techniques are exciting, do reveal, and do express care (to the parents in particular) but may not necessarily assist. Indeed they may be harmful. The (British) Association for Improvements in the Maternity Services, a consumer group, recently drew attention to 'two studies (which) have suggested an increasing rate of miscarriage following ultrasound scans'. Work in Australia has reported that 'insonated babies' are smaller. One study in Northern Ireland concluded that 46% of pregnant women at one hospital had five or more scans, and 10% had ten or more.

Ultrasound equipment, according to the Association, is becoming more powerful, and the new vaginal probes 'deliver greater ultrasound exposure' to the foetus than most abdominal scans. On occasion the family (and others) gather round the ultrasound screen, delighted – lengthily – by the scientific miracle of seeing an infant, moving, twitching and living, long before it is born. The mother is also pleased, and her pleasure might positively affect the outcome of some pregnancies, being such a psychological boost. As for routine ultrasonography, the standard advantages (detection of unsuspected foetal abnormalities being highest on the list) have to be weighed against the disadvantages – the possibility of false negative and false positive diagnoses, the concern about the effects on the foetus of exposure to ultrasound, and the tremendous cost. There is much to be said – or so it seems increasingly these days – for waiting until a birthday before attempting to examine the still-wrapped presents waiting for that time.

The threat of a malpractice suit may be having the greatest influence on whether or not to employ a particular technique. 'My child was born malformed. Should this have been detected early in pregnancy giving time for abortion? Should I sue for compensation?' The hospital, in general, tends to keep the relevant records. And it is the hospital, in general, that is being sued. Hence a current interest in (smart) memory cards, such as the recently developed Clinicard. No bigger than a credit card, it can store the equivalent of 1,000 pages of information (2.4 megabytes of formated data). It can detail whether or not a patient was given a particular course of treatment, or a particular investigation. 'The cards will provide a clear audit trail', according to the promoters, and therefore state what should have been done, what was done, and whether there is a case to answer. With malpractice being such a growth industry the cards should clarify matters, but no one expects lawyers to face penury just yet.

RUBELLA has left the headlines about as speedily as it first hit them. Only during the Second World War was it first realized that a pregnant woman catching German measles, particularly in the first three months, was at considerable risk of producing a damaged baby, perhaps with cataract, perhaps with deafness or with heart defects. In 1962, just twenty-one years after the original Australian report linking rubella with malformation, its virus was successfully cultivated in a laboratory. As if to emphasize the need for further haste, a rubella epidemic hit the United States in 1964. Among the 2 million cases were tens of thousands of pregnant women. When results were totalled four years later the disease was said to have caused 'at least'

30,000 stillbirths, miscarriages and birth defects. Fortunately, and in 1965, a one-day test was devised for accurately diagnosing German measles. For women proving positive, notably during their first trimester (and particularly the first four weeks of that three-month period), the ethical choice of whether or not to abort their foetus had been clarified. The malformation risk was 60% for the youngest foetuses, 35% for 3–6 weeks post conception, and 7% for the subsequent 4 weeks.

A rubella vaccine was first licensed in the US in 1969, namely three years after scientists had developed one. The world – or rather those parts of it able to afford such a practice – realized it had better vaccinate all girls before they had a chance of giving birth. The UK, for example, introduced rubella vaccination in 1970 for all 11- to 14-year-old schoolgirls. Later this was offered to adults who had never experienced the disease. By 1990 susceptibility to rubella among pregnant women attending ante-natal clinics had fallen to 1.8%. Two years earlier rubella had been routinely added to mumps and measles vaccines (the MMR cocktail) for children of both sexes. By 1992 rubella notifications – it is a notifiable disease – had fallen to their lowest level, but there was a disturbing outbreak in 1993. St Thomas's Hospital, London, discovered that 133 women of one entire year's batch of parturient women had apparently missed the disease, and so were vaccinated after they had delivered their infants.

Rubella is therefore, in part, ancient history; but, as historians say, those who disregard the past are thereby liable to relive it.

LITIGATION It is easy (if fruitless) to wonder, from time to time, which current practice would most astonish our ancestors. There is, for example, the present custom of employing women in the armed services (even to the extent of their fulfilling combat roles). Plainly the matter of pregnancy is an inconvenience, not least to employers, such as the military. Between 1978 and 1990 the British Ministry of Defence routinely sacked all women who became pregnant during their service careers. Some of these women (and their lawyers) then appealed to the European Court of Justice. This court ruled that the

Ministry of Defence had acted unlawfully. By September 1994 compensatory payments totalling £20 million had been paid to 2,400 women, while a further 1,700 women were waiting for their cases to be reviewed. (Payments have averaged around £7,000, but several have run to six figures, with the highest being £400,000. Those women with the brightest career prospects have, in general, received the highest awards. Many an individual thought the awards over-generous, particularly when compared with those for serious physical injury also experienced on duty.)

Then came (as further astonishment for our ancestors) the suggestion that women denied motherhood out of enforced obligation to their employers had also suffered. The Armed Forces Pregnancy Dismissal Group stated: 'Women have a right to feel aggrieved that they have lost the chance to have a family because of an unlawful government policy.' Between those years of 1978 and 1990 there had been 180,000 women, all serving in the forces, who could claim they had given up the opportunity to have a child. In *Equal Opportunities Review* it was argued that those servicewomen who chose not to have a child were discriminated against in the same way as those who left. According to the Employment Appeal Tribunal the award for injury to a dismissed servicewoman should reflect 'her frustration, anger and bitterness at having her legitimate and honourable aspirations thwarted by the MoD's unlawful behaviour'.

So too, it added, with the women who stayed. Servicewomen were either unlawfully dismissed or remained childless to avoid unlawful dismissal. But should the payments stop there? argued Michael Rubenstein, specialist in equal opportunity law. What of all the loyal husbands of loyal servicewomen who were denied the joys of paternity? As yet no male has received compensation, but the MoD is certainly changing its former cavalier approach. In October 1995 it announced that it was ordering 4,000 maternity outfits for its military personnel. Dewhirst (Uniforms), of Leeds, was both amazed and delighted by the £250,000 request.

Civilian life is also altering. A landmark victory, for one British woman in particular (and women in general), occurred in July 1994. Carole Webb was sacked two weeks after being hired because she then discovered she was pregnant. An industrial tribunal refused to promote her case. So did the Employment Appeal Tribunal. So did the Court of Appeal, but this did give permission for an appeal to the House of Lords. The ultimate authoritative body in Britain then gave its permission for her case to be referred to the European Court. Seven years after her dismissal she learned that she had indeed been unfairly treated, and was therefore well placed to claim compensation from the freight company that had employed her before, in peremptory fashion, sacking her for being pregnant. She now has two children and, no doubt, some extra money could come in handy, particularly after all that labour in the various courts of appeal. As for employers, whether MoD or others, they are being forced to realize that pregnancy, far from being some aberration (as some employers may like to think), is actually and generally a normal, reasonable and happy circumstance, for women and for men. It is a standard feature of our lives and, which cannot be forgotten, gave rise to every single one of us. Warts and boils, or toenails growing inwards, may all be aberrations, but hardly pregnancy.

INFERTILITY

TREATMENT CLINICS   IN VITRO FERTILIZATION

SURROGACY   DONATIONS   FROZEN EMBRYOS

FOETAL OOCYTES   REGULATION

All I can say, doctor, is that the boy never got anything
congenital from my side of the family.
Quoted in *Medical News-Tribune*

TREATMENT CLINICS It must be unsettling to work within a
reproduction clinic. Visitors either want contraception or they want
a child. They are desperately anxious not to conceive or, no less
desperately, long to do so. The pressure not to conceive is considerable, with an unwanted child (or an unwanted abortion) being a
frightening possibility. The pressure to produce offspring can also be
tremendous – from parents and in-laws (in particular), from fertile
siblings, from colleagues, acquaintances, society in general, and (one
assumes) from fundamental longings. Being the only form of immortality, as inherited genes are passed on and the Y-chromosome goes
unchanged from male parent to male offspring, the desire to circumvent inevitable mortality in this fashion is also strong. Breeding is not
a prerequisite of life, as if lives are worthless without descendants,
and perhaps more of the miserably infertile should read Viktor E.
Frankl, the psychotherapist: 'I observed that procreation is not the only
meaning of life, for then life in itself would become meaningless, and
something which in itself is meaningless cannot be rendered meaningful
merely by its perpetuation.' Nevertheless there is pressure, there is
misery, and there is considerable attendance at fertility centres.

Some 10–15 per cent of couples are infertile, with the problem
attributable to the male partner almost as much as the female. Some
surveys conducted for 1988 by the National Center for Health
Statistics (of the US) estimated that 8.4% of women of reproductive
age (15–44) had an 'impaired ability' to have children. Of this 4.9
million there were 2.2 million with no children and 2.7 million with
one child or more. About one-quarter of women have an 'episode of

infertility' sometime during their reproductive span. (These US figures are similar to those from Britain and elsewhere in the developed world.) In more recent years the infertility problem has increased, and for several reasons. First, there are more women in that age group. Second, more people now ask for help. Third, there are now more ways in which they can be helped (such as surgery on the uterine tubes, insemination with sperm, promotion of ovulation, and in vitro fertilization – about which much more in a moment).

About half of the couples who attend fertility clinics (2.3% of US women between 15 and 44 and 4.6% of married women, both figures being for 1988) eventually conceive. Most are treated, but about 3% of couples with unexplained infertility conceive without anyone being the wiser why this should have happened. As for the known causes of infertility there is a considerable number. The ovary may not be functioning correctly. The uterine tube(s) may be damaged. Spermatozoa may be in error (too few, too malformed, insufficiently active). Endometriosis, a wrongful form of menstruation, may be to blame. More controversial are others, such as the role of the cervix, which may (or may not) lead to infertility.

The waiting-room of a fertility clinic tends to be dominated by women, partly because failure to achieve motherhood can be more distressing than a failure of fatherhood, but mainly because, as the *New England Journal of Medicine* stated: 'male-factor infertility remains a daunting challenge, largely because spermatogenesis cannot readily be altered for therapeutic benefit'. Efforts are being made to improve the creation of sperm, their motility and their number, but success has been limited. The option of AID, of using someone else's sperm, is often effective but not always welcomed by the putative father. He, already unhappy at the state of his sperm, the very essence of his maleness, may be made unhappier by the thought that some other and unknown male will fertilize his wife. Such dismay is likely to be exacerbated by all subsequent talk, after a child has been born, about genetic likenesses and traits in general. (Some natural offspring are quite ludicrously similar to one or other parent, as if only one set of genes has been inherited, and it is easy to wonder if sperm donors

– who valiantly masturbate for a small fee – ever recognize a younger *alter ego* walking down the street.)

Stress is related to infertility, a proposition first put forward almost 200 years ago by a Lancashire doctor. His ideas were scorned (has any novel notion not been scorned?), mainly for his allegation that a change in someone's status could be influential. Work on animals has since supported Thomas Jarrold's supposition. Low-ranking female rhesus monkeys have fewer offspring than their higher counterparts. If immature marmosets are raised in groups, only the dominant females experience puberty. When such females are housed independently they all do so. Young klipspringers who stay near their parents in the third year of life do not develop sexually. If such antelopes move away they can become sexually mature even in their first year of life.

These and other examples show that severe competition (and heightened stress) leads to fewer offspring overall. This may make biological sense, by reducing competition when competition is already rife, but it becomes less straightforward when translated into human terms. Some women of South American Indian groups have ceased to menstruate when experiencing the gross trauma of being contacted (for the first time) by, for example, Brazilian gold prospectors. In quite different circumstances men produce fewer sperm, and are therefore less fertile, when undergoing vigorous training programmes. The organizers of sperm banks, who receive most of their contributions from students, allegedly know when exams are looming. Sperm counts fall dramatically.

Biologically it might seem strange that such a preponderance of infertility exists. The purpose of life is to reproduce, in so far as life can be said to have intent, and a failure to reproduce is the most fundamental failure. It is therefore even stranger that sub-fertility in men has been shown to have a familial component. For example, the form of infertility lengthily named oligoasthenoteratozoospermia, common in humans, is linked to environmental *and* genetic factors. So are various chromosome abnormalities. One answer for this strangeness is that, as a team from the University of Leeds phrased it, 'human survival is not critically dependent on maximizing the number of offspring'. There is no great pressure for all couples in a community to have as many children as they are capable of bearing. There is, and certainly was, infant mortality but couples need only create two children who will achieve adulthood for that particular population to be stable. To create three successful adults is more than adequate, assuming stability is desirable.

Such generalities are not of great interest to childless couples. Whatever the cause, or causes, of their problem (10–30% have two or more such causes), infertility is 'a major life crisis for most couples', as the *New England Journal* described it. The emotional experience begins 'with disbelief and denial' before moving on to 'frustration and anger' and eventually 'acceptance'. Friends can be wounding in their proffered thoughtlessness: 'If you want children you're always welcome at any time to look after our little monsters.'

There is also the problem of money. The cost of treatment can be high, and there is no guarantee of success. Britain has a National Health Service but, according to Cambridge's Bourn Hall Clinic: '95% of couples requiring advanced fertility care have to pay for their treatment'. Britain is one of the few countries in Europe where the government does not provide whole-hearted support for such assistance. Worse still, from the viewpoint of infertile couples, the British government actually charges a fee – this tax varying according to the intervention the couples receive – every time they receive treatment. The Human Fertilization and Embryology Authority has to license the fertility clinics, and has to pass on the fixed charge stipulated by the government for each form of licensed treatment – such as (artificial) donor insemination. Many fertility problems are caused by medical conditions, and clinics resent this extra imposition on those who happen to suffer infertility (which must be paid for) as against appendicitis or tonsillectomy (which are serviced by the state).

The good news is that unfortunate couples are increasingly seeking assistance (even if it costs). This was not always the case, such personal dilemmas being kept personal, but matters changed about forty years ago. Conception outside the body was first accomplished in 1958 – for rabbits. Twenty years later the same 'extracorporeal' trick was achieved for humans. A woman without eggs could receive them from another person, have them fertilized by her partner, and achieve the longed-for pregnancy. Men without adequate sperm could accept that their partners would be fertilized by donated sperm, and then rejoice at the pregnancy and birth. Couples with poor sperm *and* deficient eggs could still acquire offspring, with both sets of gametes being donated. If only her uterus is faulty, but her eggs and (partner's) sperm are satisfactory, a surrogate pregnancy is possible, with another woman incubating their child and then delivering it, post-partum, into their hands.

In short, infertility and its problems have entered a brand new world, with all manner of interference quite impossible at the time of the Second World War. The 'major life crisis' can be averted, time and time again, much to the joy of those who might have thought there was no hope.

IN VITRO FERTILIZATION In 1978 *People* magazine named her one of the ten most important people of the decade. She was not even an adult but a newborn baby (and not to achieve adulthood until 1996). Her arrival in the late 1970s was received differently on each side of the Atlantic. 'And here she is ... THE LOVELY LOUISE', trumpeted a British headline. 'Infant in Britain reported "Normal",' stated the *New York Times* with rather less excitement. Louise Brown had begun her life outside her mother, after sperm and ovum had been united externally, and had then been installed within her mother, Lesley Brown. She had therefore been incubated in traditional style before her ordinary birth, but the procedure of her origin had never before been successfully completed for a human being.

Part of the reason for America's lack of enthusiasm in Louise's birth was that country's simultaneous involvement in legal proceedings initiated by a New England woman, Doris Del Zio. Her involvement with IVF, with in vitro fertilization, had been nothing like so satisfactory. In 1973 Landrum Shettles had attempted to use her eggs and her husband's sperm for an external fertilization.

Raymond Vande Wiele, head of Shettles's department, only learned
of the experiment after it had been initiated. For all sorts of reasons,
not least the possible curtailment of his hospital's funding from the
National Institutes of Health, Vande Wiele effectively destroyed what
Shettles had been preparing. The destroyer was promptly branded
hero (for halting such a procedure) and also villain (for killing what
might have been an embryo). As for Mrs Del Zio, thwarted in her
desire to have another child (her first had left both uterine tubes
irreversibly blocked), she then sued Vande Wiele, and the hospital
and various others both for the embryo's destruction and for the hurt
this act had caused.

It so happened that her case reached the Federal District Court in
the same summer that Lesley Brown was quietly waiting for, and then
delivering, her famous offspring. With 'Test Tube Death Trial' as one
set of headlines, and 'Miracle Birth' for the other, the dilemma of
modern IVF was neatly demonstrated. It is novel. It is of concern,
bringing Aldous Huxley's baby manufacturers to mind. It also raises
the issue of life's sanctity, with (perhaps) several eggs being fertilized
and all but one abandoned when no longer desired. It certainly
opened up further possibilities – donor sperm (rather than the
husband's), donor eggs (rather than the wife's), selected sperm and
eggs (from the good and the great), a thorough investigation of the
developing embryo before its implantation (and a rejection of this
developing life if less than perfect). Some citizens therefore visualize
the future as a nightmare of awesome possibilities.

The principal mitigating factor, causing Huxley's images to recede,
is that external development, even in the early stages, is expensive. It
becomes more so as development proceeds, and even optimists in
external management cannot begin to visualize a nine-month arti-
ficiality. Women (and their uteruses) are better, cheaper, and readily
available. Besides, once an embryo has been initiated, most decisions
have been taken. As with an architect who designs a house, the
builders merely construct what has already been decided. So too with
human development.

Although an earlier case had been long forgotten, the arguments
of 1978 – with Louise happily on one side and Del Zio unhappily on

the other – had been enacted to some extent sixty-nine years beforehand. They were caused by revelation of a success achieved a quarter of a century even before that date. A letter from Addison Davis Hard, published in *Medical World* in 1909, affirmed that he had arranged the first AID in 1884. Artificial insemination by a donor is not in vitro fertilization, but it lies along that path. According to Hard a Philadelphia merchant had been deficient in sperm. As substitute Hard had collected semen 'from the best-looking member of his class' at Jefferson Medical College, and had inseminated the wife while she was anaesthetized, with her poorly informed about these proceedings. A son was born, and this child had reached the age of twenty-five before Hard published the relevant information. There was uproar, with bitter attacks upon the surgeon who had performed the insemination, upon the sterile father (who had been informed how his wife had become pregnant but who then kept the fact from her), upon the anonymous donor (with many an assumption that Hard himself had been the 'best-looking' member) and upon the medical school where it all occurred. Much similar recrimination was poured forth, albeit by quite a different generation, when IVF became a public issue in 1978.

The IVF procedure is reasonably straightforward. The first step is to know when ovulation is about to occur. In general it takes place a couple of weeks after the onset of menstruation, but blood samples are a firmer guide as they indicate the level of oestrogen production (and therefore of follicular development). When these tests suggest imminence it is occasion to measure urine. Samples every three hours give information about the concentration of luteinizing hormone, a yet more precise guide to ovulation. When the medical team judge that the liberation of an egg is only two hours ahead it is time for laparoscopy, for taking a look. This permits the surgeon to introduce forceps accurately, to grab hold of the ovary, and then to observe the blister-like developing follicle which contains an egg. A fine needle is inserted into the follicle, and its fluid is sucked out. Within this fluid, if all has gone well, is the precious female half of another human being.

A microscope will reveal whether the procedure has been successful.

In order to improve the chances of finding an egg the donating woman has previously been stimulated, via hormones, so that her ovaries develop more follicles (and eggs) than the customary singleton per month. It is good to have more than one egg on which to attempt fertilization. While this has been proceeding the male partner has been asked for his contribution, the sperm to perform the fertilization. A modest number, perhaps 100,000, is then added to the culture medium containing each egg some four to eight hours after extraction. The container, more of a dish than a test-tube, is then placed in a suitably warm incubator.

After twenty-four hours or so the ovum (or ova) will be checked to see that fertilization has properly occurred. After another twenty-four hours the developing embryo, having successfully reached its two or four-cell stage, is then deposited via the woman's cervix into her uterus. Normally an embryo would be sixteen cells or so before reaching the uterus, but it is difficult maintaining an embryo's development in artificial conditions. The compromise is to place the developing form in the more natural circumstance sooner rather than too late. Aldous Huxley's 'test-tubes', which achieved (in 1932) the complete creation of extra-uterine offspring, were not only ahead of their time but well ahead of current ability and the foreseeable future. (For which, says almost everyone, much thanks.)

The maternal recipient of an externally fertilized embryo rests for some hours before returning to her normal life. What she will not know, for at least twelve days but not more than fourteen, is whether the transfer has been successful and she is firmly pregnant. Samples of her blood can monitor the production of progesterone (by the corpus luteum) and of human chorionic gonadotrophin (by the developing embryo) but doubt can still exist even at this stage. Absolute certainty is not possible until an ultrasound scan at six weeks post-transfer reveals a beating heart as proof positive of a brand new life.

So much for the story of what should happen. As for the results, some useful facts were recently published in France by FIVNAT, the organization representing nearly all French IVF centres. For the period 1989–93 a total of 95,000 'ovarian punctures' were performed.

These permitted 91,695 IVF attempts to be made, and 11,405 pregnancies resulted. Various losses then occurred, such as spontaneous abortion (18.1%) and ectopic pregnancy (4.9%), but the majority of pregnancies were satisfactorily concluded. Of the 8,721 deliveries 72.7% were of singletons, 23.4% were of twins, 3.7% were of triplets, and 0.1% were of quadruplets, making a grand total of 11,423 babies.

With twins normally occurring in one in 80 European births, triplets one in 6,400, and quads one in 500,000 births, the IVF rate of multiple offspring was therefore very much higher. The French were improving the chances of implantation and pregnancy by frequently inserting more than one developing embryo into each recipient. In Britain the number of triplet births trebled in less than a decade, having been more-or-less unchanged for the previous fifty years. In 1994 there were 260 sets of triplets born, and even eight sets of quads (as against ninety sets of triplets and no quads in 1985). Twins also showed a rise, there being 6,308 in 1980 and 8,451 in 1994. As there had been consistency in multiple births beforehand, and as IVF procedures frequently put more than one egg in the basket, the various rises are entirely attributable to the new techniques.

Alison MacFarlane, of the (UK's) National Perinatal Epidemiology Unit which publishes multiple birth statistics, has said that 'from 1990 to 1992 we thought the growth might be slowing but then it picked up again. We are concerned ... The birth of triplets puts a huge strain on parents ... If one baby yells it wakes the others and the mother becomes exhausted.' One mother wrote to the unit: 'For the first twelve months I didn't go out. My husband went to the supermarket and I shopped for clothes by post.' Since 1990 IVF clinics have been limited by law to placing a maximum of three embryos in the uterus. (In early days some clinics had been inserting as many as ten to boost the chances of a birth.)

Of the grand baby total in France 1.07% had major deformities (such as cardiac defects, Down's syndrome, spina bifida), and 1.73% had minor deformities (like club feet and cleft palate). (It is important to note that such proportions are no different than for normal deliveries.) The sperm donors were aged 35 on average and the egg-receivers 33. About eight oocytes (the precursors to ova) were extracted during each ovarian 'puncture', and the number of embryos achieved after fertilization was about four (with 50% of the oocytes therefore becoming ova/eggs). Not all of these embryos were inserted within the waiting uteruses, but the average number which were inserted dropped from 2.94 to 2.73 during the four-year survey. (As confidence in the technique continues to grow, and as fewer and fewer embryos are deposited, it is likely that the number will shrink accordingly. Most parents, however desperate for offspring, are not necessarily enthusiastic for a double ration, let alone threes or fours.)

The FIVNAT report also mentioned that the direct injection of sperm into oocytes had been performed 200 times in France in 1993. This procedure sidesteps such naturalness as exists within an IVF laboratory, when hundreds of thousands of sperm are released to fertilize an egg. Instead, via micro-surgery and a system know as ICSI (intracytoplasmic sperm injection), a single sperm is used to effect fertilization. This technique was pioneered in Brussels where 350 babies were born (by mid-1994) following this most precise innovation. The Belgians report that the incidence of babies conceived with defects via ICSI is no higher than among those born naturally. Such success did not prevent France's national committee on bio-ethics calling for a curtailment of the procedure, considering that too little is known about the risks.

Also, 'contrary to all recognized rules for medical research, the first trials on humans were done even though experiments on animals were still extremely limited'. In natural conception, with sperm competing to fertilize the egg, defective sperm usually fail. The new procedure may therefore be using the wrong kind of sperm. ICSI is not likely to go away, but may be restricted until more is known. The certainty of this form of fertilization, with sperm and egg so forcibly united, has undoubted attractions over the traditional IVF procedure of extracting ova from women, sperm from men, and putting them together in the hope that fertilization will occur. (A letter to the *Lancet* in May 1996, from the Center for Human Reproduction, Chicago, stated that ICSI 'has revolutionized the management of

severe male-factor infertility ... even men thought to have azoosper-
mia due to gonadal failure can now become fathers'.)

What is additionally certain is the considerable expense of IVF
whatever techniques are used. In the United States, where all prices
are better publicized, IVF costs between $44,000 and $211,000 for
each baby born. If there are extra troubles – the woman is over 40 or
the man has a low sperm count – the price of a baby conceived after
only one cycle of treatment can be $160,000. If there is no conception
until the sixth cycle the price may have gone up to $800,000.
(Ordinary parents, who can lament the cost, year by year, of the
infant parasite in their midst, do at least achieve the product for free.)
Fortunately, for Americans and Canadians, IVF treatment is regarded
as a health act. In Massachusetts, for example, the average increment
to a family health policy that will cover all infertility services is about
$2.40 per family contract per month. Very rarely does treatment
extend to the sixth cycle. According to a reproductive centre in
Boston 'more than 95% of babies' resulting from IVF were conceived
within four cycles 'irrespective of the patient's age or diagnosis'.

Rationing of treatment in any health service is already part of the
curriculum, and will probably become increasingly so (as populations
rise in age, as available treatments become more expensive, as
aggressive life-saving policies are pursued). Currently, and within the
UK, fewer than 2,000 children are born annually to mothers whose
own eggs have been fertilized by donor sperm. Some authorities
expect this number will rise, perhaps to 2% of all births (12,000 a
year), as techniques improve.

A political history of reproductive medicine, *The Stork and the
Syringe* by Naomi Pfeffer, queried why the desire to have a child is
not considered of political importance when freedom from unwanted
pregnancy is deemed crucial. Britain's Minister of Health said in July
1994 that infertility treatment should be given only to couples (and
'ideally' these should be married) where the father is committed to
the child's upbringing. Her remarks followed the revelation that a 44-
year-old nurse had become a mother while still a virgin. 'I would
have loved a husband and family,' said this new parent, 'but I never
met Mr Right and had to create a family without a man.'

Married couples receive most support for their wish to have a child, but there are others equally keen. Lesbians can be enthusiastic, a fact highlighted by the tennis player Martina Navratilova. Her request to the Italian fertility doctor Severino Antinori was refused because he opposed donor insemination for lesbians. 'I have nothing against them ... but I am against their pregnancy because of the lack of a father figure who is necessary for the growth of a child.' Older women can also be (belatedly) keen on motherhood. Antinori achieved considerable renown when he helped a 62-year-old Italian woman to have a child, and was then dubbed leader in the field 'of granny pregnancies'.

There is no age limit on treatment in Britain but Sir Colin Campbell (chairman of the authority responsible for licensing IVF clinics) had 'strong reservations' about the Italian success. Of the 45,000 women treated in British clinics in the three years (up to July 1994) twenty-five had been over 50, and of this number five had given birth, with the oldest aged 51. 'If we found the pattern of age rising we would intervene,' said Sir Colin. Much younger women have encountered intervention. A woman of 37 was refused fertility treatment by Sheffield Health Authority because she was over 35, the reason being that 'limited resources' had led to an age limit. 'It is fundamentally wrong, fundamentally unjust, and doesn't make good economic sense,' replied IVF pioneer Robert Winston, aware that many women initially benefit the nation through their employment and then experience fertility discrimination because of their age. Most British women, whether employed or not, attend private clinics rather than the National Health Service for IVF. A greater discrimination is therefore against the poor.

In Denmark there has been considerable debate, with the Minister for Health arguing (in February 1996) that the menopause should be the upper age-limit – for government-funded donor fertilization – and with doctors (in general) proclaiming that 40 was more reasonable as a cut-off age, the expensive treatment not being worthwhile for older women. Holland announced, in May 1996 via the Royal Dutch Medical Association, that it had no objections to IVF after the menopause. The commonly used age limit had been 40 years, but the new report considered the limit could be raised to 55 years. (Interestingly there is another angle to age. The chances of success with IVF go down steeply after the maternal age of 35 but then *rise* after the menopause for the fundamental reason that donor eggs are then employed.)

No one seems to express concern about the age of fathers. There was outcry over the 62-year-old Italian woman, but many an older man sires offspring without causing unfriendly comment. As men live less long than women, and much of the resentment about women is their possible death before their child is mature, there ought to be greater resentment about elderly male parents, not less.

Many single women, as with that nurse waiting for Mr Right, welcome the chance of assisted procreation, not so much because a suitable partner has not appeared but because they have no wish for such a partner. The health authorities can also be enthusiastic, on occasion, for such assisted pregnancy, believing it may avert health problems likely to ensue if pregnancy is denied. Abortions are solutions to terrible dilemmas; so too are births, however achieved. Infertility is never solely a medical condition. It is both a social issue and an emotional crisis, each to a greater or lesser degree.

*

Once it was once the duty of women to bear children (notably after wars when manpower had been much reduced). It has now become another right, supported warmly by those in favour and contested by those anxious about present techniques (and more alarmed about future possibilities). Will women put some eggs in store for possible post-menopausal use? Can procreation be permitted, assuming surrogacy, even when both actual parents are dead? Do fathers have rights over their previously donated sperm in the event of marital breakdown? What about inheritance, of wealth or property or titles? What about the sale of sperm from named donors, the famous or the infamous? And what about sporting heroes; will they be able to trade their gametes once their active sporting days are over? Pandora, the first woman (of Greek mythology), had a name meaning all-gifted. The gift of IVF may also lead to a pack of problems, much like those linked in legend to her box.

One particular problem (which hit headlines in October 1995) was raised by an English couple named Bloor. Their daughter had been conceived artificially, with the mother supplying the egg and an anonymous donor the sperm. The father (but not of the sperm) signed a fertility clinic consent form, a legal procedure proposed by the Human Embryo and Fertilization Act of 1990. Unfortunately the couple separated. The father paid maintenance money until such time as he changed his mind, then arguing that he was not the biological father and had no need to pay. The daughter in question, aged 9 when trouble started, has allegedly 'taken it all very well', accepting the novel situation with more equanimity than the parents. Baroness Warnock, who chaired a major inquiry into human fertilization in 1984, told a newspaper that 'I don't think he's got a leg to stand on . . . his biological distance from the child has nothing to do with it.' Pandora's box has undoubtedly more such problems in store. The Bloors were not first to demonstrate one of its fractious offerings, and certainly not last.

SURROGACY, according to a Chambers dictionary, is 'a substitute, a deputy, esp. of an ecclesiastical judge: one who grants marriage licences'. To many, and certainly in the context of this chapter, it is the procedure of using another's womb for the production of a baby. In some respects it is a form of adoption, in that a third party has carried and given birth to a child before handing it over. In other respects it is quite different in that the parturient woman may not have supplied any of the genetic material. (With 'partial surrogacy' the woman who gives birth has also supplied the egg. Only with 'full surrogacy' has she supplied none of the material.) As ever the new technique is able to provoke both happiness and misery, a straight-forward result or considerable litigation.

Take happiness first. Debbie Riley, of Chester, had a hysterectomy when aged 20. She, and then her husband, were miserable at the thought of childlessness. His sister, married and with two children, offered to help. She would be a surrogate. Debbie supplied eggs, her husband gave sperm, the two sets of gametes were united via IVF and then, with six embryos developing externally, three (safety in numbers) were placed within the obliging sister. Both Debbie and her sister-in-law had been given hormones to synchronize their menstrual cycles, thus assisting all three of the embryos to become successfully implanted. Debbie was asked, when this fact had been confirmed, if she welcomed the notion of triplets. She did, and Ben, Matthew and Lauren Riley were then born successfully, a few weeks early but none the worse for their premature expulsion. The result: one very happy family, one sister proud of her altruistic service, and a £20,000 bill spread over previous years following attempts to acquire an offspring.

The outcome can also be less satisfactory. Crispina and Mark Calvert, of California, paid Anna Johnson $10,000 to act as surrogate mother. The couple supplied the gametes, a clinic arranged for fertilization and insertion, and Anna Johnson became pregnant. In time she gave birth, but then wished to keep the baby. She had become 'emotionally bonded' to the child and sought a court ruling that she was the legal mother. A court decided that Crispina Calvert held that position. In its view Anna Johnson was equivalent to a foster parent who had looked after the child for nine months. At appeal the decision was upheld, the judges stating that genes were the deciding factor, not 'birth motherhood'.

This decision therefore runs counter to guidelines issued by the American College of Obstetricians and Gynecologists. Ovum donation, in its opinion, would henceforth be in jeopardy since every donor could be regarded as the legal mother. Conversely, and in the UK, the birth mother is the legal parent. Similarly, if a couple is informed that any of their children have been sired by another man, that other man can claim legal possession – and perhaps destroy one happy family. Adoption would not be put in jeopardy because the mother customarily forsakes rights before the adopting parents formally accept the offspring. (Adoption is therefore better controlled than the standard, age-old manner of child production where no one swears anything on behalf of the newcomer.)

Surrogacy is yet another technological offering in advance of suitable legislation. Presumably this is generally the situation, with the judiciary enshrining in law the results of cases covering new ground. People, on the other hand, form opinions about such divisive issues almost the moment these hit the headlines. There was universal happiness when Debbie Riley received her triplets. There was general sympathy, but less unanimity, when the Calverts won their child (with the fact that they were white and Anna Johnson was black probably not irrelevant). A survey, published in the *Journal of Medical Ethics* (in 1994), discovered that men and women were approximately equally divided between birth motherhood and genetic motherhood. As for legislation the procedure of surrogacy has been neither outlawed nor formally approved in the US and Canada. In Britain it is illegal only as a commercial enterprise but, most importantly, no surrogacy contract is legally enforceable. (In February 1996 the British Medical Association changed its trenchant attitude. It had formerly recommended that doctors should not be associated with surrogacy, but it then considered that medical professionals *should* be involved, provided the intended parents had exhausted other means of having a child. Simultaneously the BMA expressed concern that voluntary agencies could operate with few restrictions and no system existed for dealing with complaints.)

In Israel, and until March 1996, the procedure of surrogacy had effectively been barred by the High Court of Justice. One day after its regulation had expired surrogate motherhood became *de facto* legal. Nevertheless there were half a dozen provisions which were, in essence:

- the sperm must come from the father-to-be, and the egg must not come from the surrogate mother;

- only unmarried Israeli women could act as surrogates;

- as from conception the child would be in the full custody of the surrogate mother; she may have it aborted or petition the courts to keep it after birth;

- the commissioning parents must formally request transfer of parenthood within a week of the birth;

- payment will be limited to expenses, including loss of time, pain and suffering, and temporary loss of income;

- all arrangements will be kept secret without specific court approval.

Adoption has always been popular in Hollywood, with Michael J. Fox, Michelle Pfeiffer, Nicole Kidman, Isabella Rossellini, Kirstie Alley, Jamie Lee Curtis, Linda Ronstadt (and many others) nodding in agreement. Perhaps surrogacy is next in line, it also obviating the need for a temporarily swollen figure, stretch marks, a permanently altered outline (perhaps), and an awkward interruption of a promising career. Robert de Niro and his former girlfriend used a surrogate mother successfully. And the Center for Surrogate Parenting, Beverly Hills, is happy to act as intermediary – at $12,000 per child.

Unhappiness, bewilderment and considerable regret certainly assaulted Gillian Smith, aged 38, when she offered to incubate for a couple who agreed to pay £8,000 for her labour. The deal had been arranged through COTS, the charity Childlessness Overcome Through Surrogacy. Smith received £1,000 in advance, and £7,000 on delivery but then refused to sign adoption papers. She wanted the money to assist with her four children 'but when I looked at his little face, he was just the same as my other kids when they were born . . . I felt sick inside.' The recipient couple visited the hospital one hour after the birth and, added Smith, 'I could see their delight. I went home and cried my eyes out.' Regret was paramount. 'I will be paying for the rest of my life and I don't expect anybody to understand it.' People may understand, and sympathize at this brand new style of joy for some and utter misery for others. (A new style of joy became public in April 1996 when news surfaced of a mother carrying a child for her uterus-less daughter. Grandmother and mother were therefore, in part, one and the same.)

*

As for the surrogate future, there are other possibilities. What about siblings wishing to add to their number even though both parents are dead but have left frozen embryos? Can dead people be permitted to give birth in this fashion? What if the man who donated his gametes becomes emotionally attached to the surrogate mother, and then wishes to retain the child that is his (genetically) and hers (by giving birth)? Do the two of them have more power, within a court, than the genetic mother on her own? If lesbians or would-be single mothers arrange their own insemination with donated sperm, is that a form of surrogacy subject to guidelines or even laws? Surrogacy can cause happiness. It can also bewilder those who prefer matters as they used to be, when only one procedure existed for the creation of a life.

DONATIONS of sperm and egg are most unevenly treated. In general the anonymous sperm donors (often medical students) receive modest reward – perhaps $25 or £10 to £20 – for their generosity. Egg donors frequently receive nothing. This disparity is particularly severe because the female donation takes perhaps fifty-six hours of her time whereas for males it takes, at most, one hour. The woman has to be interviewed and assessed. She has to be examined. She has to experience daily blood sampling, to be operated on at the appropriate phase of her cycle, and to recover from that operation. According to the Faulkner Center for Reproductive Medicine, Boston, an egg donor 'could expect to receive $1,400 for her time alone, exclusive of any compensation for travel, risk, or inconvenience'.

The American Fertility Society confused the situation (in 1990) by stating that 'there should be no compensation to the donor for the egg' before adding: 'This does not exclude reimbursement for expenses, time, risk, and associated inconvenience.' Within the US the donor does receive payment for her eggs (and all that inconvenience) if insurance cover for the treatment of infertility is not mandatory, but insurance companies tend not to pay for donation if already having to pay for the fertility treatment. They then (usually) pay for the male's offering but not the female's. Many women in need of donations can find a helpful relative or friend but, for those who cannot, there are 'donor programs' where anonymous women offer their services – and their oocytes.

No doubt aware of its confusing and apparently conflicting 1990 statement, the American Fertility Society issued a further guideline in 1993: 'Donors should be compensated for direct and indirect expenses associated with their participation, inconvenience, time, risk, and discomfort.' Mere expenses are not payment for the work, a point of considerable irritation when men get their $25 for the work they perform. 'Since it is standard to compensate men,' asserted the Boston clinic, 'shouldn't the policy be equal pay for equal work?' In which case, if men get $25 for one hour and women contribute fifty-six hours, their remuneration should be that figure of $1,400 – before they tack on their expenses.

In Britain, where there is a shortage of eggs, there is only modest payment for egg donations (£15), but human beings are good at circumvention. One woman was offered free IVF treatment provided she gave away half the oocytes her hospital extracted. The Bourn Hall Clinic, Cambridge, has offered free sterilization or a hysterectomy in return for eggs. The Lister Hospital, London, has put women at the head of the queue if they recruit an egg donor, either for themselves or – anonymously – for others. Such seemingly beneficial arrangements are no longer permitted in the UK, as they can have unfortunate consequences. A woman donating half her eggs for someone else may give that someone else a baby while remaining, if the fates dictate, childless herself. An infertility doctor from the Lister Hospital considered payment-for-eggs unethical 'because it lowers human dignity' and 'because of the belief that such genetic material could result in the concept of an individual who has been purchased'.

(Perhaps the egg shortage will vanish following the development of a new technique, so far only shown with mice. In February 1996 scientists at Bar Harbor, Maine, announced their creation of embryos from immature oocytes. Normal procedure for egg acquisition is to stimulate a female's ovaries with hormones, and then reap the harvest of – perhaps – a dozen egg follicles (and therefore of mature oocytes). If immature oocytes prove suitable with humans, via the mouse technique, a woman's ovaries could yield hundreds or even thousands of potential eggs, and an egg shortage would then be confined to history.)

The 'lowering of human dignity' is not a stumbling block for male donors, 'usually doing it for beer money', according to a survey by

King's College, London, but if there was no money for their services they might be less generous: 64% of 144 men who had recently donated said they only did so because cash was involved. Advertisements play upon current practice. Those requesting male donors stress the money factor; those for women mention 'the ultimate gift, the gift of life'. French clinics do not pay for sperm but acquire offerings from older married men who have at least one child and therefore a proven fertility. It would seem that the French male is more altruistic than his British counterpart.

Comparisons have been made between sperm and blood donations. In the US 90% of donated blood is paid for, but 15% is then wasted owing to the incidence of infection. The donors have strong reasons for concealing their 'blood history' as they are generally indigent, with the majority being drug addicts, educationally handicapped, prison inmates, or unemployed. In the UK almost all blood is given free. No reason therefore exists for concealing facts about, for example, hepatitis. At a relevant Ciba conference it was stated that 'in economic and medical terms the paid service is grossly inferior to the voluntary one'. Any payment for semen donors, however small the sum, could 'invite concealment of material facts in family history', so said one individual arguing 'against commercial semen donation'. Others suggested that concealment could be obviated by using 'professional' donors whose histories are well known. Those who offer themselves determinedly – 'I feel it is my duty to be a donor' – tend to be rejected.

Unfortunately there are shortages, to a small extent with sperm but in a major way with eggs. Hence the inducements, to get discounts on IVF if eggs are donated, to jump ahead in the queue. Britain's Human Fertilization and Embryology Authority, watchdog of assisted reproduction (set up in 1990), is surveying the payment-for-gametes industry. The committee is known not to support undesirable motivation, but is keen to maintain a suitable supply of material for the fertility clinics.

Not all such establishments are equal. The HFEA has published success rates for the nation's clinics, this information including the number of women treated, the birth rate, the occurrence of multiple births, and the percentage of women treated who are over 38. Successes for 1993 varied from 0% to 23%, but some clinics were most attentive to younger women, thus bolstering their likelihood of success. Couples desperately wanting children have plenty of decisions to make, without having to spend much time deciding which clinic up and down the land is likely to be most suitable for them. At least they can take comfort that success rates have been improving. In the ten years from 1985 to 1994 the UK percentage of live births per treatment cycle was, on average: 8.6, 8.6, 10.1, 9.1, 11.1, 12.5, 13.9, 12.7, 14.2 and 14.1. Typically each cycle of treatment costs £2,000 to £2,500. (The drugs alone cost £500.)

A generality is that patients are most fortunate at the large well-established centres, and should be aware that most clinics now cluster around a 14% birth-rate per IVF treatment cycle. (The 0% figure was from a private clinic which treated only twenty-two women during the year.) An HFEA official stated, after its publication of what others have called a 'league table', that it did not criticize any clinic for having a low rate. 'Our job is to inspect and register clinics to ensure every clinic is behaving properly.'

The patients will be interested in success rates, largely because IVF treatment is so rarely available on the National Health Service. Other factors are involved, such as a clinic's convenience, its waiting list, and its general atmosphere. HFEA has helpfully produced a list of queries which prospective patients should ask of fertility centres to help in the final choice:

What tests would be carried out?
What treatments are offered?
How many visits will have to be made?
What is the cost?
Who are the sperm and egg donors?
What counselling is on offer?
Is there a patient support group?
What is the live birth rate?
What is the probability of a multiple birth?
What will happen if I get pregnant?
What will happen if I don't get pregnant?

How long has the clinic been established?
How will it involve the male partner?
What other information does the clinic provide?

FROZEN EMBRYOS are not in short supply, with – in March 1995
– an estimated 30,000 in storage within the UK, most being surplus
to requirements following earlier IVF treatment. The general 'safety
in numbers' policy means more eggs are taken from the ovary than
(probably) will be needed. More of these are then fertilized than will
(probably) be needed. If implantation is successful the spare ones are
not usually destroyed, just in case of upset. Perhaps the pregnancy
will not proceed to term. Perhaps there will be a neonatal death. Or
perhaps the parents wish for more children, and are understandably
unwilling to donate further eggs and sperm when 'their' embryos are
ready and waiting.

A law, which came into force in August 1991, stated that frozen
embryos could only be stored for five years (a time increased to ten
years in 1996. Some 'disposal' of such embryos began in mid-1996).
There is uncertainty about the length of time such embryos can be
successfully maintained. There is also awareness that many a family
situation may alter dramatically – a divorce, a split, an emigration. In
theory couples have to pay about £150 a year for the cold storage. In
practice the embryos are not destroyed even if payment is not
forthcoming. The couples receive an annual letter, along with the
invoice, asking if they wish liquid nitrogen storage to continue, or if
their embryos should be donated for research or given to another
woman, or should be destroyed. About 10% of couples fail to reply
(and may not wish to be an official party to the embryo destruction).

Animal embryos have been experimentally kept for decades and
seem none the worse for their cryogenic experience. Normally there
is no reason for such lengthy storage – it is costly, it might be
injurious, and (usually) there is little point. Human babies have been
born from embryos stored for longer spans than five years but, as
with animals, there has to be extra reason for such interment. The
case of Tricia and Julian Gunther, of Chester, received publicity after

their nine unsuccessful attempts at IVF. Teresa Finlay (neither kith nor kin) heard of their plight on a radio programme, and offered herself as surrogate mother. A Gunther embryo, 4 years and 2 months old, was inserted within her, and Jennifer Gunther was born at term to the delight of her genetic parents and of their obliging benefactor.

Publicity also came the way (in 1994) of Joana and Matilda from Portugal, the one being mother of the other. Nothing strange in that, save that Joana had been diagnosed as leukaemic when aged 22. She then needed total body irradiation to survive and, in consequence, complete destruction of all her oocytes (and therefore eggs). Bone marrow transplantations were successful, and Joana was soon able to contemplate a life ahead of her, such as the conception and rearing of a child. Before the irradiation, and with this future possibility in mind, some of her eggs had been removed. After fertilization with her husband's sperm thirteen of the eggs had been stored – in case of need. When Joana was pronounced fit the embryos were unfrozen and the 'best three' were transplanted. The happy result, the birth of Matilda (at thirty-four weeks and weighing 4 lb 1 oz) occurred four years after the irradiation.

Unwanted embryos, no longer desired or claimed by their actual parents, could be used to assist infertile couples. Much like adoption (of a newborn child) an embryo could be handed over to eager parents (albeit thirty-eight weeks earlier). It could then be incubated within the recipient mother, provided her uterus was not at fault, and form a better liaison between her and her partner than traditional adoption permits. The original couple, the genetic donors, would have to sign away all parental rights, as in any adopting situation. (Unwanted foetuses, also no longer desired by actual parents, have allegedly been imported from China into Hong Kong for use in traditional remedies and health foods. The Hong Kong government took the claims sufficiently seriously to ban the trade via new legislation. A local magazine had prompted this action by referring to a doctor who said she used foetuses to preserve her complexion – 'They are wasted if we don't eat them . . .')

If freezing becomes more commonplace, if human embryos are discovered to be safely viable for longer than five years (as with animals), and if the ten-year limit is relaxed still further, the possibility arises of long-term preservation. Embryos conceived in the twentieth century need not be born until the twenty-second century, or deep into the future. Humans are already having their dead bodies frozen – in case future medicine can return them to life – and they could dispatch their embryos into the future, should they so wish and society permit. It is not the Huxley *Brave New World*. Even his imagination did not travel quite so far. (France, via its Supreme Court of Appeals, has already decided that IVF of any kind should be limited to 'living, sterile couples (male and female) of reproductive age'. It is highly likely that this rule will be broken, in different countries for different reasons, again and again.)

*

All IVF transfer work encounters issues quite different from the transplantation of organs. Gametes, however much a normal part of the human system, are not like kidneys, hearts, liver, bone. 'If I give you my kidney it dies with you,' said the chairman of the British Medical Association's ethics committee; but 'if my daughter gives her eggs or ovaries the transplant will continue on into the future generations'. Similarly, if a kidney dies after its donation, there is sadness (and perhaps a human death) but no further ethical dilemma. If two fertilized gametes die soon after development has begun there is also sadness and death – but of a growing embryo. That embryo would never have lived without artificial intervention. Hence an ethical dilemma which does not arise with organ transplantation. Such an embryo death is not reasonably defined as murder but, without human interference, it would never have occurred.

This quandary becomes more extreme when embryos survive too well. Multiple births frequently result from IVF because, hedging their bets, inseminators place faith in numbers. They stimulate the ovaries to develop more oocytes. They extract several for subsequent fertilization. They are pleased if a few actually begin development, and these reach the stage suitable for implantation. Even at this time there is uncertainty and more than one is placed in the waiting woman. She is pleased that the odds, all along the line, have been boosted in favour of success, and she may not be unduly perturbed if informed that two offspring are developing within her, rather than a conventional singleton.

However she may be aghast if told that everything has been *too* successful, and quads, quins or more are growing within her uterus. What is known as embryo reduction may then take place, a selective form of abortion. The mother will (probably) settle for twins, but not the awesome hazard of three or four or five. Her undesired stock is therefore removed and jettisoned, usually at eight weeks post-conception (by injecting potassium chloride into each unwanted embryo's chest). Those who criticize this procedure should remember the severe events which partner more numerous forms of multiple birth, the great risk of miscarriage, the near certainty of prematurity, the disadvantages of early expulsion from the uterus, and the poor survival rates of those born at such a youthful stage. Even if there is no such loss the family may not be able to cope with such a quantity of mouths. It can be better, say the embryo reductionists (and the parents), to settle for a more conventional number, however unconventional its (or their) conception.

FOETAL OOCYTES If eggs are in short supply, so runs one proposal, they could be collected from aborted female foetuses (which have millions) or from adult cadavers (which also have no further use for them). The procedure would be simple, and far simpler than having to operate on a living person. Oocytes could be collected by the thousand, and no one need think of shortage ever again. Unfortunately (for such simplicity) there are problems. Public opinion is violently opposed – one survey told of 83.2% being against, and resenting foetuses as donors even more than corpses. To a large extent the reaction is emotional, with one journalist calling it the 'yuk factor'. There is also concern for any child created in this fashion. 'My mother was an abortion' or 'I never had a mother' are two of the statements which, it is feared, may be used by such children and by the adults they become. There is no evidence whatsoever in this regard, the procedure never having been employed, but there is concern. As a correspondent to *New Scientist* wrote: 'The effects upon a child of discovering that she exists only because her grandmother chose to kill her mother before she was born are beyond calculation.' Counter-argument states that the child does indeed have a mother – the woman who gave birth to her.

There is also scientific doubt. On one hand is the possibility that foetal eggs, so recently made, are actually better than an adult's eggs. Down's syndrome, for example, occurs most frequently among offspring of older mothers. It also occurs, to a lesser degree, with younger mothers. Perhaps, so runs this thought, it would never occur with eggs from the youngest potential mothers of all, those who have never been born. It has to be remembered that an 18-year-old, at the threshold of life, is actually 18 years *old*. Her oocytes were created before she was born, and ever since have been subjected to the vicissitudes of life, to fever, to ionizing radiation, to alcohol and medicines, to unwelcome viruses. At 18 she is only half the age of those 'elderly' mothers who have a 1 per cent chance of producing a Down's baby, but is some forty times the age she was when her oocytes were being created.

On another scientific hand is the fact that no one – yet – has any idea why such a reduction of oocytes occurs between the mid-term female foetus and the adult woman, dropping from several million to several thousand. Is the loss random or selective? Are unsatisfactory cells somehow being culled, leaving only the best as survivors (as with early miscarriages which have a higher incidence than normal of malformation)? If there is this form of rejection it would be wrong to expect a high proportion of viable offspring from foetal eggs. As Anne McLaren, leading embryologist, has said: 'It would take a very long time to be satisfied that the eggs were normal enough for transfer to a woman.' This problem would not exist with adult cadavers but, as with all organs for transplantation, the ovaries would have to be promptly removed.

As yet, and as greatest scientific stumbling block of all, the use of foetal eggs is still not technically possible. In any case the British Human Fertilization and Embryology Authority banned, as from July 1994, the use of oocytes either from aborted foetuses or from cadavers in the treatment of infertile women. This committee has agreed that such material can be used for research. It appreciates that research, if ultimately suggesting that such oocytes could indeed help with infertility, might one day overturn the current ban.

REGULATION Italy has often been prominent in IVF headlines, not only for giving babies to elderly women but because of apparent freedom for fertility doctors to act without restraint. Early in 1955, for example, a girl was born following fertilization from the frozen sperm of her dead father. Many of Italy's thirty-eight (at a recent count) political parties, and numerous pressure groups, voiced opinion, both for and against. Consequently there was considerable relief in 1995 when the Ordine dei Medici, Italy's medical association, issued a firm set of guidelines, much firmer than many had expected. In essence the Ordine stipulated: no surrogate motherhood; in vitro fertilization only for heterosexual couples; no treatment for post-menopausal women (the suggested age is 50); and no IVF after the partner's death. As one of the committee's members said: 'We must avoid turning [IVF] into a supermarket where prospective parents can choose the characteristics of their offspring.'

It is easy to suspect, bearing humankind's desires and its ingenuity in mind, that choice will operate, somehow, somewhere, frequently. It is so easy to know the gender of implanted embryos, and parents paying a great quantity of money for an offspring may demand this extra benefit. They may also demand (money always ready to talk) rather more information about the donors, resenting eggs and sperm from beer-thirsty students or impoverished women part-paying for their own treatment. They may want the tall, the strong, the clever, the handsome, even if another 10% is then added to the bill. Supermarkets have a habit of appearing wherever there is demand.

As one example of confusion arising from IVF development, an Israeli circumstance caused that nation's Supreme Court to intervene. In March 1995 a panel of five justices ruled (four to one) that a man could prevent his ex-wife from using 'their' fertilized eggs in order to produce a child. He had a basic right not to be a parent, even if that meant depriving a woman of her only chance to be a parent. The woman in this case opposed the judgment, and Israel's Supreme Court ordered the case to undergo a second hearing, with nine judges rather than five. One issue was whether the original agreement (between husband and wife) 'validates its completion'. Another was whether the fertilized eggs are living beings with judicial standing and rights of their own. The Court agreed with the earlier panel, but other courts in other lands may think otherwise.

Of course IVF is a potential and actual benefit – for many. It is also a minefield, legally, emotionally, contractually. In Israel such problems are particularly acute. The country is not a theocracy, but comes closer than most to being one, with religious parties wielding considerable influence in the Israeli parliament. Fertility is a major issue for orthodox Jews, reflected in the fact that a man can divorce his wife if she does not bear offspring within ten years. The religious laws oblige Jews to reproduce, and modern techniques can promote production. In theory these should be welcomed. In practice there is considerable discussion at every turn.

As an article in *Scientific American* (in February 1994) pointed out, 'there are more than 300 IVF centers in the US, most of them private and largely unregulated'. This number is likely to grow. Of the 'more than three million couples in the US thought to be unable to reproduce [without IVF]' only 'about 20,000' go to clinics every year. When future work improves IVF's 'dismal success rate' the demand will certainly increase. But 'a far greater impact' will occur should 'uterine lavage' become feasible and readily available. (At the moment it is still a distant prospect.) This procedure would wash a naturally conceived embryo from the uterus before it has a chance to be implanted (and before six days or so post-conception). The embryo could then be examined for genetic defects and only returned to the uterus if thought to be satisfactory. Once again it is easy to be envious of earlier centuries when babies just happened, with or without ill-fortune as firm companion.

Finally, and for some levity in the matter of sperm-selection from admired donors, read *My Uncle Oswald* by Roald Dahl, first published in 1979. The story is appalling, and most enjoyable.

CONTRACEPTION

Contraception should be used on all conceivable occasions.
Spike Milligan

Most accidents are caused by a human, and most humans are
caused by an accident.
Anon

I dream of a world where every pregnancy is planned
and every child conceived is nurtured, loved, educated,
and supported.
Benazir Bhutto

The act occurs 42 billion times a year, according to the World Health
Organization. This huge total is not much easier to absorb even when
reduced to 114 million per day. Perhaps 1,300 every second is more
comprehensible, this being the number of human ejaculations which
occur in that time around the world. As many millions of spermatozoa are liberated on each of these individual orgasms (totalling
billions of egg-seeking sperm in each sixtieth of a minute) it is with
relief – rather than the customary concern – that one remembers the
actual human birth-rate of only nine new infants every second, there
having been nine successful fertilizations from those 1,300 insemina-
tions. With further relief one also remembers the three humans who,
on average, die each second. The number therefore added to the
human population is six per second, disturbing as an increment –
until one recollects what would be occurring if the human reproductive method were more akin to, say, fishes (where many tens of
thousands of eggs may be fertilized from every ejaculation).

It is also fortunate that human females only produce one egg a
month, this being viable for twelve hours or so, and none when

pregnant or recently delivered of a child. If that child is breast-fed the production of another egg is further delayed. It is also fortunate – for the over-population problem – that considerable effort is frequently employed by humans to prevent egg and sperm from successful union, thus ensuring that the success rate per ejaculation is no more than 0.7% (or one 'success' in every 143 matings). As enthusiasm for ejaculation shows no sign of lessening (*au contraire* is easier to believe) that ratio must be brought down to a third of its present level, say 0.23%, if the world is not to be destroyed by people in their billions. Contraception, and the prevention of fertility, is not just a worthy pursuit. It is, in its essence, one of the most worthy pursuits of all, if not the worthiest.

Roy Calne, transplant surgeon (and father of six), published *Too Many People* in 1995, in which he pointed out that science has bestowed the gift of death control on developing countries without birth control. Unless the two are brought into balance 'wars, pestilence and starvation seem to be the most likely way in which population numbers will be controlled in future'. He therefore argued that effective birth control should be a condition for giving aid to underdeveloped and overpopulated nations. This subject was much discussed at the UN Conference of Women, held at Beijing in 1995 (where the leader of the British delegation was Ann Widdecombe MP, a Roman Catholic and anti-abortion activist). There has been progress. In 1965 only 14% of couples in the developing world were using contraception. By 1990 this number had risen to 53%. The gain has therefore been considerable, but also means that 47% of couples were still not hindering their fertility. Wars, pestilence and starvation do therefore loom save, for example, in Rwanda where the looming days have passed and actuality has began.

HISTORY Contraception certainly has an ancient past, starting long before the current fear about overpopulation. (Earlier fears were limited to the acquisition of disease or embarrassing pregnancies, neither of these fears having vanished in the interim.) Former contraceptive methods were wide-ranging, to say the least. Ancient

Egyptians (of 1850 BC) employed crocodile dung as pessary, and honey as spermicide. Less ancient Egyptians (of 1550 BC) recommended fumes from wax burned with charcoal, presumably after rather than before intercourse. Early Chinese preferred unpleasant minerals, like mercury and lead (likely to damage both mother and possible foetus). Catch a wolf, boil its penis, and add hair, wrote St Albert the Great less than a millennium ago. Use pessaries of cabbage, colocynth, pomegranate, ear wax, elephant droppings and whitewash, wrote Rhazes of Baghdad, mixing animal, vegetable and mineral most impartially. (The Western world now favours botany for drugs, while the Orient continues to lean towards animal products, but medieval pharmacists were not averse – or so it would seem – to anything.)

Quite suddenly sheaths then enter the lists. There is allusion to such protectors in Greek, Roman and even Cretan tales, but no firm evidence. That comes first in a description by Gabriello Fallopio (he of the Fallopian/uterine tubes) published in 1564. The linen sheath he devised, intended as barrier against venereal disease, was tested formally on 1,100 men. Allegedly not one of that precise number became infected, notably against the novel scourge of syphilis (which, some argue, was the first noteworthy export from the New World to the Old, this disease being properly described first in 1494).

A similar sheath was detailed thirty-three years later by Hercules Saxsonia, a leading medical authority, but his mere words have recently been supplanted by actual relics. Some archaeologists, sifting through a heap of seventeenth-century excrement at Dudley Castle, England, recently encountered sheaths. Dudley was besieged during the mid-seventeenth-century Civil War and, along with leaves, squares of cloth, bits of brocade and tapestry to wipe Royalist backsides, were sheep-gut sheaths, the oldest such protectors ever found. Destruction of the castle, ordered one year after its surrender, entombed 100 cubic feet of faeces, wipes and sheaths for excited researchers to explore 350 years afterwards.

Writers, diarists and others were mentioning sheaths as 'protective armour' almost casually during the eighteenth century. Whereas Casanova was describing (and being reluctantly grateful for) 'English overcoats', and the English shifted this fame on to 'French letters', no one ever called them 'condoms' – in print – until 1706. Lord Belhaven then used this six-letter word in a famous poem. Daniel Turner followed suit in 1717, and the term – now in wider circulation than ever before – had solidly arrived. Far less certain is its origin. It does not come from Latin (*condus* = receptacle), as some allege. There was no doctor (or colonel) of that name, as others prefer. The Gascon community of Condom has been investigated as possible eponym, but rejected. The Persian *kendu* (or *kondu*), a storage vessel for grain etc., made from animal intestine and much like Scottish haggis, is another – and most reasonable – contender, but no longer finds favour among etymologists.

It would seem, sadly, as if the condom's literal origin must remain a mystery.

> Oh Mary Mother, I believe, Thou did'st not sin but did conceive;
> Oh Mary Mother, still believing, teach me to sin without conceiving

wrote Lord Byron.

One suspects he, almost two centuries ago, could have known, or discovered, how the word originated. Better still Casanova, half a century earlier, had he chosen to discuss the point in his memoirs. He certainly told how he inflated the things for the amusement of his ladies (and to check for holes), and how these friends often disapproved of the overcoats, thinking them scandalous, nasty or merely unwelcome for covering '*le petit personnage*', but he never delved into their linguistic source. He mentioned sizes, qualities, virtues and drawbacks, but never origin.

Once condom had entered literature, with James Boswell – for example – describing how he used 'armour' on Westminster Bridge (of all places), it might be expected that the subject of contraception would steadily gain universal acceptance, but not a bit of it. Indeed, as contraception improved, discussion of the procedure began to wane. The vulcanization of rubber, developed in 1844, permitted technical advance – but also retreat, with the subject almost disappearing from view. The final thirty years of the nineteenth century witnessed a gradual bettering of rubber sheaths – they became seamless, for example – and therefore they gained in popularity, but social attitude towards their manufacture, their sale and distribution became entrenched. There were then no clinics to disseminate contraceptive information, even to married women. The medical profession, by and large, was antagonistic to the whole idea, with many textbooks of the time not even mentioning reproduction, let alone the effrontery of contraception. Politicians were also not in favour, liking neither the subject, nor its airing in public, nor the dread thought of a shrinking population.

Contraception did not go away because so many backs were turned; on the contrary. In the seven decades which lasted from 1861 to 1931 the mean completed family size in Britain diminished steadily, the respective figures being 5.9, 5.3, 4.1, 3.3, 2.8, 2.4 and 2.1 surviving children. This fall did not also cause population to fall, partly because people were living longer but because a drop in birth-rate is not immediately followed by a drop in total numbers. Between 1871 and 1931 the population (of England and Wales) actually rose from 22,713,000 to 37,886,000. During this period the number of live births per 1,000 inhabitants (of the UK as a whole) dropped from 35.5 to 32.5 to 29.9 to 27.3 to 23.6 to 20.0 to 18.3. This steadfast reduction in fertility was occurring despite the antagonism from so many quarters concerning contraception, despite the medical resentment, the restrictive legislation, and the back-street aspect of the contraceptive trade.

The opening years of the twentieth century were steeped in hypocrisy, with public statements entirely different from private happenings, but even more recent years can seem embedded in the past. In Connecticut, for example, a 1965 amendment to the old Comstock law meant that married women living in that state could finally use contraception without fear of arrest. (The law had been named after Anthony Comstock who, in 1873, had initiated legislation forbidding the distribution of contraceptive literature through the US mails. He had also established the New York Society for the Suppression of Vice, and was rigorous in punishing – through the courts – those who disagreed with him.)

A CONTRACEPTIVE CALENDAR can best clarify the recentness of many key events, and therefore the changes in attitude throughout this century that, oh so slowly, brought legislation into line with actuality, and reflected the general wish of people concerning, at long last, what they could and could not do.

1900 – Emma Goldman travelled from the US to Europe to acquire contraceptive devices and information, these being unobtainable in America. Thereafter her pamphlet *Why The Poor Should Not Have Many Children* was widely disseminated. She recommended cervical caps, diaphragms, condoms.

1912 – Margaret Sanger, principal US pioneer, watched a woman die from an attempted abortion before dedicating her life to birth control. She too travelled across the Atlantic for information and supplies (only returning home after Comstock had died).

1914 – Sanger published (surreptitiously) *Family Limitation*. This contained an illustration of a 'French pessary' but nothing about diaphragms.

1915 – The UK Notification of Births (Extension) Act stated that contraceptive advice could be given 'to cases where further pregnancies would be detrimental to health'. Public money could be spent on this.

1916 – Sanger arrested for opening a birth control clinic – at 46 Amboy Street, Brooklyn. There was then no such clinic in the UK, but also no law – such as Comstock's – for preventing one. It was argued by some that contraceptive literature could be seized under the Obscene Publications Act of 1857, but the government disagreed.

1918 – Marie Stopes published *Married Love*, an instant success – 2,000 copies in first fourteen days. It made only the briefest reference to contraception, but that subject was covered by her *Wise Parenthood*, also published in 1918. Her own marriage, and possible parenthood, was proving difficult. She was already PhD, and had been – 1905 – the youngest UK DSc but had to read extensively in the British Museum before appreciating that her first marriage had not properly occurred, and could be annulled on the grounds of non-consummation, she still being *virgo intacta*. It was formally annulled in 1916.

(On reflection Marie Stopes was a bizarre promoter of her cause, and of a book which sold 1 million copies in its first twenty-six editions. She was a virgin when it was published and 44 years old when her first child was born, her second and more successful marriage having taken place in the same year as *Married Love* appeared. No single girls were ever treated in her clinics. *Enduring Passion* was another of her books but her second marriage ended after twenty years, her husband having been ostracized by her for a considerable period of that time. When her son became engaged – to a daughter of the brilliant aeronautical engineer, Barnes Wallis – she tried to prevent the marriage, and did not attend the ceremony, mainly because the bride-to-be was a 'handicapped' person by virtue of wearing glasses. After abandoning birth control, when she lost control of it, this first woman ever to sue for libel then became a poet and took a lover thirty-three years her junior when she was seventy-two. She died in 1958, and to her son – on whom she had doted violently for all his early years – she left one dictionary.)

Most flatteringly to her memory Marie Stopes's *Married Love* was re-published – by Gollancz – in 1996 after being out of print for over forty years and seventy-eight years after its first appearance.

1920 – France introduced a law which totally forbade birth control literature. (The First World War's slaughter had prompted its arrival.)

1921 – The *Lancet* reported that no part of the medical curriculum (for students) dealt adequately with the human reproductive system, with many anatomical textbooks still omitting this topic.

Marie Stopes opened the 'first birth control clinic in the British Empire' on 17 March at 61 Marlborough Road, Holloway, London. (By the end of 1921 it had been attended by 518 women, all of whom had to be married to obtain advice.)

Lord Dawson, the King's physician, supported birth control in a speech to the Church Congress, and was widely condemned.

The Malthusian League opened its first birth control clinic on 9 November at 153a East Street, Walworth, London. Fewer than twenty attended in its first three months. (Stopes resented the competition, and resigned that year from the League.)

1925 – The Marie Stopes clinic moved to better premises at 108 Whitfield Street, some 5,000 women having attended Marlborough Road. (It has stayed ever since at Whitfield Street, although run by Population Services since 1976.)

1930 – An encyclical from Pope Pius XI stated that artificial contraception (everything except abstinence or the rhythm method – also known as Roman Roulette) was shameful, intrinsically immoral, and an unspeakable crime.

The advisory council on Catholic marriage considered it was clarifying rhythm when it stated that the 'first fertile day is found by subtracting 19 from the number of days in the shortest recorded cycle, and the last fertile day is found by subtracting 10 days from the longest cycle the woman has ever recorded. This formula gives at least 10 possible fertile days in the middle of the menstrual cycle. And the greater difference between the shortest and longest cycles, the greater the number of possible fertile days.' Couples in possession of their necessary cycle facts (and sound in arithmetic) could still be confounded: about one in four women are extremely erratic in their cycles, notably after birth.

At the Lambeth Conference (of Anglicans) a resolution approved of birth control 'when there is a clearly felt moral obligation' to limit parenthood, and a 'morally sound reason' for avoiding abstinence.

The UK Ministry of Health permitted – for the first time – local authorities to give birth control advice to married women who needed it. (The Stopes clinics – there were now several – still gave no advice to the unmarried.)

Most women in the UK were now using contraception 'at some time' during their married life, the proportions having been 15% before 1910, 40% up to 1919, 61% up to 1929, and 66% up to 1939.

1937 – A *fatwa* issued by the Grand Mufti permitted Muslims to take any measures by mutual consent to prevent conception.

1940–45 – Live births per 1,000 of population (in England and Wales) were still on a downward trend, although figures were complicated by wartime changes. The number had been 18.3 in 1931, and became 14.1, 13.9, 15.6, 16.2, 17.7, and 15.9 during the six war years. A baby boom then followed (in many countries) as couples were united and more optimism reigned.

1951 – Helen Brook, British birth control enthusiast, joined Islington Family Planning Association. Seven years later she directed the Stopes Whitfield Street clinic.

1958 – Marie Stopes died, having abandoned the cause of birth control and stipulated that her clinics were not to be taken over by the FPA. Brook was then able to start 'thinking independently'. In 1962 she 'secretly' opened a session at one Marie Stopes clinic for the unmarried. (Conventional FPA practice at this time was only to see either married women or women with proof of forthcoming marriage. 'Imposters' were turned away.)

Hans Lehfeldt, a cervical cap enthusiast, originally from Germany, opened the first birth control clinic in a US municipal hospital (at New York City's Bellevue).

1960 – 20% of couples in developing countries now using contraception. Average age of first intercourse in the UK had dropped from 21 to 19 (according to Brook).

1961 – In many rural areas of the world, said Dr S. Chandrasekhar (of California University), the cost of having a baby was cheaper than the price of birth control equipment.

A birth control clinic in New Haven, Connecticut, was forced to close after only nine days. (It was not re-opened for another four years.)

1964 – Brook Advisory Centres were founded in London with a gift from a 'generous donor'. These were specifically for young unmarried people.

1965 – Only one British medical school compelled students, at some time during their six-year course, to attend an FPA clinic for instruction.

There were still twenty-eight states in the US which had laws on their books limiting contraceptive advertisements or sales outlets or the spread of information. (Garages and tobacconists still sold more contraceptives than did pharmacies.)

1967 – France revoked its forty-seven-year-old law totally forbidding birth control literature. (It was then believed that 50% of all French conceptions were ending in abortion.)

1968 – Pope Paul VI issued an encyclical reaffirming faith in the 'safe period' technique of contraception. At Dallas, Texas, a huge gathering of scientists called his message 'repugnant to mankind'. At that time the eleven Catholic countries of Europe had an average birth-rate slightly lower than the fifteen non-Catholic European nations.

A poll taken among British Roman Catholics showed that 54% disapproved of the Pope's ban on birth control. (The president of an American Catholic feminist group later asked: 'How come a so-called country, that is in essence 800 acres of office space in the middle of Rome, that has citizenry that excludes women and children, seems to attract the most attention in talking about public policy that deals with women and children?' One answer is that it doesn't. Another is that only lip-service is paid – in general – to the Vatican's pronouncements.)

1969 – The UK's Family Law Reform Act established 'medical age of majority' to be 16, namely two years less than general majority age (for voting etc.). Brook centres extend help to under 16s. (Brook centres had been visited by a total of 10,000 clients, 85% of them requesting the birth control pill.)

The pill could be sold in France for the first time. (Advertising contraceptives was not permitted there until 1987.)

The first Italian birth control clinic opened, in defiance of both church and law. (The country then had over a million abortions a year.)

1970 – The Family Planning Association decided that advice should be for all, irrespective of marital status. A 500-school survey showed that the majority of secondary school heads disapproved of giving contraceptive information to their students.

1971 – British Medical Association stated that confidentiality should be offered to under 16s.

The UK Minister for Health paid first-ever official visit to a Brook centre; it was seen by others that 'advising the unmarried' had become respectable. (The first royal visit was to occur in 1989 to celebrate twenty-fifth anniversary.)

1972 – The first contraceptive commercial for cinemas was screened. A thousand FPA clinics were transferred to the UK's National Health Service.

1973 – The FPA stated that doctors must themselves decide whether to inform parents of under-16s. Doctors would be 'prudent' to seek the patient's consent to tell parents; failure to do so would 'put them in jeopardy'.

1974 – Contraception advice in the UK became free for all (after considerable wrangling by the medical profession over who should get how much for what).

1975 – The UK teenage pregnancy rate, which had been steadily rising, started to fall. (It started rising again in 1983.)

1976 – Some 90% of British doctors had registered to provide family planning services (but only a few thousand had had any training).

1978 – Brook centres received their first government grant.

1980 – In England and Wales men make up only 1.2% of the 1.5 million annual users of birth control clinics.

1984 – An encyclical from Pope John Paul II, even more restrictive than those of 1930 and 1968, stated that Catholic couples should on occasion not even use the rhythm period. 'The use of infertile periods in married life can become the source of abuses, if the couples seek in such a way to avoid, without just reasons, procreation, [so] lowering procreation below the morally correct level of births for their family.'

US aid to family planning programmes limited by a 'gag rule' – no funds were available for any organization that counselled on abortions, even if US money was not involved in that counselling. (This ruling was overturned in 1993.)

Appeal Court in UK (*Gillick* v. *DHSS*) prohibited doctors from giving contraceptive assistance to under-16s without parental consent.

1985 – The UK Law Lords reversed the 1984 (*Gillick*) ruling, and restored rights of under-16s to confidential medical services.

1995 – The Alan Guttmacher Institute, New York, stated that 'the proportion of births that are actually wanted at the time they happen is only about 40% in Kenya, 50% in Japan, Mexico and the Philippines; and 60% in Egypt, Jordan and the United States ... Of the 190 million pregnancies that occur each year, 51 million are ended by abortion, with 21 million of these being in countries where the procedure is illegal.'

TODAY'S DILEMMA Plainly there is still progress to be made in the matter of contraception; but, equally plainly, the twentieth century has witnessed a greater shift, in attitude and practice, than all previous centuries put together. The subject of contraception has moved from the back street to the front page of every form of publication. And so it should have done, in the opinion of all those who consider that our zeal for reproduction (and intercourse) could destroy our planet, via over-population, as effectively as war.

Nevertheless there are still contradictory elements, of legislation, of attitude, of fact as against intention. Like a pair of youngsters, who take no precautions and dread the thought of pregnancy, the world is worried about increasing numbers and yet knows of 300 million women (according to the WHO and Population Concern) who have no access to contraception. Even the lower figure of 120 million, favoured by the UN Population Fund and the International Planned Parenthood Federation, is cause for anxiety in a world adding to its people numbers by 90 million every year. Half of all pregnancies overall are unplanned and a quarter 'certainly unwanted', says the WHO. The carefree pair of youngsters acting without precaution can seem almost responsible when compared with the planet as a whole.

In 1992 a Dublin court found the Irish Family Planning Association guilty of selling a condom. (It was fined £500, but still sells condoms.) The California penal code prohibits a male of any age from intercourse with a female under 18 to whom he is not married (a law without much meaning in, of all places, California). In the US there are some 3.5 million unintended pregnancies each year, 800,000 of them among teenagers. The US has fewer birth control options than many other countries, and 1.6 million abortions annually. In developing nations there are, according to *World Health*, 'at least' 12 million illegal abortions every year, leading to 200,000 maternal deaths. There are half a million maternal deaths overall from causes related to pregnancy (according to the IPPF), with an African woman having a 1 in 22 likelihood of dying from a pregnancy-related cause. Cutting down on conceptions would not only reduce births but keep more women alive. (The figure of 500,000 maternal deaths a year, frequently quoted, is a similar degree of mortality to one airliner – of Pan Am's Lockerbie kind – crashing every four and a half hours. In February 1996 the WHO increased its estimate of pregnancy-related deaths to 585,000 a year, 99% of them occurring in developing countries. The aircraft crash rate therefore goes up to one every 3.8 hours.)

The WHO published (in February 1996) a country-by-country breakdown of maternal deaths per 100,000 live births, the first time it had ever done so. It hoped, via this novel policy, to stimulate all countries to do better, particularly those at the bottom end. The world average is 430, a figure which conceals considerable disparity around the world. Switzerland, Norway, Canada, Cyprus, Luxembourg and Iceland all scored six or fewer. At the wrong end were Mozambique (1,500), Bhutan, Somalia, Guinea (each 1,600), Afghanistan (1,700), and Sierra Leone (1,800). 'There is no mystery to how these figures could be reduced,' said Carla AbouZahr, of the WHO's maternal health programme; 'the priority is to improve the economic and social status of women.' A similar, but broader-based survey, conducted by Population Action International, investigated more than maternal mortality to reach the 'reproductive health status' of women in 118 countries. This organization also took note of contraceptive use, abortion policy, pre-natal care, attendance at birth by trained individuals, and HIV and AIDS rates. In this 'overall ranking of reproductive risks' Italy was top country, with Denmark a close second. The UK was thirteenth and the US eighteenth, but all such countries at this end of the index were considered 'very low risk'. Each one of the twenty-one 'very high risk' countries was in sub-Saharan Africa, with the exception of Haiti and Afghanistan. Zaïre was the nation to score least well.

No wonder there is argument that Britain, for example, should spend more of its overseas budget on contraception, with the amount for population programmes (in 1994) forming only 1.5% of total aid. This proportion, it is argued, should be raised immediately to 2%, rising to 4% by the year 2000. If population growth is *the* dominant concern, notably in developing nations, that improved donated percentage can still seem modest.

The United States, so frequently a country of extremes, shifted gear dramatically between 1995 and 1996. USAID, its agency for inter-national development, had been a leader in promoting worldwide family planning, and gave $546 million for this purpose in 1995. For 1996 the equivalent funding was slashed to $72 million, with inevi-table consequences to planning procedures in developing countries. 'Is it too much to hope,' wrote Malcolm Potts, fervent lobbyist for such endeavour, 'that the cuts will generate sufficient anger to put international family planning and reproductive health back on track?'

Kenya, a leading African country for birth rate, received about £40 million in annual aid from Britain. Euro-parliamentarians have stated that such a sum is spent in Kenya on job creation and on teachers for every 12,000 births, the number this single country produces in each four days. Developing countries in general, despite the frequent lack of contraceptive availability (and the total lack in many areas), have successfully reduced family size. This number was an average 6.1 children per woman within such countries in the early 1960s. It is now 3.9, and therefore only (if that is the right word) 1.8 children per woman above the level for zero population growth.

Something similar, by way of reduction, happened in the US during the nineteenth century. Between 1842 and 1900 average family size shrank from 6.5 to 3.5 children, despite the moral antagonism, the lack of contraceptive assistance, the stringent legislation, and a total absence of official support. It is therefore arguable that contra-ception is only of secondary importance in determining family size, the prime force being a wish to have fewer children; but no one argues that contraceptive advice, provision, etc. should therefore be abandoned. The wish will come first – and then the contraception, thus permitting a greater quantity of that wish to become reality.

The longing for fewer offspring is now gaining even in those countries traditionally regarded as big breeders. One comparative UNFPA survey, carried out first in 1970 and again in 1980, discovered a noticeable reduction within that single decade. Senegal's women, for example, dropped their wish for children numbers from 8.1 to 6.5, Ghana's from 5.9 to 5.1, Kenya's from 7.4 to 4.5 (the greatest drop), Mexico's from 4.2 to 3.2, Indonesia's from 4.2 to 3, and Egypt's from 3.9 to 2.6. The figures are still substantially higher than the zero growth of 2.1, but are moving – speedily – in the right direction.

Unfortunately (for the birth rate) girls are becoming sexually developed earlier, in many regions of the world, than in recent centuries, with 13 now the average age of first menstruation. Adult hesitation about sexual advice for extreme youngsters can seem archaic, notably to such youngsters. Many a teenager is now sexually active throughout all her teenage years. In Britain there are conceptions for 10-year-olds almost every year, that average age of menarche at 13 only being the *average*. Sexarche, or earliest sex, has been falling along with puberty, and changing morals, and increased permissiveness – and, of course, for increased abortion if things go wrong. (Some 30–50% of all women presenting for abortion in Britain were not using contraception when they became pregnant.) A survey by *Nineteen* magazine in 1995 suggested that one in two British girls has sex before the age of consent, and one in ten at the age of 13 or younger. The survey was only of 1,000 girls and also self-selected, in that they were asked to respond to a questionnaire within the magazine, but no one is denying that sexual relations are starting earlier these days.

ADOLESCENCE Society, particularly its legislators, welcomes the solidity of an individual's age. If someone has lived for 5,844 days, and is therefore just 16, certain activities are permitted, but forbidden if the total is less. Society also finds it difficult to accept that a

16-year-old of today may be older – in physique, in maturity, in general development – than the 16-year-old of yesterday when poor living conditions, poor food, and more disease tended to inhibit growth in all its forms. Arguably, therefore, the 14-year-old of today *is* the 16-year-old of yesterday. The current crop of youngsters are, in short, older than they used to be. Perhaps the term 'teenager' is at fault, implying not so much youth as someone under age. In consequence it has the wrong associations. Youngsters are becoming adult (and therefore ready for adultery, as the saying goes) earlier than before.

There is concern in Britain that its teenage pregnancy rate is the highest in Europe, being seven times greater than in Holland (where contraceptive advice and assistance have always been more available – and which was visited by both Sanger and Goldman from the US during their search for information). Those in favour of more liberal abortion laws, established (within the UK) in 1968, have been saddened that the annual total (over 100,000) remains so high. Perhaps it is the lack of counselling and general education that is at fault rather than – as easily alleged – the girls themselves.

Figures from Western Europe are markedly different, with the UK and Holland topping and tailing the list. Teenage pregnancies per 1,000 females aged 15–19 were (for 1992): UK 32.9, Portugal 22.7, Greece 19.9, Ireland 16.1, Germany 12.4, Spain 12.1, Luxembourg 11.8, France 9.3, Belgium 9.3, Italy 8.9, Denmark 8.8, and Holland 4.8. Very few teenagers wish to become pregnant (and abortions often follow) but something must be happening to cause such differences. How odd, for example, that Italy and Denmark, so dissimilar in religion, prosperity, climate and tradition, should be so equal when Holland and Britain, so much nearer in so much, are quite so separate.

In the world as a whole, more than 15 million girls aged between 15 and 19 annually give birth. Waiting in the wings, as it were, is the one-third of the world's current population who are under 15. Particularly disturbing are those of them who live in Africa. *World Health* has stated that 95% of couples in East Asia have good access to contraceptive services, 55% in South and South-East Asia and Latin America, around 20% in North Africa and West Asia, and less than 10% in sub-Saharan Africa. If the world's population in 2000 is not to exceed the UN projection of 6.3 billion the global rate of contraception use will have to rise from the current mean of 51% to 59%. That change demands the provision of contraception to an additional 86 million couples, a paltry task – say all those who know full well what the world can do, or even a part of it can do (if it really tries), such as forcing Iraq from Kuwait.

*

A final general point. The battle for contraception used to be called Birth Control. It then became Family Planning. A move is now afoot for it to be known as Sexual Welfare or even Sexual Health. With so many unwanted pregnancies around the world, so many unwelcome babies, so much abortion, and so much infant and maternal death, contraception is truly one more medicament aimed at preserving health, along with all the others targeted at malaria, tuberculosis, diarrhoea and every other major killer still rampant on this Earth. Sexual health is as crucial as every other form.

METHODS The ideal contraceptive does not yet exist. It should be cheap and simple. Its effects should be speedily (and reliably) reversible, preventing conception only until conception is required. It should certainly forbid the transmission of sexual disease. In no way should it diminish the sexual act. There should be no side effects, either immediately or later. There should be no storage problems, no individual problems (such as allergy reaction), no packaging problems, no need for medical expertise (or involvement) – in short, no problems of any kind. And, as final (or prime) point, it should be 100% effective in preventing pregnancy.

Perhaps no such ideal form of contraception will ever be developed, save (it should be remembered) for celibacy and total abstinence. Each generation has always had to make do with whatever has been available. The blessing of now, as each generation must also have said, is that more options exist than ever before.

THE SHEATH has to take pride of place, having such an ancient history. Most of that history, involving animal bladders, leather and the like, can be swiftly bypassed. So can the early, crude, vulcanized objects made from crêpe rubber, which were even thicker at their seams. There was a slow but gradual improvement in their quality during the latter half of the nineteenth century, and they became seamless by the end. The first teat-ended products (for holding the ejaculate) appeared in 1901. The introduction of liquid latex, in the 1930s, permitted not only thinner sheaths but a longer shelf-life (up to five years from three months).

Today's products are 0.065 to 0.075 millimetres thick, save where they are deliberately thickened (with ribs and bumps for extra effect). They can be coloured, lubricated with spermicide and – most modern trick of all – formed from plastic. They are airtight (think of toy balloons, often made by the same companies) and watertight. Hence their ability to act as barrier to micro-organisms. The smallest sexual disease transmitter, the herpes virus, has a diameter of a tenth of a thousandth of a millimetre, about 1,000 times larger than air or water molecules.

The modern sheath does embody a good many of the criteria for an ideal contraceptive. It is light, compact, and easily disposable (too easy, say some, who see it 'everywhere'). In the literature it is either 'relatively inexpensive' or 'not cheap', depending upon each author's perspective. It requires no medical examination, either before or after use, and has no side effects. (There is, one can assume, disagreement concerning every topic. A letter from Safat, Kuwait, in the *Journal of the Royal Society of Medicine*, September 1994, stated that the 'condom will eventually prove to have been one of the greatest, albeit overlooked, killers in twentieth-century medical history'.)

The modern sheath provides visible evidence of its effectiveness, it is easy to use, and permits the male to share in the primarily female concern of contraception. Since the arrival of AIDS awareness (in 1986) the sheath's preventative abilities have not only encouraged its popularity but have opened up the subject of contraception quite extraordinarily (when all those years of reticence are remembered). Nevertheless there are also disadvantages to the sheath. Sex may not be so enjoyable. Users can discover they are allergic to its rubber. The objects may break, or slip off unintentionally, notably when the penis is flaccid once again. They can be sabotaged – for a varied set of reasons – by either partner. But, if pregnancy is not intended, the failure rate is, according to one textbook, two to four pregnancies per 100 woman-years (the usual, if somewhat blunt, formula for assessing contraceptive risk).

Enthusiasm for the sheath has altered over the years, notably as other methods have appeared or it has been discredited, and also – to a large degree – among different countries. Before 1910, there being so little choice, the sheath was overwhelmingly the most favoured 'appliance' method, the 'non-appliance' system of withdrawal (*interruptus*) being the major rival. Free sheath supplies issued to servicemen in the First World War boosted the condom's popularity.

Despite the subsequent arrival, and proliferation, of contraceptive clinics, very few women (out of the total population), and virtually no men, visited them between the two world wars. Contraceptive advice came from friends or, for men, from over-loud mutterings at the barber's shop about the forthcoming weekend (with no equivalent outlet for women). During the decade leading up to the Second World War 46% of British couples (using any contraception) employed the sheath, 50% withdrawal, 12% the safe period, 11% the pessary, 8% the cap, 6% the douche, and a few were using other methods, such as foam tablets and creams, all these percentages totalling more than 100 because many couples employed more than one system. The sheath gained a little in the decade embracing the Second World War. So did the safe period and the cap, whereas withdrawal waned. The important point – for women, when sheath and withdrawal occupied pride of place, was that contraception remained male-orientated, with failure – as ever – landing squarely in the female court.

That fact shifted gear when the contraceptive pill (about which more soon) arrived in the 1960s. So did IUDs, the loops, bows, spirals, rings and coils that had been investigated since the 1920s, and which then advanced as rival for the pill. Sterilization also gained favour, in partnership with better surgery. Sir Alan Parkes, distinguished biologist, said at the time that contraceptive methods had, until then, 'been so crude as to disgrace science in an age of spectacular achievement'.

This crudity was steadfastly eroded, and on several fronts, in the second half of the twentieth century; but the sheath, principal barrier method, still marched on, its sales viewed anxiously by its manufacturers. The Genetic Study Unit was active in demoting the pill, and busy spreading news about possible drawbacks, until the *Sunday Times* revealed (in 1965) that this unit was supported by the London Rubber Company, principal manufacturer of sheaths in the UK. The GSU promptly disappeared from view. So did some of LRC's 'obscene profits', as they were described by a British MP following a report by the Monopolies Commission. Leo Abse, that politician, regretted the soaring cost of living but was pleased at his part in 'reducing the cost of loving'.

Somewhat inevitably there are price differences, with the more expensive sheaths putting up that loving cost by selling at the top end of the range; but size is less of a variable. One or two brands are larger but, generally speaking, one size fits all men (despite locker-room talk about big ones marked 'Medium' before their dispatch to Russia/Ireland/Poland). The European Union announced in March 1996 that it had agreed a common standard for condoms. They have to be at least 170 millimetres long but can be any width. If they depart from the usual width of 52 millimetres the package must state how wide the condom is, and whether wider or narrower than the regular size. Italy had been lead country in requiring a range of widths, but eventually accepted the general ruling.

In 1975 the sheath was being used by 25% of British couples (as against 58% for the pill). Other countries, such as Japan (top country in this regard), much preferred the sheath; but a number, such as the United States and France (the French almost bottom in this league), were less enthusiastic. In Europe, where prices vary as much as attitude (but not directly), sheaths cost 60 pence on average, with 1,500 million being purchased annually (or 3.8 per person). In Britain each one costs about 35 pence, and 152 million of them are sold (or 2.7 per person). In France the cost is slightly higher (44 pence) but use of the *préservatif* is much lower, being two per person per year. Holland is most logical, with the cost highest (71 pence) and the use lowest (1.7 per person). The United States has never been outstandingly enthusiastic about the sheath, often preferring (and more adventurous about) the newest forms of contraception. Global production (with Japan using 58% of it) is said to be 2.8 billion annually, or slightly less than half the total of every man, woman and child alive today. To put that figure yet another way round, the sheath total is one-fifteenth of the ejaculation total. (The Japanese dominance is caused partly, if not mainly, by the fact that hormonal contraception and IUDs are not licensed for contraceptive use in that country.)

AIDS has been a great boost to the barrier industry and stopped a downward trend, with the British percentage of sheath use being 14 in 1976, 13 in 1983, 13 in 1986, 16 in 1989, and 17 in 1990. In those five years, according to the same surveys, the pill's percentages were 29, 28, 26, 25 and 32; the IUDs stayed virtually constant; withdrawal went down; sterilization doubled (for both sexes); and the percentage of those using no method whatsoever dropped from 31 to 25 to 25 to 28 to 22.

Although safe sex is most possible via a condom, many males refuse to wear one and many women, notably those in the business of sex, have to comply if they are not to lose their partners. They are then at risk from HIV, but there is current talk about a 'chemical condom'. Vaginal virucides, as they are more formally known, will permit women to guard against infection and thereby care for their sexual health. Testing of such agents is now in progress, and could be beneficial for both sexes. Women would be less likely to acquire HIV, and therefore less likely to transmit it to their men.

Giovanni Jacopo Casanova de Seingalt, whose memoirs describe his sexual enthusiasms during a good many of his 73 years, might have changed his critical attitude to sheaths had he been blessed with the modern, seamless, colourful, extra-thin protectors, available at modest cost from every pharmacy. And perhaps he would be even more interested in plastic sheaths, fashioned from polyurethane and half as thin as the latex varieties, which went on sale for the first time in 1994. Or, conversely (and quite possibly), he might have preferred all his ladies to be themselves prepared, and on the pill.

THE PILL arrived, in opportune fashion, at the start of the so-called 'swinging sixties'. (Indeed it assisted them to swing.) Only in 1930 had it been properly discovered that the female's ovary released a ripened egg 12–16 days before the start of menstruation – although ordinary mortals (and religions) had long suspected such a happening. Science also learned that changing hormone levels caused the egg to mature, the uterus to prepare for possible implantation, and then for its endometrial inner layer to be sloughed off when fertilization failed to take place, causing the bloody flow. Tinkering with the hormone levels could therefore – it was reasoned – disrupt the normal process, prevent the liberation of an egg, and forbid conception.

It was Gregory Pincus who, in 1955, told a planned parenthood conference about the inhibition of ovulation in women who had taken progesterone or norethynodrel. His first large-scale experiments were performed in 1957 on Puerto Rican women. The demand (and need) was high in that country, the educational level low, and governmental restriction lower still. His act was later criticized 'as one of the most risky and foolhardy measures ever to be put into general use', this same critic then adding that the pill's introduction 'will turn out to be one of the most responsible things done by any one individual'.

The remarks are not inconsistent. It was a risk. Puerto Rico was (and is) an undeveloped nation. The women there did not ask too many questions, and were desperate for contraception. The colossal trial permitted the new product to be ready for sale within an extremely short time – by 1960 in the US and 1961 in the UK. By 1962, according to figures from the FPA, 3,536 British women were 'on the pill', by 1963 13,760, and by 1964 44,000. At the time of Pincus's death (in 1967) the estimates were 1,000,000 in Britain, 7,000,000 in the US, and 2,000,000 in the rest of the world. His pill had fulfilled so many of the criteria for an ideal contraceptive, being simple to use, extremely effective (if the instructions were followed correctly) and in no way disruptive to the sexual act. Unfortunately it did not hinder transmission of sexual disease; indeed its very attractiveness and simplicity probably helped disease to spread.

Such a burgeoning business inevitably appealed to the pharmaceutical industry. By 1968 there were twenty different oral contraceptives on the market. Medical journals were flooded with advertisements, these sometimes consuming a quarter of the total advertising space. The big business did indeed become bigger. By 1980 over 50 million women were on the pill (of one sort or another) and 150 million had tried it. Popularity varied tremendously between countries, with Japan always low and Holland high (current figures for married women in those two countries being 2% and 40%). Enthusiasm has also varied as the pill itself has varied (with new combinations of constituents) and as various forms of adverse publicity have surfaced.

The first report of venous thromboembolism (clotting) with possible linkage to the pill appeared in the *Lancet* in 1961, and the subject has never gone away. Without doubt there is risk, whether from extra clotting, and breast or other cancers. This risk is extremely small, and varies with the type of pill, the age of the woman, and with her life-style (notably whether she smokes or ever has done). Behaviour of human beings is so complex that comparative studies – concerning the health of pill-users v. non-pill-users – are so fraught with problems that many results are suspect from the start. (Science loves experiments where everything is equal among two cohorts, save for a single difference. Humans cannot be paraded in this fashion.)

In one study of pill-users compared with non-users it was discovered (to quote from John Guillebaud, leading expert in contraception) that users were 'on average taller, more physically active, more likely to smoke or drink in moderation (though not more likely to be heavy smokers or drinkers), to sunbathe more frequently, to have initiated sex earlier, and to have had more partners than non-users'. It would seem, from this single quote, that pill-using women were either more attractive than non-users or more in favour of sex, both these attributes demanding sound contraception. They – as a group of people – were certainly not equal to the non-users, and their list of differences – more sunburn, more smoke, more drink, more lovers – was possibly of more consequence than whether or not they took the pill.

To try and initiate scientific method, to take a large population, to divide it at random into two halves, to instruct one half to take the pill while the other half stays pill-free, and then to tell both halves to behave equally is immediately problematical. Large numbers of the pill-free cohort will soon become pregnant – unless it is using some other form of contraception, or abstaining from all sex. If so – pregnant or puritanical – the two halves of this experiment will no longer be comparable.

A further complication concerning risk is rendered by the pill itself. It does prevent pregnancy. Therefore it prevents all drawbacks associated with pregnancy, such as – not least – the possibility of death during childbirth. The pill also obviates need for other forms of contraception, such as – most importantly – the sheath. Sexual diseases are therefore more likely to be transmitted. Cervical cancer, for example, has been implicated, notably in younger women who have been infected during sex. This association is rare, as is the incidence of other cancers thought, somehow, to be encouraged by the pill. Travel by bus, and you are more likely to be killed in one. Stay at home, and a home accident is more probable. Overall, as John Guillebaud has asserted (and taking the good news with the bad), the pill does not significantly increase the risk of cancer for women taking it.

*

There is no single pill, however much this word is used as if a solitary entity is involved. The combined pill, the COC, the kind most commonly used today (of which there are at least thirty varieties), consists of oestrogen and progesterone, both hormones produced in the ovary. Between them, and via the pituitary gland located near the brain, they control the menstrual cycle, including the preparation of the uterus for implantation, and then the abandonment of that preparation when no pre-embryo is ready to implant. The pill's great alteration to normality is that it prevents ovulation, the production each month of a single egg (or sometimes, as dissimilar twins will tell you, more than one). Nothing therefore exists for the sperm to fertilize, just as there is nothing during pregnancy. The two conditions, being pregnant or on the combined oral contraceptive, are therefore similar.

Some of the pill's occasional side effects emphasize this fact, such as (to quote from *Sexuality and Birth Control*) 'nausea, weight gain, dizziness, sore breasts, absent periods, breakthrough bleeding, vaginal discharge, and depression'. Women react differently to pregnancy, and women react differently to the pill, or to one pill (with slightly varied constituents) rather than another. More serious side effects include a rise in blood pressure, affecting 2.5–5% of pill-takers. Women should not be on the pill if already with circulatory problems, or cancer or liver problems, or if they smoke a lot. The pill's oestrogen is responsible for thrombosis, and no contraceptive pill now contains more than 0.05 milligrams of the hormone.

The *Biologist* regularly produces lists of inventive errors spotted by school examiners, such as sperm 'causing fermentation', menstruation lasting 'about sixty years' and how 'You don't have to remember to take a condom every day.' Indeed the pill can be forgotten. If it is taken more than twelve hours late there is risk of pregnancy (and therefore of uncertainty and misery among those involved). The woman should take the next pill immediately she remembers *and* use other contraceptive measures for the following seven days. However unintended and unplanned the actual sexual act may be, the pill as contraceptive does demand daily vigilance. As one more child reported: 'There is no danger of addiction to the pill.' The need for regularity, particularly among the highly fertile, often casual, generally impulsive young, can be construed as the pill's major snag.

VASECTOMY

'Pick up the vas in its covering sheath of fibrous tissue . . . lift it up out of the incision with a tenaculum forceps . . . Then incise longitudinally

along the course of the sheath ... Gently get hold of the vas ... It is
important not to be too rough at this stage ... make sure you have
got the vas, and not a thrombosed vein ... Having freed up about 5
to 6 cm of vas, place a mosquito artery forceps at each end and cut
away the segment in between ... Tie off each severed end with 3–0
black silk ... Inspect the incision for haemostasis, and take care to
stop any bleeding ... Then close the skin with one vertical mattress
suture of 4–0 plain catgut and cut the knot short ... A spray of
Nobecutane is enough dressing ... '

Thus one particular text for a surgical procedure becoming increasingly common as a form of birth control, with more than 40% of
Chinese, 30% of American and 23% of British couples having one
sterilized partner. In the US this means 500,000 men annually
choosing vasectomy – the cutting of their vas (or rather of both vasa,
each testicle being supplied with one vas deferens to conduct sperm,
via a most tortuous route, to the penis's urethra). The surgical
operation takes ten to fifteen minutes, is performed under general or
local anaesthetic (there are different reasons for each), and prevents
sperm from travelling anywhere – as into the seminal vesicles, where
they are customarily stored, or into the ejaculate. Vasectomy does not
immediately cause sterility. The sperm on the wrong side, as it were,
of the cut can still be effective. Therefore time must pass – perhaps
two to three months – before a man (and his lady) can be confident
of infertility. Sometimes, very rarely, conceptions occur on later dates,
the sperm having unhelpfully found a way around the surgery.

Vasectomy does not, like the sheath, have an ancient history, unless
its most brutal method – castration – is considered a form of it. The
operation of vasectomy is (relative to castration) a minor affair,
diminishing neither sexual appetite nor performance. It maintains the
quantity of ejaculate (for only the sperm are missing), preserves the
sensation of orgasm, and sometimes actually improves sexual enthusiasm (in each partner) because the dread of possible pregnancy has
been removed. Its disadvantage lies in its irreversibility. As yet the
procedure for reuniting the severed ends of each cut is generally
ineffective. Even if the reunion looks good, having been well performed, the sperm – almost always – have lost their potency.

Counsellors have to make certain that males (and couples) requesting vasectomy will not subsequently wish to change their minds. The operation is therefore inadvisable for the young (whose future is obscure) and unnecessary for the old whose partners are already infertile.

Nevertheless sterilization is growing in popularity. In 1964 the US estimate was of 100,000 'doctored' males, and is now five-fold that number every year. India, always grateful for any form of contraception, possesses some clinics more like production lines. The actual surgeon performs his cutting and removal in forty-five seconds, provided that each patient has been adequately prepared beforehand and is then properly treated afterwards. Such formidable haste means – assuming unremitting labour – a mere eighty men an hour. Shirley Green, in *The Curious History of Contraception*, quotes a disturbing statistic. 'To sterilize all Indian men who have *already* fathered three or more children would take 1,000 surgeons performing 20 vasectomies a day at the rate of 5 days a week, a total of 8 years.' While these 41,600,000 men are being treated another tremendous quantity of individuals will have been busy siring their three offspring – and have therefore become eligible for the operation. Vasectomy *is* effective, but needs to be partnered by every other possible form of contraception if an expanding population is to be controlled.

India's statistics, whether involving vasectomy or not, make both happy and sad reading. Since 1951, for example, general mortality has fallen by two-thirds, fertility has dropped by about two-fifths, and life expectancy at birth has almost doubled. Indians are indeed having fewer children and are much less likely to die prematurely. Unfortunately the country's population has doubled even since 1961 (according to figures released in 1995). By 2025 the total is expected to reach 1.5–1.9 billion, and therefore half as much again as China today. Indian families are still too large for the (steadfastly elusive) goal of zero population growth. They have shrunk commendably, from an average of six births in 1951–61 to 3.4 in the 1990s, but 3.4 is still a long way from 2.1, the level of replacement fertility.

Vasectomy's history only really began in the second half of the twentieth century. There are some very early claims but, knowing the power of sepsis, it is easy to discount them until sterile surgery could safely be used to make sterile men. All manner of obstacles loomed when the procedure did become possible, such as the UK acts of 1861 concerning offences and assaults against the person. (Surgeons could, it was agreed, always remove damaged or diseased portions of a body, but not any parts still in sound working order.) Once again, as with other forms of contraception, ancient laws – and their interpreters – became out of step with practice. In Britain the matter was most actively debated in 1972. About 30,000 operations were then being performed, but only within private clinics. Not until a new Bill had had a bumpy ride in both Houses of Parliament was the National Health Service (Family Planning) Amendment Act finally given the royal assent at the end of that stormy year. Local health services could then provide vasectomies on the same basis as other forms of contraception. (Strictly speaking, for pedants, vasectomy is not a form of contraception, there being no sperm available to be forbidden from achieving conception.)

The United States, so much behind Europe in so much of its contraceptive legislation, was ahead in permitting vasectomy to be established. In January 1965 *Time* magazine assessed that 1.5 million American men had already been voluntarily sterilized. The equivalent US total, a third of a century later (and with an annual 100,000 rising to today's yearly 500,000), must now be well in excess of 10 million.

Brazil has possibly led the world in its contradictions, having a tremendous but illegal abortion rate and having a huge need for contraception in the face of governmental stricture, but the law suddenly switched concerning sterilization. Up to 1994 the operation had been illegal (with all manner of penalties for those involved) but it became abruptly legal in 1995. The country's largest city, with some 20 million inhabitants, promptly ordered that the operation could be free on the public health system. The Church had opposed the new legislation, and continued to oppose sterilization as a contraceptive method. São Paulo's Physicians' Union have also criticized the new law mainly because it does not alter the situation for other contraceptive methods. The city's mayor, seemingly caught in the middle, stated that the procedure will not be encouraged among the poor (who need it so desperately); it will merely be made available to them. The citizens of Brazil often cry that no other country has so many laws, but are also happy to state that no law has been passed which demands obedience to all the legislation. A vibrant sexual life-style coupled with conflicting beliefs, laws and proclamations concerning contraception have combined to give Brazil a burgeoning population, particularly among the very poor.

In most contrary fashion, although without a family planning programme, Brazil has done better in reducing its population growth than, say, India which does have such a programme. Its growth rate was 2.99% a year between 1951 and 1960, and this proportion fell to 1.93% a year between 1981 and 1990. Some observers have attributed this descent to television. Globo, the most popular TV network, shows three hours of 'soaps' – *novelas* – for six nights a week. Such programmes do not banish the thought of sex – far from it – but they tend to show wealthy characters living a high life in big cities, such families unburdened by children and doing very nicely. Despite the fall – and the television – Brazil does not expect its population (currently 150 million) to stabilize until it reaches 250 million a quarter of a century or so from now.

ENFORCEMENT For a time the United States, leader in vasectomy, was also a leader in the creation of involuntary sterilization. It certainly led most other countries in the introduction of legislation permitting the operation on those it deemed unsuitable for breeding, an issue focused sharply by Justice Oliver Wendell Holmes (appointed to the Supreme Court by Theodore Roosevelt) with his famous remark, made during a rape case, that 'three generations of imbeciles are enough'. The victim of the rape had been feeble-minded, and she needed – so some said – to be sterilized.

The campaign to sterilize the 'unfit' started in the 1890s. Michigan, for example, discussed in its legislature the 'asexualization' of both the mentally deficient and certain criminals in 1897, but did not pass a bill to that effect. Sterilization was performed in Kansas (on forty-four boys), even without legislative consent, before public disapproval called a halt. With the start of the twentieth century, and following the rediscovery of Gregor Mendel's laws (which so simplified genetic inheritance in many minds – and whose publication had scarcely been noticed thirty-five years earlier) the campaign gained new impetus.

Indiana became the first US state to pass a sterilization measure based on eugenic principles. By the 1920s many others were following suit, with Harry H. Laughlin (prominent American eugenicist) most active in promoting the new laws. There was considerable opposition from the Roman Catholic Church, but by 1931 sterilization could be performed in thirty states on, for example, 'hereditary defectives' such as 'sexual perverts', 'drug fiends', 'drunkards', 'epileptics', and 'diseased and degenerate persons'. (Kenneth M. Ludmerer has written an excellent account of this particular chapter in American history. So too has Stephen Trombley in his documentary, *The Lynchburg Story*, thus titled because Lynchburg's Colony, in Virginia, sterilized 8,300 – some as late as 1972 – of the 70,000 people thought to have experienced compulsory 'asexualization' within the United States.)

Other countries were not against the notion of preventing procreation by those considered less worthy of producing the next generation, with Finland, Sweden, Norway, Iceland and a couple of Canadian provinces bringing such legislation into effect. Germany then instituted its Hereditary Health Law (on 14 July 1933), only a few months after the National Socialists had gained power. Germany's extreme implementation of eugenic thinking caused other countries to retreat, despite hesitant (and not so hesitant) steps having been taken along that path.

Harry Laughlin's invitation from Heidelberg University (in 1936) to receive an honorary MD degree, and his acceptance of it, showed that blatantly racist Germany and well-intentioned (as was claimed) eugenic America were not so far apart. Since he had been appointed secretary of a 'Committee to Study and to Report on the Best Practical Means for Cutting Off the Defective Germ-Plasm in the American Population', Laughlin had laboured to eliminate, within two generations, inheritance lines of 'the most worthless one-tenth' of Americans. Earlier he had written that the principal end of research on human heredity, although 'not stated in so many words', was to 'enable predictions in hereditary behaviour to be made'. Copies of his writing had plainly reached Heidelberg, and Laughlin was duly rewarded for his thinking – even though Nazi legislation never went so far as he had done in defining 'socially inadequate persons'.

In Britain J. B. S. Haldane and Lancelot Hogben, both popularizers of science, were consistently adamant in condemning Germany's eugenic practices, with Hogben publicizing 'the strain of haemophilia in the Royal Houses of Europe' before admitting that 'no eugenicist had actually proposed sterilization as a remedy for kingship'. A Bill put before Parliament in the 1930s would, if passed, have permitted the sterilization 'of persons who are deemed likely to transmit a mental defect or a grave physical disability to subsequent generations'. It was never passed, but haemophilia is undoubtedly a 'grave physical disability', whoever happens to carry, and perhaps pass on, its defective gene. (In truth the proposed British law was even more severe than those passed in Germany. The Germans were sterilizing 'defectives' but the British law-makers were also concerned with those who might sire offspring with a defect or disability. Mere carriers, as right as rain themselves, would therefore have been subject to mutilation.)

The German situation altered, year by year, as the National Socialists entrenched their power, as the war loomed, and then as defeat became a possibility; but it is important to remember eugenic thinking in other areas, as in Britain. There were, for example, annual 'Galton Lectures' delivered under the auspices of the Eugenics Society. Their titles are quietly revealing: 'What Nations and Classes Will Prevail?' asked Dean Inge, in 1919. Other speakers were to address similar topics: 'The Ruin of Rome and its Lessons for Us', 'Some Reflections on Eugenics and Religion', 'Causes of Racial Decay', and 'The Social Problem Group, as Illustrated by a Series of East London Pedigrees'. Many individuals in Britain would not have been in violent disagreement with some of Hitler's racial statements, expressed for all to see in *Mein Kampf*: 'The mixing of higher and lower races is clearly against the intent of Nature ... The stronger must dominate and not mix with the weaker ... Only the born weakling can feel this to be cruel.' Hitler was a politician who actually acted upon his earlier declarations once power had come his way. (He even wrote, in 1923, 'The greater the lie, the greater the chance it will be believed' but no one seemed to absorb this awful truth until it came to pass.)

The excesses (if strong enough a word) made others in other countries realize how close they had been to travelling along the self-same path. Sterilization of the hopelessly incapable (whose procreation would be a disaster for everyone concerned, not least the parents) could lead on to the fairly incapable and then to those, believed by society at the time, to be less than average in their capability. Germany's slaughter, which accelerated as the war progressed (and degenerated), meant that eugenic policies became confused with retribution and revenge. Until that happened quite a few American scientists, for example Paul Popenoe, were saying that Germany was 'proceeding toward a policy that will accord with the best thought of eugenists in all civilized countries'. Or, as *Eugenical News* reported: 'It is difficult to see how the new German Sterilization Law could, as some have suggested, be deflected from its purely eugenical purpose, and be made an "instrument of tyranny" for the sterilization of non-Nordic races.' Well, it was deflected, and no one spoke or wrote like that after Germany's procedures had become more widely known. Even today many prefer the term vasectomy to that of sterilization, so tainted is the latter by its awesome history.

*

As for vasectomy's future, the apple-cart may next be turned completely upside-down. Currently the whole purpose of every form of contraception is, as the word states, to prevent conception. Perhaps, so say some futurists, fertility should be viewed contrarily, and for it to exist *only* when fertility is desired. At present each man's inborn fertility is (often) an unwelcome trait.

Therefore, so this argument runs, why not render him infertile as soon as possible? Perform vasectomy early (much as circumcision has been routinely performed in many societies). Simultaneously remove a quantity of sperm, store it in liquid nitrogen (at minus 196 degrees Centigrade), extract a sample when need be (should fertility be desired), use artificial insemination to create the pregnancy, and forget entirely about the archaic procedure of contraception. *Infertility* would therefore be the normal state for a male, with fertility – however contrived – the rarer situation.

Human pregnancies resulting from frozen-thawed sperm are becoming commoner all the time. France, a leader, has achieved 17,000. With animals the procedure is often commoner than normal mating. About 46 million head of cattle, for example, are annually created by this method. Women have been receiving frozen sperm for a shorter period than have animals, but some ten-year-old human sperm has been used to produce normal human offspring. Bovine sperm has been successful after thrity-seven years, and it does not seem to matter greatly how long the sperm has been stored. The lambing rate, in one major study, was equal for ewes artificially inseminated with frozen sperm, whether this had been stored for five years or two weeks. As for artificial insemination of humans, about 170,000 American women undergo this treatment every year (at a cost of $164 million, and making use of 11,000 physicians), but most of this insemination is with fresh sperm rather than the frozen kind.

If all goes well with the proposed technique, and if cryogenic storage proves entirely satisfactory, vasectomy can then be considered reversible. Contraception will have been left behind in the medieval twentieth century. Save, of course, in all those poorer regions where even babies cost less than their prevention, but that poverty will not stop wealthier areas from adopting infertility as the norm rather than the fertility which is, so often, a form of scourge in every region of the world, both rich and poor.

SALPINGECTOMY A salpinx was an ancient Greek instrument, resembling a trumpet. Hence salpingian, of the trumpet. Hence also a variety of medical terms associated with tubes thought to be (modestly) trumpet-shaped, like the Eustachian tubes of the ear, and the Fallopian tubes between uterus and ovary. Examples of the Greek inheritance are salpingitis, when the Fallopian tubes are inflamed, and salpingectomy – for the sterilization of females.

This too, by whatever name, has an ancient history which, as with all claims about early surgery, needs to be viewed sceptically. The King of Lydia is reported to have 'castrated women'. Ancient Egyptians, a little like early Chinese in having done *everything* long before everyone else, are said to have removed ovaries. The purpose, allegedly, was maintenance of the sylph-like female form (which could be lost during repeated pregnancy). A few native Australians were said, by some nineteenth-century anthropologists, to have been like 'spayed cows', the purpose of their mutilation being for them to serve as prostitutes who could never become mothers. Perhaps some women, at different times in different lands, did have their ovaries removed, or they did suffer some other form of surgical damage to render pregnancy impossible, but such procedures cannot have been common – if, indeed, they occurred at all.

The procedure is certainly common today. Some 37 per cent of American women of reproductive age who do use contraception (almost all at some time) rely on sterilization. In the UK about 90,000 women are sterilized each year. A third of Brazilian women have undergone surgical sterilization, with the proportion being 70% in some areas. Many of these operations are performed at the same time as Caesarean births (which, in Brazil, account for a third of all births). These sterilizations have undoubtedly helped to bring down the number of children per woman in that country from 5.7 in 1970 to 2.5 in 1991.

The major disadvantage of salpingectomy/female sterilization is its irreversibility. The procedure is final, as with male vasectomy, and should be considered so, even though attempts have been made, with both males and females, to undo what has been done – or rather to do up what has been undone. Two major reasons for a change of mind (with women) are either a new marriage and relationship or the death of an offspring. Plainly, according to various authorities, young women should not be sterilized, nor should those who might (later) want children. A formula is even mentioned: if the woman's age plus the number of her children totals 35 or more the operation can be considered.

Other disadvantages (apart from irreversibility) exist, such as failure (the figure of two per 1,000 is often quoted, or more if the operation is made soon after a pregnancy), 'heavier' periods (said to occur with a third of all women), weight gain (which may only be a relaxation of previous slimming regimes or so-called 'compensation-eating' following the important loss), and a temptation to blame all subsequent events and emotions, such as depression, upon the operation, it being handy as a scapegoat for every ill.

The major sterilization difference between men and women relates to the age at which the two sexes become infertile. A woman of, say, 35 who requests the operation has only a few more years before natural infertility is imposed upon her. A man of 35 who requests vasectomy is denying his fertility for the remainder of his days. If a couple, both aged 35, decide upon sterilization there are many years ahead which may witness a breakdown of that relationship (and perhaps the finding of new friends). For the woman it is likely to be too late for further procreation. For the man, if he has acquired a younger mate, he (and she) may greatly regret his imposed sterility. Counselling and care are therefore vital before irreversibility is undertaken, particularly for younger women and most men.

Vasectomy can be considered somewhat crude – a portion of each sperm-conducting tube is severed and removed – and female sterilization is equally basic. With salpingectomy a portion of each Fallopian tube is removed, and the severed ends tied so that eggs can no longer travel. (Partisans and others who remove one short length of railway track are no less subtle in their approach.)

Two principal methods are used for female sterilization. Minilaparotomy, requiring a general anaesthetic and a couple of days in hospital, involves a small cut (just below the bikini line, says one text) and then a reaching for the Fallopian tubes. A portion of each is cut and tied (tubal ligated) or cut and sealed by burning (diathermy). Laparoscopy is the commoner method, and can be done with a local anaesthetic. An incision is made just below the navel. A laparoscope (for looking into the abdomen) is then inserted. This guides the surgeon as he finds and then blocks the Fallopian tubes with rings or clips. (Future generations may consider current sterilizing techniques almost as basic as we view former agricultural processes for castrating domestic animals, using bricks or even teeth.)

Failure can and does occur. There may even be death, allegedly with one in 5,000 abdominal operations (as risk and operations are always partnered to some degree). More probable is a failure of the operation itself, with a subsequent unplanned (and unwanted) pregnancy. Some unhappily gravid individuals then sue for negligence, often affirming they were not informed that success was never assured. According to the Medical Protection Society (of the UK) such claims peaked in 1986, and have since diminished. Both the counselling – and the technique – have improved in recent years.

As for sterilization of the mentally unfit, a subject so tainted by its history, the subject will never be easy. It is currently impossible, according to an excellent summary of the situation (by C. Orr), to make an adult a ward of court 'in order that any sterilization, or for that matter termination, may be ordered by the court'. The surgeon, and those involved, 'must be in a position to show that the actions taken were in the best interests of the patient concerned on a strictly individual basis'. This summary concludes that 'the courts seem now to be tending to the view that an assault, or more properly a battery, cannot be established without ill-intent, which manifestly would not be the case when the mentally-handicapped patient is under the care of reputable practioners ... The actual application to the courts for leave to operate should be made by the care authorities and not by the surgeon in question.'

OTHER DEVICES When illustrated collectively a group of modern IUDs resemble a batch of modern earrings (and maybe the things *are* used as earrings in these confusing days). There are twirls, whirls, loops, bows, circles, rings, wings, and they all make an observer wonder, not so much at the variety but whether any shape is not included. According to dubious legend various objects were first inserted into camels, but facts are also scarce concerning their use with/by/on humans until the 1920s. Ernst Gräfenberg, of Berlin, put coils into a large number of patients, but some of these women conceived, some bled badly, and some aborted foetuses. His devices therefore fell from favour.

Then, when the pill was arriving in the early 1960s (and contraception of all kinds was leaping forward), intrauterine devices reappeared to do rather better than forty years beforehand. By 1970 some 3 million had been distributed in the United States. Today the global figure is nearer 90 million, with China using the lion's share. There has never been the publicity over IUDs that was accorded to the pill, or even to sheaths, but the devices have co-existed, quietly, less obtrusively and – save for China – less importantly. In Britain before 1951 some 6% of women (who were using contraception) had a loop or bow or some other earring look-alike as their contraceptive. Currently the proportion is nearer 5%.

Despite China's enthusiasm for IUDs, and for its home-made steel ring – usually inserted after the birth of a first child – this form of contraception has not been outstandingly successful for the Chinese participants. Of the country's 10 million annual abortions about 3 million are performed on behalf of women who have been using the cheap, local steel ring, the nation's most favoured contraceptive. No one (yet) knows for sure how IUDs work, and how they (usually) prevent embryos from implanting, but it is known that copper is a better material for them than iron. China has realized, even if copper is more expensive, that about $19 million could be saved each year from abortions occurring as a result of the cheaper alternative. A recent UN Population Fund report concluded that, if China switched completely in the next ten years to the so-called Copper-T (six times more effective than the traditional ring), 36 million abortions could be prevented. China has understood, and four Copper-T factories were operating by the end of 1994.

That country's well-televised suppression of democracy has been one impediment to foreign aid. Its vigorous one-child programme of birth control, with a preponderance of males among those single children, has been another. The presumption of many female murders, and China's policy in general, caused President Reagan in 1985 to cut off aid to the UN Population Fund. Only in 1994 was that aid resumed – but still with the proviso that none should be spent in China.

There is also antagonism, from various quarters, concerning IUDs in general. They cause, it is argued, a form of abortion by preventing the fertilized egg and subsequent pre-embryo from implantation. More modern thinking (convenient for the encouragement of IUDs) is that – somehow – they block fertilization. Perhaps there is a bit of both – stopping fertilization and stopping implantation – because the devices can be effective even if inserted post-coitally. Individuals who wish their contraception to resemble abortion as little as possible, who like sperm and egg never to meet (as they fail to do with sheaths or sterilization), can object to IUDs on the grounds that egg and sperm may meet, if unproductively. John Guillebaud, steadfast devotee of infertility (already mentioned as a defender of the pill), lists the 'wonderful advantages' of IUDs in his best-seller *Contraception*, such as 'effective', 'no known systemic effects', 'nearly always reversible', 'independent of intercourse', no 'day-to-day actions', and 'easy to distribute'. The main disadvantage, which he also publicizes, is an increased risk of pelvic infection. He reminds his readers that IUD-linked infections are caught sexually. The chances of infection rise with the number of partners, whether of the user or of the user's partner, however single-minded and faithful the user herself may be.

In 1995, causing Guillebaud to be yet more enthusiastic about IUDs, Mirena was officially launched, with him calling it 'the most significant advance in reversible contraception since the invention of "the Pill"'. In essence it is both IUD *and* hormonal agent, in that it has a T-shaped frame whose upright, the 'hormone sleeve', contains a supply of progestogen, this substance also present in a combined contraceptive pill. It leaks out at the rate of 20 micrograms a day, and does so for five years. The device works, so its promoters state, by keeping the womb lining thin, by thickening the cervix mucus (thus impeding sperm) and by hindering ovulation. The sleeve is about one inch long, and the two arms of the T together total about one inch. There are a couple of removal threads, also cut to about one inch, that lead from the cervix into the vagina. The whole device is known as an IUS (for hormonal intrauterine system).

Advantages are numerous, with enthusiasts claiming it fulfils more of the requirements of a 'perfect' contraceptive than any predecessor. It does not cause infection, or heavy periods, or ectopics. Periods are either reduced or absent. There is no oestrogen (and therefore no risk of thrombosis). It stays in place for three years (or maybe five if current licensing is extended). Its contraceptive effects are completely, and speedily, reversed when it is removed. Little time is needed for fitting, being five minutes or so for preparation and seconds for the actual insertion. In the largest study (up to 1995) the failure rate – unwanted pregnancies – was 0.1 per 100 woman-years.

Such a battery of benefits is mitigated by one major deficit: it affords no protection against HIV. The IUS is not cheap, being about £120 in 1996, or £200 including the fitting, but these sums appear more appealing when the single outlay is spread over five years. It then totals little more than the pill over a similar period. A big disadvantage in Britain is that IUDs in general are not much favoured. Therefore any new one, however good, has an uphill struggle to compete with, for example, the well-established condom and the pill.

FUTURE POSSIBILITIES Sir Alan Parkes's remark that contraceptive methods were, until the 1960s, a 'disgrace to science in an age of spectacular achievement' has already been noted. A few years later (and still master of the pithy phrase) he said that 'no woman should be kept on the Pill for twenty years until, in fact, a sufficient number of women have been kept on the Pill for twenty years'. Both statements help to explain the current situation. Those great advances of the 1960s – the astonishing pill, the range of IUDs, the skilled severing of ducts derailing sperm and egg – took contraception, if belatedly, into the most scientific century of all, at least for its second half. They also took some steam out of the situation. Modern scientific methods do now exist. Women no longer have to rely upon men, either for timely withdrawal or remembering the sheath. The quantity of devices, pills, caps, and other options available are considerable, relative to the first half of the century.

But, as Sir Alan indicated, it is now hard to introduce new techniques. The Puerto Rican experiment would not be tolerated today, and there is less enthusiasm for such risky work. Carl Djerassi, of Stanford University (proclaimed 'inventor' of the *modern* contraceptive pill), says that every barrier preventing new contraceptives from reaching the market is still firmly in place: the high development costs, the poor profit potential, and the fear of litigation. Of the dozen major pharmaceutical companies pursuing contraception research in the 1960s, according to a Rockefeller Foundation report, only a handful remain. 'Contraception is not AIDS,' said a molecular biologist of Colorado University. 'It's not a big issue in the USA.' 'When you throw everything into the balance, companies decide to work on other things, such as developing new drugs against cancer or heart disease,' said the vice-president of Genzyme, a Massachusetts-based biotechnology firm.

One more problem, affecting the United States in particular, is the fourteen-year-old ban on federal funding for research involving the special creation of human embryos. 'If you want to prevent fertilization, you have to study fertilization,' says John Eppig, of the Jackson Laboratory, Bar Harbor, Maine. And so say a good many others who liken the problem to designing aircraft without benefit of flight. The opposition's strength was demonstrated when the (admittedly controversial) abortion pill RU486 (or mifepristone) failed to find a pharmaceutical company in the US prepared to test and market the drug. They all turned it down, reported Wayne Bardin, of the Population Council in New York City, for fear of being targeted by political backlash. Anything which smacks of abortion, however much it is intended to reduce abortion (as with all contraception), is in for a rugged ride.

*

None of this means that contraceptive research has come to a halt. For example, there is Norplant, first used in Finland in 1983, approved by the US in 1990 and tried out by 900,000 women in its first four years. The matchstick-sized capsules have to be inserted beneath the skin where they liberate a synthetic hormone. In theory they should remain active for five years. In practice 15% of the American women asked for removal, either because they wished for pregnancy or because they resented Norplant's wide-ranging side effects (irregular and abnormal bleeding, headaches, nausea, weight increase, arm pain, general discomfort, acne, depression) or the implants had caused scarring. The principal reasons given for removal (in one Belgian study of 749 women) were excessive bleeding and a wish to become pregnant. Removal should take only thirty minutes under local anaesthesia, but badly implanted capsules can cause more trouble. About 400 women have – so far – filed suits against the manufacturer.

There are also vaccines. In 1994 a paper in the *Proceedings of the National Academy of Sciences* reported on trials in India which had immunized women against human chorionic gonadotropin (abounding in the blood – and urine – of those who are pregnant). It is created by the pre-embryo, and is essential for implantation to occur. Four-fifths of the Indian women produced large quantities of antibody to the hormone but, as an article in *Science* reported, the possibility that such vaccines 'act after fertilization has kept many funding agencies out of the picture'. Moreover 'no vaccine that has yet reached trials on primates matches the efficacy rate of oral contraceptives'.

Some contraceptive pills can be effective even if taken after unprotected sex. These must be taken at the right time and in the right dosage, but this information is not (in general) included with their labelling. It should be, according to the US Food and Drug Administration. Similar oestrogen/progestogen preparations are used elsewhere as 'morning after' pills, and could be used more widely – if instructions were available. This more general use could, it has been reported, bring down the number of unwanted US pregnancies by 1.7 million to 2.2 million a year. And this fact could, in its turn, reduce the country's abortions by an annual 1 million.

No one is absolutely certain how such pills operate contraceptively – perhaps by impairing movement of the egg down the Fallopian tube or by interfering with implantation – but there is universal agreement that 'morning after' is not the term for them. It suggests that the subsequent morning is the only occasion for their use. Instructions could (and would) point out that their taking is more complex and prolonged. Their wrongful taking nowadays (without advice for guidance) can even be dangerous, word having spread around that such post-coital pill-taking is a feasibility but with this word having failed to include proper information. As the saying goes – 'If all else fails, read the instructions' – but there must be instructions to read.

As for a new male contraceptive (to supplant, or supplement male methods of sterilization – the condom and *interruptus*), there is little on the immediate horizon. Carl Djerassi (and S. P. Leibo, of Guelph University, Ontario) stated in *Nature* (July 1994) that the prognosis 'by the year 2000 is nil'. Indeed, because the 'development, testing and regulatory approval of a truly novel, systemic male contraceptive requires 15–20 years ... the expectation for a "male pill" even after 2010 is dismal'. Others have been less dismal in their forecasts, such as those promoting a sugar which allegedly cripples sperm attempting to break through each egg's natural coating. The substance has worked effectively on rats, bulls and rams, and has the extra blessing of not interfering with sperm production or hormone levels.

In April 1996 results were announced of two-year trials based upon the injection of a hormone (testosterone) into human males who had been willing to try this new form of contraception. Weekly doses significantly reduced their sperm count – one man's production dropped from 40 million to 200,000 – and thereby rendered the men less fertile. (It is thought that the considerable quantities of additional hormone cause the testes to be less diligent in producing sperm.) The men's sex life was apparently unaffected, but ordinary life – with those weekly injections into buttocks – was presumably impaired. Fred Wu, British doctor on the research team, claimed the results as a 'very significant' advance towards 'permanent contraception for men'. The failure rate was 1.4 pregnancies per 100 couples, including that man whose sperm count had dropped to one two-hundredth of its former level. Further trials are now being conducted using pills, skin patches and longer-term injections, with results expected by 2004.

Some women, when asked about the blessing of this new form of male contraception, were a touch sceptical. Trust would be involved and they, as ever, would be the victims if it had been misplaced.

*

It would seem that the revolution of the early 1960s, when contraception moved so swiftly on so many fronts (after years in the doldrums), will not be overtaken by great change for quite a while. (Perhaps that is the trouble with revolutions: having, as it were, leaped ahead, there has to be a quieter phase after their arrival.) The planet still has a tremendous need for contraception. Unfortunately its people have become extremely aware of possible (or even certain) drawbacks for every method, whether old or new. Human beings frequently embrace novelty, as they do for so-called wonder drugs always so loudly heralded, until realizing that all silver linings surround some form of cloud.

As for the objectors, who can want contraception without the immorality – in their view – of killing a human, however small and ill-defined, they certainly have power to obstruct. It may diminish their concern (and perhaps lessen their argument) to remember that less than 0.1% of an embryo twelve days post-fertilization will actually be transformed into a human being. The rest is preparing the nest, assisting with implantation, and forming the amnion, the umbilicus and much of the placenta. Failure to destroy that 0.1% (and its accompanying 99.9%) can lead on to the greater destruction – so often performed – of a three-month foetus or, no less welcome, the creation of yet another infant, unwanted first by its family but also by the world.

In the Philippines, where abortion is rife but illegal, and where the government has been accused of 'promoting pornography, abortion, homosexuality, incest and sodomy' under the guise of sex education and birth control, the principal (and most famous) Catholic spokesman in this most Catholic of eastern nations is Cardinal Sin, of Manila. His memorable name can make us wonder what future generations will think of their inheritance, after our errors have donated to them a less satisfactory world, a planet polluted, heated and overpopulated. Knowing that pollution, heat and other errors are all people-based, they might consider the most cardinal fault was overpopulation. As excuse we cannot blame our lack of means. The existing forms of contraception could be effective in reducing the ejaculation strike-rate from its current level of 0.7%. We have the technology, as the saying goes. We have sufficient wealth. We have – most of us – the wish. The future may indeed wonder why that triumvirate proved so inadequate, and why several billion people became several billion more people, notably as the twentieth century yielded to the twenty-first.

POPULATION The precise number of future billions, horrendously relevant, is both difficult to envisage and – for the forecasters – tricky to estimate. From a current 5.75 billion the total is thought (by the UN) likely to rise to 7.1–7.8 billion by 2015, and perhaps as high as 11.9 billion by 2050. The extremely reputable International Institute for Applied Systems Analysis (in Vienna) prefers 10 billion by 2075. This institute does not claim certainty for these projections, only a level of probability, such as a 64% chance that a doubling from the present level will not occur. The thought of a peak, a cessation to the inexorable rise, is comforting, despite that formidable total of 11 billion.

Such modest comfort can be swiftly eroded by the expectation, promised by the WHO, that 20–30 cities will have populations in excess of 20 million even by 2011. Already 45% of the world's population lives in urban areas, and half will do so by 2005. Tokyo now has the largest conglomeration, with 26.5 million inhabitants (in 1994). Second largest is New York with 16.3 million. (Some argue that São Paulo should lead if amalgamated with neighbouring conurbations, such as Santos.) As for the United Kingdom, and according to the Office of Population Censuses and Surveys, it is expected to reach 62.3 million by 2027 and then return to its present quantity (of 57.6 million) by 2062.

Pakistan on its own can kill much surviving optimism. It has the fastest rate of population growth of any large Asian nation. It is now the world's seventh largest country but, in half a century, will become third largest, having overtaken the US, Russia, Indonesia and Brazil. Pakistani women are still having 5.9 children (as against 3.4 for India and Bangladesh). Only 18% of Pakistani couples use contraception (with India 41% and Bangladesh 45%). Only a fifth of Pakistani girls learn to read and write, and only a third attend primary school. By 2035, according to present trends, Pakistan's population will be growing faster than China's, in both percentage and absolute terms. Until 1996 Pakistan had a female prime minister ruling a country where each woman has a 3.5% chance of dying while giving birth, where young girls die more frequently than young boys (in a ratio of 5 to 3) and where girls are more likely to suffer malnutrition and receive less health care than their brothers.

*

How Many People Can the Earth Support? is the intriguing title of a book by Joel E. Cohen (published in 1995). So how many can it cater for? Is the projected 11 billion of the twenty-first century likely to be excessive, creating gross starvation and consequent warfare? Unfortunately (for any peace of mind) Cohen gives no easy answer. It all depends what is meant by support. He lists eleven simple questions by way of elaboration, such as: 'How many at what average level of well-being?' 'How many with what distribution of material well-being?' and 'How many with what values, taste and fashion?' Or, to phrase such questions in even simpler manner: 'How many North American and European life-styles?' 'How many peasants of, say, the Indian variety?' and 'How many indigenous people, living traditional lives on traditional lands?' (a number now estimated as 300 million in seventy countries).

For a final thought, the children now being conceived are likely to live to 2070 or so. They will know absolutely what it is like to have twice as many people on this planet. They will certainly know if warfare, inequality and premature death continue as relentlessly as during the twentieth century. And they will undoubtedly wonder why contraception was not better employed when, compared with the nineteenth century, it was all so easy.

ABORTION

HISTORY   IRELAND   POLAND   ELSEWHERE

UNITED STATES   RIGHTS   MIFEPRISTONE

SOME ALTERNATIVES

Abortion; premature delivery, or the procuring of it,
esp. (*med.*) in the first three months of pregnancy: arrest of
development: the product of such arrest: anything that fails
in course of coming into being: a misshapen being or
monster.

Chambers Dictionary

The law should not forbid what it is not necessary to forbid
and, so far as is possible, it ought to authorize what people
feel they want to do.

Lord Kilbrandon

Abortion is a new procedure; it is as old as the hills. Women have a right to choose; doctors have a right not to assist them. Abortion is a form of contraception; it is a form of murder. It is the killing of a human being; therefore the killing of abortionists who carry out such acts is acceptable. It is unacceptable in a modern world; the modern world would be appalled if the 50 million abortions carried out each year were allowed to grow as human beings, as unwanted human beings adding to the score. The subject has no place in a book on reproduction; the subject cannot be denied for the part it plays in the human story.

Every issue raised by abortion finds fierce enthusiasm on either side. Many new medical procedures – anaesthesia, vaccination, asepsis – have encountered opposition, but resentment against abortion continues long after it might have settled down. Unfortunately, abortion numbers have not diminished despite earlier assurances that better contraception would reduce the need, that better education would do likewise, that free facilities and advice would render obsolete (almost) all unwanted pregnancies. In England and Wales it was expected that abortions would rise once the procedure had been legalized, and then – or so it was hoped – would fall. They did rise, but did not fall: 21,000 in 1968 (legalization year), 73,000 in 1970, 105,000 in 1972, 117,000 in 1979, and 127,000 in 1983. For the United Kingdom as a whole the total has stabilized at about 175,000 (this number including non-residents).

Japan, frequently providing exception to any rule, has witnessed a steady downward trend and no stable plateau. It is the only industrialized country where steroid oral contraceptives are (still) illegal, and it was the first industrialized nation after the Second World War to sanction abortion as a method of birth control. The increasing use of other means of control – condoms and the Ogino calendar rhythm method – has brought down the abortion rate. In 1955 there were 1,170,000 induced abortions. By 1965 the number was 843,000, by 1975 671,000, by 1983 567,000, by 1992 410,000, and the number has continued to fall. There was a slight drop in the number of registered births during those years, but the proportion of Japanese abortions to births fell from 67% to 37%.

Conversely, in the United States, the number of abortions has been rising (to well in excess of a million per year) and the proportion has also been increasing to 25% or so (indicating one abortion for every three live births). In Britain about one in five pregnancies is terminated, and the ratio is one in two for women under 20. In countries of the former Soviet Union recourse to abortion varies from 2.5 to 4 per woman (according to the WHO), as against 1.5 for Eastern Europe and 0.63 for Westerners. The 'fifth freedom', as it has been called, will plainly continue to be exercised for many years to come. It has also been called 'the greatest epidemic of all time'. 'Even in the best organized society,' as David Baird (of Edinburgh University's Centre for Reproductive Biology) has written, 'there will be a continuing demand.'

What is not known is the number of illegal abortions occurring before 1968, or the number that would have occurred had the new laws not arrived, but Madeleine Simms (long-time expert in this subject) believes it was 'probably near 15,000' annually for the UK. In which case approximately 4 million Britons are not alive today –

half London's population – because abortion freedom has arrived. If the global annual figure is indeed 50 million, the world has been spared the arrival of China's current population in the past thirty years. Abortion is certainly a contentious issue, but its form of killing undoubtedly reduces population (unless, as further thought, the procedure encourages less concern over fornication).

HISTORY Until 1803 (in Britain) the law was tolerant of abortion, provided 'quickening' – usually in the fifth month – had not begun. Increasingly restrictive legislation was then initiated, culminating in the 'Offences Against the Person Act' of 1861 with the maximum penalty for both abortionist and aborted mother being imprisonment for life. This effectively forbade the procedure legally (and condemned it to back streets) until 1968 when, after tremendous debate, the modern situation had its origin. The medical profession was then divided. Most gynaecologists disliked the act, contravening – as it did – their prime purpose of assisting in the creation of healthy infants. Most psychiatrists welcomed it, being more aware of the aftermath of unwelcome pregnancies. Nurses, torn every way, had (and have) a terrible time. Loyalty to their superiors, their patients, their religion and their own sensibilities can be most divisive. The patients themselves, however relieved that an unwelcome happening can be banished, do not enjoy the procedure. Some even say the pain is therapeutic: 'I wish to suffer as this is a wrong act.' Abortion, as has been said about so much else, is not good but is better than alternatives.

One reason, to quote Baird again, why most societies and cultures tolerate the act is that 'the rate of abortion which occurs naturally in our species is very high'. Only about one fertilized egg in four results in a viable offspring. Most of these failures occur before the woman realizes she is (or was) pregnant. Much of this expulsion is a filtering process, or so it is believed, much like a factory's rejection of faulty components. Over 60% of all abortions up to six weeks of age are associated with chromosome abnormalities. As the gestation weeks pass by the rate of natural abortion decreases, along with the incidence of abnormal foetuses. The later a medical abortion is performed the more likely that the expelled foetus, having passed natural rejection procedures, is normal, and would have been normal had it proceeded to term.

One poll, taken shortly after the 1968 Act had come into force, revealed that 63% of those questioned thought abortion should be permitted in particular circumstances, and a minority – 18% – believed it should be available on demand. Certain cases helped to polarize opinion, such as *Roe* v. *Wade* in the United States. This famous trial (spread over three years) assisted the Supreme Court to legalize abortion in 1973. Norma McCorvey, who revealed in the late eighties that she had been the case's Jane Roe, told of an appalling relationship with her mother, of her mother's attempts to abort her, of robbing a filling station and being sent to a Catholic Reform school (before she was 12 where she was sexually assaulted by a nun), and of being repeatedly raped (when 16) by a relative. She also spoke of being beaten unconscious by him, of her ignorance of life's facts, of not even realizing she was pregnant on the first occasion, and how she was deceived into forsaking her maternal rights (and had her baby adopted). As a review of her book in *Nature* stated: 'To say to such a woman that she simply should have been more careful is nonsense.'

Norma McCorvey, a lesbian, later became an active advocate of a woman's right to choose. Anti-abortionists then attempted to shoot her. They shot up her car, threw baby clothes over her lawn, and accosted her in the supermarkets. Later still, and in 1995, she sided with the anti-abortionists. This switch was a blow to those in favour of abortion, but McCorvey had become a born-again Christian. 'The pro-choice poster girl has jumped off the poster, into the arms of Jesus Christ,' reported Phillip Benham, co-founder of Operation Rescue, a radical anti-abortion group. 'Miss Norma has answered our prayers,' he added.

The 1968 British law on abortion remained virtually intact for twenty-two years, despite many attempts to amend it. Eventually, in 1990, a new law was implemented. This took note of some of the objections and also protected practitioners who might otherwise have been found guilty under the Infant Life Preservation Act of 1929. It listed four grounds under which termination of pregnancy could be undertaken, provided:

a) that the pregnancy has not exceeded its twenty-fourth week and that the continuance of the pregnancy would involve risk, greater than if the pregnancy were terminated, of injury to the physical or mental health of the pregnant woman or any existing children of her family; or

b) that the termination is necessary to prevent grave permanent injury to the physical or mental health of the pregnant woman; or

c) that the continuance of the pregnancy would involve risk to the life of the pregnant woman, greater than if the pregnancy were terminated; or

d) that there is a substantial risk that if the child were born it would suffer from such physical or mental abnormalities as to be seriously handicapped.

The twenty-fourth week is therefore the great barrier, save when the terminations involve b, c or d. That week represents quite an advanced state of pregnancy, being five and a half months since conception. Foetuses delivered at this time have survived – about half do so if the facilities are good. They weigh 600 grams and measure 12 inches from crown to heel. Their eyelids can open, and they can grasp with their hands. Hence much of the controversy. They are not bundles of cells, waiting to be fashioned more correctly. They are very small people (and one wonders how many of those colourful photographs, extolling the 'miracle of life' by portraying pre-term foetuses, are in fact of abortions before these miracles were jettisoned). Plainly, for all concerned, the sooner the better whenever abortions are the preferred alternative.

IRELAND Anyone who understands Ireland, it has been said, can understand the world; it possesses all the contradictions. Only in 1979 did the sale of contraceptives become legal in that country. Only in 1993 was homosexuality decriminalized. The Irish family is generally considered sacrosanct, but one in five births is to an unwed mother.

'Marriage has been a disaster for Ireland,' said a nun recently. A national plebiscite in the 1980s voted to forbid abortion (and essentially maintain the situation since Britain drew up the Offences Against the Person Act of 1861).

Even Northern Ireland, a province of the United Kingdom, does not adhere to Britain's Abortion Act of 1968. Its unhappily pregnant women (and girls) have to cross the Irish Sea to achieve abortions – and half a dozen do so every day. At an international abortion conference (held in Newcastle, Co. Down, in 1994) it was reported that 35,000 women had visited England for abortions since the 1968 Act, these women from both the Roman Catholic and Protestant communities. Madeleine Simms, in reporting on this conference for the *Galton Institute Newsletter*, wrote that: 'The state of the abortion law in Northern Ireland is much as it was in the rest of the United Kingdom in the early sixties. The evidence shows that there exists a substantial majority in Northern Ireland among both the public and the professionals that favours early reform to bring Northern Ireland into the modern world alongside the rest of Western Europe.'

To the south, and in the Republic of Ireland, such gravid journeying is much more commonplace. The generally quoted figure is 5,000 annually, a probable under-estimate as many of the travellers do not register with their home address. Following a national plebiscite, Ireland passed the Eighth Amendment to its Constitution: 'The state acknowledges the right to life of the unborn and, with due regard to the equal right to life of the mother, guarantees in its laws to protect, and as far as practicable, by its laws, to defend and vindicate that right.' Some have argued that this enactment can mean almost anything; others that a constitution is no place to be determining abortion policies. The discussion certainly raged, in newspaper columns in particular, with sex, gender, promiscuity, infidelity and unwanted pregnancy all topics for debate. A *National Geographic* correspondent, made aware of Ireland's particular obsessions, wrote: 'I sometimes felt that people were more concerned about whether girls should be allowed as altar servers at Mass than about children being murdered in Northern Ireland.'

Confusion became yet more complex in early 1992 when the

Dublin High Court prevented a raped 14-year-old from travelling to Britain for an abortion. The Irish Supreme Court then overruled that travel ban. It argued that possible suicide by the young girl posed a 'real and substantial threat' to her life, thus affirming that abortions were tolerable (and should be made legal) in certain instances. Nevertheless the ruling still existed that travelling for abortion was unlawful, and doctors were acting illegally if they provided information about suitable locations for abortion. Therefore another referendum was held. The vote was then 60:40 in favour of giving women information about foreign abortion services. Doctors are now permitted (as from May 1995) to give the addresses of such clinics, but they cannot make referrals for women seeking abortion. The 44-year-old man who had caused the 14-year-old's pregnancy had first 'indecently assaulted' her when she was 12 (her parents then on a pilgrimage to Lourdes). The court sentenced him to fourteen years' imprisonment, the lengthiest such penalty in living memory. There was no further referendum to ask the people if they thought the punishment reasonable. In the same month, but in England, a boy of 15 who caused a 10-year-old to have an abortion was told to pay £100 and receive counsel.

POLAND is as confusing as Ireland, or more so. Only 8% of Polish women use the pill, diaphragm or IUD, according to the Federation for Women and Family Planning, and 40% have never used any form of contraception. For many decades there was easy access to abortion, as was the procedure in all Eastern European nations. That old system was overturned in 1993. A new law stipulated that abortions were only permissible when the life or health of the mother was threatened, when rape or incest was involved, or when the foetus was seriously malformed. For those in favour of liberal abortion the clock had been put back. It was probably not irrelevant that the Pope was of Polish origin (and former Archbishop of Cracow). It was certainly not irrelevant that Lech Wałesa, Poland's president, was a devout Catholic. Poland's 1993 law became the only one of its kind in Eastern Europe.

As with Ireland's 14-year-old girl, whose case brought that country's dilemma to the fore, so has Poland agonized over Barbara Pawliczak. A divorcee with a 10-year-old daughter to support, and surviving on a poor salary from her job as clerk, she had an abortion in March 1994. Her former lover then informed the authorities, allegedly because he had wanted the baby (although offering only £6 a month for its upkeep). The officiating doctor could be imprisoned for up to two years and lose his medical licence for ten years. The lover also faces two years in gaol, but Pawliczak herself faces no punishment. The case, although well publicized, may never reach the courts as Poland's justice ministry has received innumerable other reports of illegal abortions. The former situation of poor (or non-existent) contraception leading inexorably to unwanted pregnancies has not been altered merely because a new law has been written. Abortion used to take care of such unwantedness. It will continue to do so, illegally, because human beings cannot adjust their ways as speedily as legislation can suddenly be pronounced.

ELSEWHERE The abortion conflict is only partly a division of opinion. There is also disagreement between actual occurrence and official statistics. In the Philippines abortion is illegal – and common. In Germany, according to its latest law (passed in 1995), abortion is still illegal, but neither the women nor the doctors will be prosecuted if various stipulations of the law have been followed. (For example, the women must undergo counselling, by a professional organization and *not* by the gynaecologist, at least three days before the abortion takes place.)

In Brazil there are phenomenal sales for Cytotec, a drug marketed since 1986 for ulcer treatment. Everyone knows it can induce abortion, and it frequently does so. Pregnant women taking Cytotec are doing so without any form of medical supervision. Abortion is a criminal offence in Brazil, and 1.5 million abortions are thought to take place annually in that land. Mexico is similar, with a tremendous population problem and a government forbidding abortion, but it has recently made an exception concerning AIDS. This disease now justifies an abortion to save a mother's life, it being the only ground for abortion under Mexico's legislation.

There are, says the WHO, 4–6 million abortions annually in Latin America and the Caribbean. Almost all are clandestine, and 'most are performed using inadequate techniques under unhygienic conditions and without medical supervision'. One in eight of all maternal deaths, or 65,000 a year, are believed to be due to unsafe abortion, with 95% of these unsafe abortions occurring in the developing world. The US death-rate per 100,000 abortions at less than eight weeks is 0.4. In India it is ten times higher. India has the extra problem of female feticide, the selective form of abortion favouring masculine birth. Dowries are costly when girls are married, and males care for parents in old age; hence a double enthusiasm for boy offspring. The recent passing of a Prenatal Diagnostic Techniques Bill means that mothers undergoing sex-detection tests can face three years in prison and a fine of £200. Their participating doctors can be similarly punished and then struck from the register. Once again practice can make a mockery of official intent, with inspection difficult to administer. There are over 2,000 private clinics in Delhi alone, the city where 70% of all abortions are thought to be female feticides. In India there are 927 women for every 1,000 men, a ratio contrary to most other communities where women's numbers are consistently superior.

Even in countries with legalized (and sexually impartial) abortion many women, as the WHO affirms, resort to either unapproved facilities or unskilled providers. Abortion has been prohibited in Argentina since 1921, save for exceptional circumstances (extreme danger to the mother's life or where a rape victim is mentally incapacitated), but a figure of 365,000 abortions annually has been estimated. The risks are high, with patients facing gaol sentences of four years, and their abortionists three times longer. In South Australia nurses refused to care for patients whose abortions were in their second trimester (foetuses older than thirteen weeks). State law permits abortions up to twenty-eight weeks (within the third trimester). Eventually, according to the *Journal of Medical Ethics*, it was

concluded that 'the actions of the South Australian nurses, which have over the last few years both terminated and disrupted second trimester services, are morally impermissible'.

Whether abortions themselves are thought morally impermissible, or positively illegal, or against the will of a local majority, the single certainty is that expulsions will continue to take place in colossal numbers. Societies may be against, or in favour, but individuals of those societies react most independently when unwillingly pregnant. Whether self-inflicted injury is employed, or abortifacient, charlatan or skilled practitioner, the wish for termination will seek out any means. One way or another some 50 million women a year achieve the end they desire so determinedly, rather than the birth that would otherwise occur.

THE UNITED STATES is a conundrum entirely by itself. Abortion has been legal since 1973, but many gynaecologists will not terminate any pregnancies. Some 80% of counties have no provision for second trimester abortions. Termination techniques are an optional subject for medical students. More than 2,500 'events', violent or gentle, were carried out against abortion clinics in 1993 (according to the National Abortion Federation). Fifteen of the fifty states require a waiting period between asking for an abortion and it being performed. Only fifteen states provide full abortion services for poor women. Pregnant minors in thirty-six states must obtain consent from their parents before receiving abortions. Anti-abortionists have a majority (or near majority) in Congress. The Supreme Court, following its 1973 ruling, added that each state may regulate abortions provided an 'undue burden' is not imposed on pregnant women. (State legislatures, and lawyers, have had a field day in defining those words.)

American abortionists have found themselves in the front line. John Bayard Britton, a 'circuit' abortionist (serving many clinics within Florida and Alabama), would strap on a bulletproof vest before arrival. At Pensacola he was more than justified with his precautions, the town's clinics having been 'bombed, torched, invaded, picketed, burgled and ransacked'. In March 1993 his protective clothing was strapped on more securely after he had heard that a fellow (and unprotected) circuit doctor, David Gunn, had been shot three times in the back. Sixteen months later the 69-year-old John Britton suffered a similar fate, along with a 74-year-old ex-Air Force colonel, a supporter of women's rights and volunteer escort for visiting doctors. Paul Hill, the 40-year-old assassin, was quickly arrested. This former Presbyterian minister knew about Britton's vest, and had therefore aimed at the doctor's head. Hill attempted to justify his murder by arguing that the killing of unborn children was a more serious offence. The court found him guilty after a three-day trial. 'I think I acted nobly,' he said during ABC's *Good Morning America* programme when transmitted from 'Death Row' in a Florida prison.

Many anti-abortionists disapproved of Hill's violent act, but the picketing of abortion clinics, the threat of further violence, and the fact that a Canadian gynaecologist was shot four months after Britton's death, have had effect. According to the Alan Guttmacher Institute, which monitors trends, the 2,500 hospitals, clinics and doctors' offices offering abortion services in 1994 were 12% fewer than ten years earlier. In 1986 almost a quarter of medical schools and residency programmes offered instruction in abortions. Five years later only an eighth did so. After Gunn's death 33,000 medical students received lengthy pamphlets from a Texas publisher containing heavy-handed jokes. 'Question: What would you do if you found yourself in a room with Hitler, Mussolini and an abortionist, and you had a gun with only two bullets? Answer: Shoot the abortionist twice.'

Not only the clinics but other medical establishments are also in danger. In September 1995 the *New England Journal of Medicine* received a menacingly worded fax from the founder of Operation Rescue. It said, in part: 'Let Richard Hausknecht and every chemical assassin who follows him be forewarned: when abortion is made illegal again, you will be hunted down and tried for genocide.' The journal had earlier published a study by Hausknecht, New York gynaecologist, which had found that combined treatment with methotrexate and misoprostol was effective in terminating early pregnancies. These two drugs can readily be obtained from a physician's surgery, thus enabling women to circumvent any restrictive legislation by finding a sympathetic doctor, acquiring the drugs, and achieving their abortions independently. After receiving the fax, the journal's editor immediately stepped up security at his offices.

The Freedom of Access to Clinic Entrances law, drawn up in 1994 to curb the violent actions of groups seeking to close down abortion clinics, has caused further antagonism. The national director of Operation Rescue, always keen on organizing demonstrations outside clinics, has called the law 'ungodly, unconstitutional legislation'. Paul Hill, condemned to die in the electric chair for his double killing (and a wounding of the escort's wife), has said that what he did would 'become commonplace and generally accepted as normal'. The Pope has consistently repeated that 'the direct and willing elimination of an innocent human life is always gravely immoral ... no law in the world can ever make licit an act which is intrinsically illicit'.

The fact that anti-abortion forces have had a comfortable majority in the US House of Representatives and a slight edge in the Senate (in recent years) means that various changes have been enacted. Since the 104th Congress began (in January 1995) it has voted to:

- bar abortions at US military hospitals;

- bar abortions in federal prisons;

- bar health insurance plans from offering abortion coverage to federal employees;

- bar the use of Medicaid funds for most poor women seeking abortion.

These alterations run counter to the fact that most Americans (and the law and the courts) support a woman's right to choose. The abortion issue has not been making life easier for President Clinton. He vetoed (in April 1996) a proposed ban on 'partial birth' (very late) abortions, this intent considered by the Lancet as 'yet another excess by Congressional members'. A couple of months earlier, and scarcely noticed, the President had signed a law making it a criminal offence to disseminate certain abortion-related information over the

Internet. No longer can tranmission be made concerning how to obtain any 'drug, medicine, article or thing designed, adapted or intended for producing abortion'.

Meanwhile abortions in the United States steadfastly continue – at about 1,400,000 every year, or 5,300 every working day.

RIGHTS There are plenty in the abortion sphere. The pregnant woman, it is said, has a right to expel her foetus. Gynaecologists have a right not to perform this act. Nurses and secretaries can believe they have a right to refuse assistance. Fathers sometimes think they have rights concerning their child, and whether or not it should be born. People have a right to healthy infants, it is also claimed, and a right to send them permanently to an institution if imperfect. Parents have a right to sue if an abortion, recommended when a foetus appears abnormal, proves to have been quite normal. Conversely parents can sue if a foetus, said to be normal, is discovered at birth to be otherwise. Finally, and by no means least, the foetus has a right to life, either from conception or from implantation or after the first trimester or merely when labour is initiated. Never before has society been suffused with so many rights.

At a lesser level society can think it reasonable that a man wears a condom so that a possible birth does not conflict with next year's holiday, but thinks it highly unreasonable that an abortion should be undertaken for such a minor inconvenience. At a grosser level society can approve of legislation clarifying the issues, but be appalled by some governmental restriction. Nicolae Ceauşescu of Romania gave his country little birth control, less sex education, a high maternal mortality (ten times the European average) and no abortion. There is also the dilemma of handicap. Many an abortion is recommended, or even demanded following amniocentesis, if the foetus is demonstrably faulty. As one contributor said at an abortion conference: 'There seems to be an implicit assumption that it is not socially desirable to have handicapped people about, an assumption which must be carefully examined.'

A gynaecologist's right to refuse has both lost and gained ground in recent years. Initially, after the more liberal abortion laws were implemented, it was considered appropriate that gynaecologists, having had abortion imposed upon them, should be able to refuse this extra duty. Individuals who chose this branch of the medical profession after the legislation knew they were entering a world where abortion occurred. They, it has been argued, are less able to choose which aspects of their work are more acceptable. In general they are still able to refuse. The British House of Lords has ruled that conscientious objection can only apply to those performing the abortions. It does not relate to assistants, such as nurses.

The objectors and promoters will probably continue their form of warfare for all time. Killing a foetus *is* distressing. Preserving an unwanted life is also mortifying. John Britton, killed for his work around the clinics, was – an extra irony – disdainful of abortion, severely questioning the women whether they were adamant. Law suits against some of the violent protestations (such as an $8 million award in Portland, Oregon) have helped to turn the tide in favour of the abortionists and their often desperate clients. It turned again when much publicity followed a professor's announcement that twenty-three-week foetuses could feel pain. Large amounts of the stress hormone had been detected after some foetuses were given blood transfusions, a possibly painful procedure. Unconscious adults, incapable of feeling pain, can also produce stress hormone if stimulated in a fashion that would normally be painful. Moreover, as others pointed out after the professor's announcement, very few abortions are of twenty-three-week foetuses. In the UK almost 90% are carried out before twelve weeks, and only 1.2% after twenty weeks.

The pros and the cons are important, but so is that awesome WHO estimate of 50 million abortions a year. The awareness of a pregnancy ought to be a blessing. It plainly is nothing of the sort at least 50 million times a year, and probably is nothing of the sort on millions more occasions when the pregnancy continues until ending with the birth of one more unwanted child.

MIFEPRISTONE   It may well be that the surgical ending of a
foetus's life, the moral dilemma involved among those taking part,
the reaction of so many outsiders, the cost, the lack of facilities, and
much of the restrictive legislation may all be altered dramatically if
RU486 takes the stage as all-embracingly as the pill has done. Indeed
it has been said (by Sir Malcolm Macnaughton, gynaecologist) that
this drug's development 'is an advance in reproductive medicine of
the same magnitude as the development of the hormonal contraceptive pill, ... [it being] the first effective "medical" method that can
be used for the termination of early pregnancy'.

RU486, also known as mifepristone, received its first approval for
use by the French government in 1988. As the maintenance of
pregnancy is entirely dependent upon the secretion of progesterone,
it was inevitable that a substance to block the effects of this hormone,
an antiprogestin, would be sought. RU486 is a steroid hormone
similar in its structure to the natural hormone progesterone. It is not
yet 'fully understood' how the steroid works, but it is effective,
particularly if given with prostaglandins (that are normally produced
by the uterus). If both are taken within forty-nine days of the last
menstrual period an abortion will be induced in 95% of cases. Not
only is there no need for any form of surgery but the drugs can be
consumed privately. In the US a large number (some say 83%) of
counties do not possess abortion providers, and those that do are
often subject to harassment. Therefore RU486, or some similar
abortifacient, could become tremendously popular – as soon as it is
given the go-ahead. For most countries of the world that permission
has not yet been granted. In the US the group which holds patent
rights to the French drug, the Population Council, filed an application
to the Food and Drug Administration for its use in April 1996.
(Normally the FDA takes six to twelve months to process an
application.)

Roussel Uclaf, holders of the patent for RU486, have been criticized
for 'their conservative marketing policy'. The drug has proved
popular in trials but is expensive. However, there is also methotrexate
which is cheap, off-patent, readily available in many parts of the
world and, when combined with prostaglandins, works almost as well

as mifepristone, according to the *Lancet* (in September 1995). Both drugs take more time than surgery – sometimes (according to the same article) as long as fourteen days, and the process is 'messy' – but '9 out of 10 women reported they would choose the method again if they needed it'. Methotrexate costs less than $7.00, and is registered in many countries for various other reasons, such as cancer therapy and psoriasis (the skin disorder). The FDA does not have jurisdiction over the 'off-label' use of a drug in this way, as the *Lancet* explained, but 'it is both possible and probably advantageous to seek FDA approval for this new use'.

SOME ALTERNATIVES It is as well to remember, when discussing abortion clinics, artificial steroids, hormone blocks, dead gynae-cologists and foetal rights, that another world exists where such subjects are of lesser consequence. At one London abortion conference a delegate from Thailand spoke of the induction methods used in rural areas he had visited. These were (in brief):

- Insertion of vegetation. Certain types of shrubs, grasses, stems are inserted into the vagina, and left there until bleeding occurs.

- Traditional emmenagogues (medicines to bring on the menses), such as chewing and swallowing betel-nut paste.

- Insertion of hard objects. Elongated and narrow probes are used to puncture the foetus.

- Curettage. Standard D and C used by practitioners. The contents of the uterus are scraped out.

- Injection of solutions. A catheter or drip tube is inserted into the uterus, and any one of a wide variety of solutions is infused.

- Massage or chiropractic technique. 'Probably the most widely used among rural practitioners ... An estimated 80 per cent of all abortions are done by this method.'

There may be 1.4 million abortions in the US every year, 175,000 in the UK, and many more in the rest of Europe, notably in its eastern areas, but there is another world beyond all their borders where most people live. They inhabit simpler dwellings, and are short of possessions, but they too have a crying need for abortions. They, and their methods, make up most of the WHO's estimate of 50 million every year. RU486 and its successors may never come their way, but their need for abortions will remorselessly continue, just as it has done for countless centuries.

BIRTH

PRIMATE BIRTHS   MIDWIFERY   SOME STATISTICS

CAESAREANS   MATERNAL AGE   PREMATURITY

WATER BIRTHS   PARTOGRAPH   BIRTHRATE

BIRTH DEFECTS

Birth, and copulation, and death. That's all the facts when
you come to brass tacks.
T. S. Eliot (in 'Sweeney Agonistes')

If men had to have babies they would only ever have one
each.
Diana, Princess of Wales

He does not realize that, instead of conceiving him,
his parents might have conceived any one of a hundred
thousand other children, all unlike each other and unlike
himself.
Peter Medawar

PRIMATE BIRTHS 'Humans do any number of things better than animals, but giving birth is not one of them,' wrote Joshua Fischman (in *Science*). They do seem to be markedly inept concerning the birth procedure. Apes bring (admittedly smaller) infants into the world through a roomy birth canal with little fuss. 'In contrast,' added Fischman, 'human beings often spend hours corkscrewing their way down a narrow birth canal, finally emerging head down, away from the mother – the only primates to do so.' 'It's not the type of system you would invent,' anthropologist Karen Rosenberg has said, 'if you were designing it today.'

Of course it was not invented today any more than the rest of the system, but there are reasons for the lengthy, twisting, backward approach. The maternal pelvis, which surrounds the birth canal, constricts it lower down, but even at the top (where it is widest) there are problems. As Fischman has explained, 'the longest dimension of a baby's head is from the nose to the back of the skull, and so the baby enters the canal facing sideways. But lower down, the canal changes its shape so that its longest dimension is from front to back. As a result the infant must rotate 90 degrees.' There then follows one more twist to this event. The lower part of the canal is a bit broader at the front, whereas the baby's head is broader at the back. Therefore, most unhelpfully for any help the mother might wish to give, the baby finally emerges facing downwards.

With an orang-utan, chimpanzee or gorilla the infant's head is smaller than the area of the birth canal. Consequently, for mother and infant, there is no problem. With humans, and with their bigger heads at birth, the distended canal area is about the same as the head area – causing the tight squeeze – but the shape of canal and head do not coincide. Like two similarly sized ovals they can only be superimposed when aligned correctly; hence that 90-degree rotation. This very awkwardness, and a need for assistance, has – it is said – been important in human evolution. Chimpanzees hide at birth time, and produce their infants independently. Humans, as anthropologist Wenda Trevathan has explained, are different, as were their pre-human ancestors. Birth was 'so painful and risky' that mothers needed 'help from others to deliver a baby successfully'. Cooperation and communication were therefore encouraged, these skills integral to humanity. Precisely when this happened, and when the ape ancestor became a human-ape (being born backwards, or even sideways) rather than an ape-ape is unknown. Further fossil evidence would help resolve the problem, but pelvic fossil material is (and has always been) in short supply.

Human brain size and head size, so much larger than with ape infants, can seem to cause the problem (particularly to parturient mothers), but it was gait that upset earlier simplicity. Human ancestors acquired their narrower pelvis to facilitate upright, two-legged walking. Simultaneously, more or less, their heads enlarged. Therefore the canal shrank just when – for the head – it ought to have increased. 'Lucy', the three-million-year-old fossil female from Ethiopia discovered in 1986, had a pelvis more oval than with apes. Therefore, or so it has been suggested, such an australopithecine would have given birth to infants not facing upwards (as with modern apes) nor facing downwards (as with modern humans) but sideways. Lucy is in consequence a link between the old style and the new. It is further argued, after examination of later fossil material, that sideways delivery was the norm until quite recently, perhaps 'a few hundred thousand years ago'. Bipedalism and a narrow pelvis were then entrenched, but the head was still enlarging. The resulting compromise was today's tight-fit, painful, sometimes damaging, face-down, corkscrew delivery, imperfectly catering for a varied set of needs.

*

Human birth happens after 266 days, on average. The big apes are similar, with gorillas averaging 250–270 days (before producing their 4–5 lb offspring), orang-utans 260–270 days, and chimpanzees 230–240 days. In general these infants are more competent than newborn humans (who, it has to be admitted, are not much good at anything). Once again there has been evolutionary compromise. To delay delivery would make for greater competence but a tighter fit at birth. To advance delivery would mean an easier birth but less maturity (and premature human babies make the point that less womb-time is rarely advantageous).

MIDWIFERY If any individual fathered midwifery it must be Nicholas Culpeper. He died in 1654 and his *Directory for Midwives* was last published in 1777 after repeated reissues. His book *The English Physitian* went through 100 editions and became the first medical text to be published in America. Originally he had been herbalist and astrologer but he switched to medicine following family tragedy. 'Myself having buried many of my children young, caused me to fix my thoughts intently upon this business', this business being the birth and rearing of children. When aged 24 he had married the 15-year-old Alice Fields but, of their seven children in fourteen years (for he died at 38), only one survived even to young boyhood.

Culpeper was extraordinarily ahead of his time. He believed ovaries were the source of human eggs. He considered that 'in the act of copulation the woman spends her seed as well as the man, and both are united to make conception'. To describe sex in Puritan days was one astonishment. To describe it accurately, when medical opinion still adhered (largely) to early Greek teaching, and when even his contemporary William Harvey thought the female role to be passive with male semen the sole origin of life, was yet more remarkable. He also credited the foetus with an active role during birth (an opinion now finding more favour than ever before): 'How much the foetus contributes to the acceleration and facilitation of its own birth is plainly to be seen in the hatching of oviparous creatures, for it is apparent that the foetus itself breaks the eggshell and not its mother.'

He is forever a joy to read, having a turn of phrase and sprightliness that often pleases. In his *Genitals of Men* (a remarkable title for the seventeenth century) he, not unreasonably, turns his attention to the penis, and a name for it that has, alas, become extinct: 'The Latins have invented very many names for the yard, I suppose done by venerious people ... I intend not to spend time in rehearsing the names, and as little about its form and situation, which are both well known, it being the least part of my intent to tell people what they know, but teach them what they know not...' He certainly had much to teach. For example: 'A little medical knowledge about cleanliness and care can do more good than many costly potions from the apothecary.' That was written 200 years before Ignaz Semmelweiss suffered derision in Vienna for suggesting that medical students should wash their hands between the mortuary and the maternity ward. Unfortunately Culpeper, the father of so much, save for children, died young, and his unhappy wife followed him aged 34. The world then took its time to catch up with him.

No European medical college was interested in the subject of birth until 1726, when Edinburgh appointed a professor of midwifery. A century later there was still no formal teaching of obstetrics anywhere in Britain, save at Edinburgh. Not until 1886 was midwifery considered a necessary subject for British medical students, even though non-medicals had long been active. The Chamberlen family, for example, had been most vigilant. Peter Chamberlen, Huguenot immigrant to England, developed birth forceps towards the end of the sixteenth century. Together with his brother-in-law, also named

Peter, they travelled the country, charging considerable fees in advance, keeping their instruments locked securely out of sight, blindfolding the labouring women, and being famously successful. Many an awkward birth was assisted, despite bells, chains and hammers rattled and struck to confound all listeners.

The first Peter died in 1626. His brother-in-law continued before passing the secret to a son, also called Peter. By this time royalty were among the patients, but the happy situation then began to deteriorate. Female midwives successfully petitioned Parliament, thus reacquiring some of their ancient privileges. A difficult birth in Paris, attended by a Chamberlen descendant, left both mother and infant dead. (Louis XIV's jealous doctor, after cunningly offering a 38-year-old dwarf with a narrow pelvis having her first child, was not displeased at the outcome.) The family therefore sold the instrument. It went through several hands before arriving at the Amsterdam College of Medicine. This institution, most unethically, ordered no one to practise midwifery unless the forceps secret had first been bought from the college (for 2,500 guilders). In 1753 two Dutch doctors revealed what they had purchased. It was no more than an iron bar. Someone along the line had been guilty of even grosser misconduct.

Despite ups and downs the Chamberlen story promoted the notion of forceps. Suddenly (in the eighteenth century) everyone was using them. Midwives always had a pair. Nature, said one critic, had apparently abandoned her work of propagation, and had left it to forceps. The craze for interference then diminished and, as is the way of pendulums, was replaced by a policy of *laissez faire*. Nature, and mother and child, were determinedly left to their own devices. Sir Richard Croft, royal physician, watched for fifty-two hours while Charlotte, only child of – soon to be – George IV, struggled to give birth. The child was stillborn. The mother died six hours later. Sir Richard then died, after putting a bullet through his brain.

Whether to interfere or not to do so oscillated throughout the nineteenth century, with advocates for and against. Difficult births, with protracted labour, caused many assistants to reach for forceps. The instruments were most successful when the birth was well advanced, but least so when the operation known as high forceps was employed. By 1964 between one-quarter and one-third of all births were assisted by forceps (according to twenty-two US Navy hospitals), but thereafter the tide turned as Caesarians became more commonplace.

SOME STATISTICS   The forceps story illustrates fluctuating enthusiasm for one approach to birth, but changes about the ancient procedure are rife, with practically everything altering to some degree. Total birth numbers are certainly going up – to today's figure of 540 a minute. Conversely the number of offspring per woman has been going down. Home births versus those in hospital are certainly altering. Sixty years ago only 15% of births (in Britain) occurred in some form of maternity institution. This proportion rose dramatically after the Second World War to 70% and to over 90% by the 1980s. It is now falling again, with many mothers expressing a preference for home.

The push to have babies in hospitals, said Alison Macfarlane, a medical statistician, was based on faulty assumptions. Between 1964 and 1992 two trends coincided: declining deaths of newborns (including stillbirths) and a drop in home births. Facts from Denmark and Holland have given the lie to this alleged (and frequently believed) cause and effect. The decline in death was similar in both countries during those years, but Denmark almost eliminated home birth whereas a third of women in Holland persisted in using their homes. 'Correlation does not mean causality,' Macfarlane concluded, when addressing the Royal Statistical Society in May 1996; 'That's the first thing they teach you in statistics.'

A British survey (of 1995) stated that one-fifth of British women would never have babies, a larger proportion than ever before. Previously over 90% had given birth some time in their lives. Initially fathers were firmly excluded from obstetric wards as the hospitals took charge. Even twenty-five years ago only 10% of fathers were present when their women gave birth. The current figure is nearer 95%, fatherly attendance not quite compulsory but expected (much to the dismay of those who are unattracted by blood, pain, discomfort and even, on occasion, the cutting of flesh). Medical personnel can see fathers as mere spectators, without a proper role.

For stable communities (as with England and Wales) the numbers of births have not altered greatly in recent years, despite the Second World War and a lengthy time of peace. In 1930, depression time, there were 648,811 births. In 1941, mid-war, there were 579,091. The peak year of 1947 witnessed 881,026 births after much postwar reunion. Ten years later the number was down to 723,381. Ten years further on, in 1967, there were 832,164. Birth quantities were therefore altering, but not extraordinarily. By 1987 the number was 681,500, and by 1990 706,100. During the six decades from 1930 to 1990, with that peak of 881,000 (in 1947) and the low of 590,000 (in 1941), the lowest figure was only two-thirds of the highest. In short, there has been a surprising consistency (in those two regions) over giving birth, despite the alarming arrival of a six-year war, the freedom of a National Health Service, the provision of simple and effective contraception, and a greater shift in life-style than in any earlier span of sixty years.

CAESAREANS The name implies antiquity, alleging that Julius Caesar was born this way, but such operations performed successfully in pre-antisepsis days are difficult to believe, particularly when there was a wish to save the mother. (Julius's mother lived long after her famous son's arrival.) Even at the turn of this current century they were uncommon. Somerset Maugham, medical man before becoming writer, described one such operation: 'I went into the theatre to see a Caesarean. Because it's rarely done it was full . . . I seem to remember [Dr C] saying that the operation so far was seldom successful . . . He'd explained the danger [to the mother] and said that it was only an even chance she'd come through . . . Her husband wanted it too, and that seemed to weigh with her. The operation appeared to go very well and Dr C's face beamed when he extracted the baby. This morning I was in the ward and asked one of the nurses how she was getting on. She told me she'd died in the night . . . I suppose what affected me was the passion of that woman to have a baby, a passion so intense that she was willing to incur the frightful risk . . . The nurse told me that the baby was doing well.'

In the 1960s about 2% of all deliveries in the United States were by Caesarean section. In 1964 one Scottish hospital, enthusiastic in this regard, reported a rate of 5.5% (among its 18,102 deliveries). This Glasgow hospital also reported twelve successes with one woman. By the 1980s the average figure (for England and Wales) was 7.2%. In the United States, according to Public Citizen, a consumer group, the average rate peaked in 1988 with 24.7% of deliveries being Caesarean. Some southern hospitals were then operating on more than half their pregnant women. In Brazil, and in the town of Pelotas (which has been particularly enthusiastic), almost a third of births have been by Caesarean section. In its richest families this proportion was over one-half. Even among the poorest families the incidence was still one-quarter. Much of this impetus arose from the belief that normal births could not follow a Caesarean. Whereas a third of nulliparous women (no previous births) had Caesareans this Pelotas ratio increased to four-fifths for second births, and to nine-tenths for third births.

In Britain there has never been such enthusiasm, although the national number has been rising, from 13% in 1992 to 14.6% in 1993. Middlesex University studied 90,000 births in eighty-five hospitals, and encountered (almost) uniformity, with the peak hospital reaching 19.6% and only one hospital below 10%. The United States' national percentage fell to 22.6 by 1992. This slight drop is attributed mainly to increasing denial of 'once a Caesarean, always a Caesarean'. It had formerly been feared that the uterine scar would rupture during natural labour, but there is a growing movement for trying such labour after a Caesarean. In the United States obstetricians are now being actively encouraged by relevant organizations 'to counsel patients that the benefits of a trial of labour outweigh the risks'.

Swimming against this particular tide (and according to a *Lancet* report in February 1996), one in three women obstetricians in London would choose a Caesarean delivery for herself. The proportion of male obstetricians who would favour this unnatural method of childbirth (for their consorts, one assumes) was slightly under one-fifth. If such experts are recommending one method for others and a different method for themselves it would seem as if not all the truth is changing hands. As the report concluded: 'In this era of patient choice, should information regarding the potential benefits of elective Caesarean delivery be given to women?'

It has been said, by some, that Caesarean babies grow up differently, being 'dependent and impatient' (according to American physicist Jan English, who has studied the 'subtle' alterations). The Vaginal Birth After Caesarean group considers 'it might be a good idea if Caesarean babies were given some massage after birth to simulate the contractions they missed'.

(As we all know, ever since Anon wrote the verses, 'Friday's child is loving and giving; Saturday's child works hard for its living; but the child that is born on the Sabbath Day is blithe and bonny, good and gay.' Unfortunately, according to a Pregnancy Outcome Unit in Adelaide, Caesareans have deprived that southern Australian state of 741 hard workers and, worse still, 1,030 blithe and bonny individuals. The existence of weekends, and a reluctance then to operate, has led to an inequitable birth distribution. Similarly 'We always induce on a Wednesday', reported by a British hospital registrar, is a policy that further diminishes society of the loving, giving, and hard working, as well as the good and gay (that word in its earlier connotation). Such loss is not to be accepted lightly. As the Australians pointed out: 'The consequences [of increased Caesareans] are dire . . . How can we ever restore our balance of joy and goodness that the Sunday children bring?')

MATERNAL AGE, and age of the foetus when born, are both changing dramatically. One early English contender for oldest mother was Winifred Wilson, of Cheshire. She was 55 when giving birth to her ninth child in 1937. The oldest American mother, beating Mrs Wilson, was Ruth Kistler, of Portland, Oregon. She was 57 when giving birth in 1956. Then came artificial insemination plus the preliminary hormone treatment able to create fertile women when, if nature had been in charge, their child-rearing days had been concluded. In general AI staff do not encourage old motherhood, but exceptions have been made. Rosanna Della Corte, of Italy, lost her only son in a motorcycle accident when he was 17. She was desperate and longed for another child. Severino Antinori obliged. He collected an egg from a donor, some sperm from Rosanna's husband, and enabled Rosanna to give birth again (another boy) when she was 62, becoming (it is believed) the world's oldest mother.

British women are, in general, becoming older when they have their babies. In 1992 their average age at childbirth was 27.9 years, the highest for thirty years. The fertility rate for women in their early thirties exceeded, for the first time, the rate for women in their early twenties. As for fathers there is little concern about either their fertility rate or their maximum age for siring offspring, but an English 93-year-old must come close. He died in 1994 five weeks after the birth of his child when about to undergo sight surgery so that he could better see his offspring.

PREMATURITY has undoubtedly been altering in recent years. Normal babies, in the womb for 266 days (thirty-eight weeks), weigh about 7½ lb (3,400 grams) when born, but a proportion are born before the proper date. In North America and Europe the small percentage of pre-term births account for 65–70% of early neonatal death. Premature infants, often casually known as preemies, are in a different category to pre-term, being (usually) defined as weighing less than 5½ lb (2,500 grams). About 5% of all births are premature, if judged by weight alone. Their degree of development, which is not necessarily matched with weight, is more significant for survival than their weight. And so is the time since conception.

Whether the premature infant is born at twenty-four weeks or twenty-eight weeks is of considerable importance, with the former less than two-thirds of the standard time. The weight of the smallest survivors may only be 1.3 lb (600 grams), but their gestational age should be twenty-four weeks if they are to have much chance of survival. One American hospital (which reported its findings to the *New England Journal of Medicine* in 1993) had studied all of its 142 infants born at twenty-two to twenty-five weeks' gestation. Not one of the 29 born at twenty-two weeks survived, 6 of the 40 born at twenty-three weeks did so, as did 19 of the 34 at twenty-four weeks, and 31 of the 39 at twenty-five weeks. The survival percentages therefore rose from 0% to 15% to 56% to 79% in that critical period of four weeks.

Making early arrival yet more dangerous is the possibility of stillbirth. In that American hospital there were seven at twenty-two weeks, and four at twenty-three weeks. The fact of survival for prematures is not the total blessing it might be. Only one individual of those born at twenty-three weeks was without 'severe abnormalities on cranial ultrasonography', only one-fifth at twenty-four weeks, and two-thirds at twenty-five weeks. Bernard Shaw's dictum of taking care to get born well is frequently quoted. Part of that care is not to arrive too soon. To arrive sixteen weeks early is, at present, likely to be lethal.

To arrive early, and then to survive, certainly has drawbacks. Another *New England Journal* article reported on a follow-up study of sixty-eight American children born during the early 1980s, all having weighed less than 750 grams at birth. These were compared with another premature group, all of whom had weighed between 750 and 1,499 grams at birth. In a *BMJ* summary of this study 'the smaller children scored less well in tests of cognitive ability, psycho-motor skills, and academic achievement; around one-fifth had one or more severe disabilities, and as many as half had less serious but functionally important disabilities'. This blunt statement concluded: 'While it is true that the management of very small babies continues to improve, the message is plain: prevention of premature delivery is the real answer.'

It is not anyone's wish that a foetus is so prematurely expelled from the uterus. Any choice in the matter comes afterwards, involving what is known as aggressive delivery-room care. Such care may occasionally be rewarding for infants born at twenty-three or twenty-four weeks but, as the *New England Journal* article affirmed, 'the considerable mortality and morbidity of the majority is a question that should be discussed by parents, health care providers, and society'.

In short, should a line be drawn? The New York State Task Force on Life and the Law has stated that there may well be technological advances in the future, further improving survival of those born at twenty-three or twenty-four weeks (or more), but a threshold for survival will continue. At present that cut-off point is thought (with very few exceptions) to be 600 grams and twenty-four weeks. Below that weight, and before that time, the pregnancy has failed. The Fetus and Newborn Committee of the Canadian Paediatric Society has suggested that only 'comfort care' should be given to infants born at twenty-two weeks and weighing less than 500 grams. For those of twenty-three and twenty-four weeks there should be flexibility concerning resuscitation, taking into account the views of the family. ('Go home and try again'?) For those of twenty-five and twenty-six weeks (and over) there should be full resuscitation.

Everyone is agreed that prematurity should be discouraged. The womb is a far better environment for the foetus than any incubator, however skilfully managed. Unfortunately it is not known what triggers normal births, let alone the others occurring before their proper time. It is believed the foetus chooses the moment (much as Culpeper suggested, all those years ago). The foetus, after all, maintains its own daily routine before birth, exercising and resting as it thinks fit rather than following the mother's schedule, and it may also be making the ultimate choice concerning when to end its foetal life. As for premature births, most of these occur spontaneously, without anyone the wiser (at present) about their cause. The hormone oxytocin, produced in the mother's pituitary gland, is known to be relevant to early and normal births, but not known to what degree. It is used to initiate uterine contractions artificially when need be. Consequently a drug that interferes with the action of oxytocin has been developed (and is being used in trials). The hope is that premature labour can be halted, or at least delayed, by its administration. Every week gained in the uterus will be of benefit.

One other hormone, known as CRH, has also been implicated, but in contrary fashion. It seems, according to some recent work (published in *Nature Medicine*), that high levels of this hormone in early pregnancy, rather than later, were linked to early labour. If this proves to be the case, thus giving early warning of trouble ahead, the threatened pregnancies could be treated, perhaps by giving the mothers steroids to mature their developing foetus's lungs more speedily, all immature lungs being so relevant (and often disastrous) to premature welfare.

*

The switch from a liquid to a gaseous environment is particularly difficult for the premature, with 'respiratory distress' being the most common cause of death. This arises because the small babies lack a surfactant chemical. Such a chemical lowers the surface tension on the air side of the lungs, causing the alveoli (air sacs) to open up. If these do not open correctly the necessary exchange of oxygen cannot occur and the baby is 'distressed'. More to the point, it dies from the damage done by low oxygen levels in its blood. To counteract this unfortunate chain of events an artificial surfactant can be introduced. Ventilators can also be used to assist in the breathing. But the surfactants do not always help as they should, and the pumping of air can be damaging. Trials have begun in America which promote a watery rather than a gaseous approach. The baby's lungs are filled, for at least twenty-four hours but no more than ninety-six hours, with a liquid possessing a high affinity for both oxygen and carbon dioxide. The situation is then much more akin to life inside the uterus, where the foetus's lungs are filled with amniotic fluid that is saturated with oxygen.

Meconium asphyxiation is another lung problem of the newborn, a major cause of neonatal death and possible origin of severe respiratory problems in those that survive the meconium assault. This greenish substance, mostly bile and mucus, is expelled from the baby's bowels, usually within minutes of being born. In about one in ten births, generally when these are delayed, it may be expelled earlier and then find its way, damagingly, into the baby's lungs, thus preventing normal breathing. This mishap can be prevented by flushing the uterus with saline water when meconium is abundant, a procedure bringing to mind the former enthusiasm for baptism even *in utero*. If the foetus was not emerging, and its life might be lost, its soul could at least be saved if holy water was applied. Saline in liberal quantities should do better, by saving both body and soul.

WATER BIRTHS Liquid outside the parturient body is also becoming commoner via so-called water births. Advocates say that water's buoyancy and warmth promotes natural labour while providing 'a non-invasive, safe, and effective form of pain management'. Critics have questioned possible infection and inadequate information. The Royal College of Obstetricians and Gynaecologists achieved widespread publicity (and opprobrium from the water advocates) in 1993 by being less than immediately enthusiastic. The Royal College of Midwives had been warmer in its approach, welcoming the caution being applied. (London's *Daily Telegraph*, making a point about such deliveries but spreading unique confusion, wrote (on 15 October 1993) that 'not even rhinos – who live their life in water – give birth in water'.)

Oxford's John Radcliffe Hospital has been the location for a considerable number of water births, and is wholly in favour of caution. 'Until [its] studies and trials are completed,' wrote an enthusiast from the Royal College of Midwives Trust, 'no one can judge the effectiveness or safety of the procedure.' By then more than 600 women had used the pool during labour, and over 300 had actually given birth under water. Risk of infection was the safety feature that had caused most concern.

The Northern General Hospital Trust (based in Sheffield) has reviewed its own experience. A total of 122 mothers had used the pool in labour, and forty-one had actually delivered in the pool. The babies had all received ear swabs (to check for infection) and three of these had proved positive. The three were then successfully treated with antibiotics. Of the total of 122 offspring there had been five 'with a raised respiratory rate or grunting', four of whom had been delivered under water. These five were also given antibiotics, even though no bacterial tests on them had proved positive. Among the 122 there was one 'unexplained birth asphyxia'. In short the new procedure had been reasonably satisfactory. This Sheffield authority was adamant about 'bacteriological surveillance'. It recommended that a sample of pool water should be taken (for testing) before each woman's immersion as well as afterwards. The taps should be opened for five minutes each day whether or not the pool needed to be filled. Water should also flow for two minutes before filling the pool which should be well cleaned and dried after use.

A team from the Middlesex Hospital, London, gave details in 1994 of one infection, adding that it 'did not know of any [other] reports of a baby becoming infected owing to a water birth'. The 3,600 gram boy had been born in its birthing tub. The mother's membranes had been ruptured for less than twelve hours, and she had had no fever. At eleven hours of age the boy had become poorly 'with two episodes of cyanosis' (going blue and lacking oxygen). 'Probable septicaemia' was diagnosed. He was given antibiotics, and soon recovered. Cultures made from his urine, blood and cerebrospinal fluid proved sterile, but those from the tub, its disposable lining, the filling hose, the taps and exit hose all grew *Pseudomonas aeruginosa*. The contamination had therefore occurred even though there had been 'meticulous washing with hot water and detergent and drying' after each birth. This Middlesex team concluded that 'despite the increase in popularity in water births ... reliable evidence is lacking about the benefits and hazards with such births'. *P. aeruginosa* is a virulent organism, and had managed to invade the system despite all that scrupulous care.

PARTOGRAPH Scrupulous care sounds time consuming, costly and unlikely to be relevant to most of the 250–300 million births around the world each year. Instead of the tub it is the partograph that will be much more influential, being cheap, simple and designed to improve birth for the great majority of women. Currently around 40,000 die annually in childbirth and, as the WHO reports, 'huge numbers are left with painful injuries as a result of obstructed or prolonged labour'. Such prolongation can produce 'obstetric fistulae, a condition causing incontinence which often leads to a woman's rejection by her partner, family and society; leaving her destitute and ashamed'. The WHO added that more than a million women in Africa suffered in this fashion. Hence enthusiasm for the partograph, designed to prevent much unnecessary suffering and loss of life. A fifteen-month study on 35,000 women delivering in eight hospitals in Indonesia, Thailand and Malaysia had produced encouraging results.

Essentially the partograph is a chart to record the progress of labour, such as the rate of dilatation (opening) of the mother's cervix, and the heart rate of mother and baby. On the graph are 'alert' and 'action' lines. These show whether the labour is progressing normally or whether some form of intervention is necessary, such as a Caesarean or drug administration. As the 250 million or so babies born annually arrive in every sort of circumstance, from mud-hut to shed to cabin to cottage to fancy hospital or to the open air, and as the attendants vary from relatives to frightened partners to traditional deliverers or passers-by or uniformed staff, the mothers in question either do or do not receive appropriate treatment and advice. The partograph is almost a user's manual, telling what to do and how and when and, no less important, what not to do when there is no need.

The fifteen-month Asian study was most revealing. Partographs had been introduced into each hospital part-way through the study so that 'before' and 'after' could be compared. As a result of their introduction 'the number of prolonged labours was halved, the rate of postpartum infection (sepsis) was cut by over half and the number of stillbirths fell from 5 to 3 per 1,000 babies'. In addition, the number of unnecessary interventions was reduced – fewer drugs were needed, and Caesarean sections carried out on women without complications were avoided 'with no adverse impact on the condition of the foetus'. According to the World Bank it would cost $2.00 per head to bring about 'a substantial cut in maternal deaths and illness'. The annual slaughter of 40,000 women, labouring to give birth, and the terrible destruction of so many of their offspring, do seem set to be reduced by the 'cheap and simple chart' that has been named the partograph.

BIRTHRATE Fifteen years ago Britain's fertility (babies per 1,000 women) was close to the European average. Most of the European birth rates have since collapsed, leaving the UK, France, Norway, Sweden, and Finland as leaders. In all other areas the UK now scores lower than the European average, as with the average age at first birth, the average age at all births, the percentage of births to mothers over 30, and the average age at first marriage (for both males and females). UK mothers have been getting older, but nothing like so much as in Europe. The average age at first marriage has also been rising, but is lowest in Western Europe, save for Belgium, Greece and Portugal. Conversely Britain's divorce rates are currently the highest in Western Europe, and 25% of British children will either be brought up (for part of their lives) by one parent or will have one parent not their own. UK births outside marriage are high, but beaten by Norway, Sweden and Denmark. 'We are still a long way from an adequate explanation of fertility trends and patterns even within individual countries,' according to an article in the *Galton Institute Newsletter*, 'let alone understanding the reasons behind the differences in fertility between countries and their regions.' Even on such a fundamental issue as birth there seem to be no comprehensible rules governing human behaviour.

It is also not known why, when mating is popular throughout the year, there are more births in certain seasons. Thirty years ago the peak delivery period was the second quarter, April to June. More recently it has become the third, July to September. In 1990, and for the United Kingdom as a whole, the four quarter totals were 191,000, 202,000, 207,000, and 197,000. The provinces of England, Wales and Scotland independently obeyed this trend, but Northern Ireland behaved differently, with the year's first two quarters each providing a greater number of births than either of the final two. Perhaps there is a prosaic reason for the major shift from second quarter deliveries to the third. In earlier years the August holiday was paramount for days off work – leading to more births eight and a half months later. The Christmas break has more recently been extended from a couple of days to – in many occupations – a couple of weeks. Hence more births in the third portion of the year. Or maybe there is some other and subtler factor at work than increased opportunity for cohabitation – and for sex.

BIRTH DEFECTS There is definitely a relationship between certain psychiatric disorders and the birth season. This does not mean that anyone born in a certain month will be prone to some malfunction, but a certain month may consistently be accompanied by a higher than average figure for that complaint. Schizophrenia, for example, peaks for August and November births. Dyslexia is most evident among those born in the summer months of May, June and July. It is thought that influenza, suffered during the second three-month period of pregnancy (in the previous winter), may explain these and other findings. If so dyslexia might be reduced by greater immunization of women of child-bearing age. Conversely the immunization itself might trigger some other error.

Many researchers have 'discovered' associations between birth-month and disorder. Such relationships do not always agree with similar work in other areas, at other hospitals, at other times, but some alleged relationships are stronger than others. One form of cretinism was investigated via all Scottish hospital registers, and the number of cases did show a strong seasonal pattern. For births in the twelve months – from January to December – the numbers were: 2, 4, 9, 6, 7, 4, 1, 3, 2, 2, 2, and 1. The three peak months (March to May) therefore witnessed more cases of this cretinism than the other nine. Something must be happening to cause this incidence, and maybe many other somethings are relevant to the time of year when we are conceived, and eventually are born.

The unfortunate and always unwelcome sting in the tail to the event of birth is that serious defects, often genetically determined, 'complicate and threaten the lives of 3 per cent of newborn infants', as the *New England Journal of Medicine* phrased it recently. They account for one-fifth of deaths during the newborn period and an even greater fraction of serious morbidity in infancy and childhood. The cost of such error is colossal, both initially and subsequently. So

too the consequences for affected families. Cerebral palsy, the commonest cause of severe physical disability in children, is even on the
increase owing to the improved survival rate of low birth-weight
babies (and most children with cerebral palsy now survive to
adulthood). Various other malformations, such as Down's syndrome,
are not so much diminishing as frequently being rejected (and
aborted) before their actual birth. It is not decreed that pre-natal
diagnosis is obligatory but the procedure is increasingly the norm.

Amniocentesis, usually performed during the second trimester (at
sixteen weeks), is the most widely used method. Chromosomal studies
are carried out by culturing and examining cells present in the
amniotic fluid surrounding the foetus. The risk of 'foetal loss' is 0.5
to 1% and culture failure occurs in less than 1% of cases. A second
form of pre-natal diagnosis – chorionic villus sampling (or CVS) –
has both advantages and drawbacks. It can be performed earlier than
amniocentesis – at eight to ten weeks – and produces its results within
twenty-four hours rather than the two weeks (or more) for amniocentesis. Unfortunately it is not only less accurate but a proportion of
CVS procedures need to be repeated for a variety of reasons and,
according to one report (in the *Journal of Medical Ethics* by Judith A.
Boss of the University of Rhode Island), there is 'a 3.2 per cent
procedure-related foetal loss'. This means, putting benefits and
drawbacks within the same equation, that 'as many as 66 normal
fetuses (will be) lost for every 100 abnormalities detected'.

Women in developed countries, as already mentioned, are having
their babies later than was traditionally the case. Therefore Down's
babies and chromosomal abnormality in general are becoming
increasingly more probable with that rise in years. At age 20 the
Down's risk has been calculated as 1,667 to 1 against. For chromosomal abnormality the equivalent figure is 526 to 1. These odds are
then steadfastly lowered with rising maternal age, being 1 to 1,250
and 1 to 476 at age 25, 1 to 952 and 1 to 385 at age 30, 1 to 227 and
1 to 129 at age 37, 1 to 106 and 1 to 65 at age 40, 1 to 30 and 1 to 20
at age 45, and 1 to 11 and 1 to 7 at age 49. The likelihood has
therefore risen 150-fold for Down's and 75-fold for chromosomal

abnormalities in general during that span of twenty-nine years. Delaying pregnancy may have its advantages in modern society, but adjustments to earlier customs are not always of benefit in every area.

*

It is true, as this chapter began, that 'humans do any number of things better than animals, but giving birth is not one of them'. That in no way diminishes, and possibly enhances, the astonishment of this event. Even hardened obstetricians, accustomed to a production line, will speak of birth's miracle if nudged in that direction. Therefore it is small wonder we are disturbed, for example, on reading (in 1994) that a British woman prisoner, serving a three-year sentence for a modest crime, was handcuffed during her delivery. (On hearing this news the president of Royal College of Midwives said she was 'appalled' by the act – it being 'positively barbaric'.) Similarly we are aghast, when reading of the Warsaw ghetto in the Second World War, to learn of a woman who, hiding in an attic, gave birth without uttering a sound while German soldiers searched the floor below. Birth can be the greatest miracle we will ever see. It can also be quite horrifying, should the fates choose otherwise.

It can even, on occasion, be troubling when all should be going well. Two scientists from Pisa, Italy, wrote to the *Lancet* in 1995 about a distressing incident:

> We would like to bring to your attention a patent violation of animal rights, paradoxically caused by extreme human care. The episode took place last January in an Italian medical institute, where a precious primate was kept in temporary custody. The specimen was an adult female, in the latest stage of pregnancy. Because of the considerable value of both mother and fetus, labour was monitored throughout by the use of a cardiotacograph. For this purpose, the mother was restrained in an uncomfortable, supine position. For greater safety, uterine dilation was monitored by manual inspection, about six times an hour, each time causing evident pain to the animal. Finally, just before delivery, episiotomy was performed to reduce the risks of delivery.

Both mother and baby are now perfectly well, but we believe that this and similar cases should be brought to the attention of whoever cares for animal welfare. We certify that what is reported is true: one of us is the primate in question.

GROWTH

DEVELOPMENT   WEIGHT   HEAD SIZE

DENTITION   BREASTS AND GENITALIA

VARIATION IN DEVELOPMENT   ADOLESCENCE

PUBERTY   INEQUALITY   ACNE

OVERWEIGHT   WHO MAKES WHOM?

By the time the new human being is 15 or so, we are
left with . . . a half-crazed creature more or less adjusted
to a mad world.
R. D. Laing

Our DNA travels down the generations, and death can
achieve no victory over our genetic pool.
Liam Farrell

The human infant, having spent a long time in the uterus (compared
with most mammals, such as rodents, marsupials, carnivores, other
primates), is even more leisurely in its post-birth development. A year
must pass before it can stand. More years are necessary before it
could fend for itself in any environment. It will not be sexually
mature for thirteen years or so, and not reach adult stature – let alone
strength – until in its twenties. Nevertheless that is its virtue. It does
take time, even to grow – whatever parents may think as the 7 lb
baby puts on 14 lb in one year and weighs 38 lb at six years.

There are some surprising swiftnesses to contradict the laggardly
approach. A mere fifteen minutes after birth an offspring can imitate
– by putting out its tongue, opening its mouth (a fact generally
disputed until filmed sequences clinched the matter). At forty-five
hours a baby has learned its mother's smell, and vice versa. (Such
cleverness cannot compare with, for example, many a herbivore
where scent distinction happens virtually instantaneously.)

The subsequent process of rearing is practised almost entirely by
amateurs. Humans are less competent than animals, or so it would
seem, in that there is more conspicuous problem within human
nurseries. Theories about parental procedure are entrenched, often
before the first experience. Such theories become either more rigid,

as that experience is gained, or are abandoned in tune with lessons learned. Most important fact of all is not knowing the precise intent of this endeavour. That has to be the biggest conundrum of them all.

The practice, the procedure, the endeavour, is the raising of another generation. Human parents *are* amateurs, in that they perform the task only once or twice (in the modern age) and rarely more than ten times (in much earlier years). The entrenched theories either mimic each parent's own upbringing or repudiate them. Policies are also based on local culture, with the practice of swaddling, for example, held to be anathema in many nations and yet obligatory in others. Africans can be appalled by the Western pram or baby carriage, wondering how infants respond to such enforced separation from maternal contact. Theories are often amended when the second child is born. It behaves so differently from the first, being quite a different individual, not so much newborn as born with a (different) personality intact. Each newcomer is a unique event, a human being whose set of genes have never previously been combined.

As for the purpose of parental duty, apart from the production of one more individual, there is enigma. Is it to create an excellent baby (of no trouble to anyone), a splendid child (equally free of unwelcome traits) or a supreme adult (of the highest quality – generous to its parents and its community)? The three phases do not necessarily harmonize, with repellent offspring often becoming excellent adults (or vice versa). Parents, with theories, aspirations, prejudices and yearnings, do not really know what it is they have in mind. They have the basic longing to reproduce. They may expect their offspring will eventually repay the care bestowed upon them and care for their creators. In the meantime, having been presented with a mewling, puking child, they make the best of the unique consignment they have received. They feed it. They love it. They train it and reprimand it, praise it and observe it, but do not truly comprehend what they are attempting to achieve. 'I'm just doing my best,' they may say, without an inkling what is best. They – probably – want their child to be better than most, to be happy, to be healthy, and to be rewarding; but, as to method in acquiring such excellence, they do not really know. They simply do their best.

The offspring have to make do with whatever circumstance into which they have been born. Perhaps they are Inuit and there is blizzard beyond that icy wall. Or they are African with excessive heat as instant concern. Their parents may be wealthy or outrageously poor. Their mother may be young, and inexperienced, or middle-aged and firmly set about what or not to do. The infants are like first-time parents, never having experienced anything of the kind, but they also do their best, come whatever may. Upbringing, on the face of it, is an extraordinary system, the blind caring for the blind, but with a set of instincts and programmed development as built-in factors for survival. In essence parents need only feed the gaping mouths, protect the vulnerable forms, keep them free from excessive want, and watch as nature takes her course. All a child need do is live and learn, and gradually understand what kind of world is its environment. In time, and having learned, it too will become a parent, one more amateur, one more procreator of another generation.

DEVELOPMENT Given food, air, warmth and water a child will grow, more or less, as already planned. That fact has not stopped a plethora of books on how best to follow this simple dictum. Truby King, for example, was much followed in the 1920s, his stern discipline forcing children (and parents) into a similar mould. Reaction was inevitable, and Benjamin Spock sold 17 million copies in 145 printings before the mid-1960s of his *Baby and Child Care*, a more relaxed manual which appealed when so many barriers were tumbling down. The plethora then came into its own, with volumes to suit every pre-conviction. There has never been such a clamour concerning how to achieve what has been achieved ever since life began, how small human beings should become big human beings, and how the results of nine months of pregnancy should continue to grow and reach the stage when they too can embark upon parent-hood. What policy should parents pursue, other than to provide the basic needs? 'Before I got married I had six theories about bringing up children; now I have six children and no theories,' said the Earl of Rochester.

The Holy Bible, never averse to giving opinion, can claim priority in proffering child care advice. 'He who spares the rod hates his son' – Proverbs 13. 'Do not withhold discipline from a child. If you beat him with a rod he will not die.' – Proverbs 23. 'There came forth little children ... and mocked him ... And he turned back ... and cursed them in the name of the Lord. And there came forth two she-bears out of the wood, and tare forty and two children of them.' – II Kings 2, 23–4. 'If a man have a stubborn and rebellious son ... then shall his father and mother ... bring him unto the elders of his city ... and all the men of his city shall stone him with stones, that he die.' – Deuteronomy 21, 18–21. Truby King's insistence that a child may have insufficient air, even if placed directly between open window and chimney, can suddenly seem most tame.

Punishment has frequently been strict, with intriguing deviations. Whipping boys, for instance, received the blows which should rightly have landed upon princes following royal misdemeanour. (Henry VIII and French kings of that time favoured this approach.) Rods were not spared much in nineteenth-century schools. They were even used in anticipation of misdeed, with likely troublemakers flogged at the beginning of each day to save time. Many a current juvenile court, sentencing after a pack of crimes, might not wonder whether such a procedure still has advantages.

Ronald S. Illingworth (who died in 1990, aged 80, shortly before the tenth edition of his famous book went to press) first published *The Normal Child* in 1953. This landmark in paediatric literature has been consistently reprinted, translated and updated ever since. He too has words about stricture. 'Every child must experience discipline ... Lack of discipline in the first years is a major factor in juvenile delinquency, accident proneness and other undesirable traits.' But 'excessive discipline is hardly less harmful'. Punishment must, in any case, 'be immediate', so that the cause is related to the effect. The child must understand, which means that 'no punishment is ever justifiable in the first year'. He/she can learn on reaching three, and 'somewhere in between' one and three is the age at which such teaching can begin.

While parents are learning how best to treat the child, that same child is also learning how best to treat its parents. As Ogden Nash proclaimed: 'The wise child handles father and mother, by playing one against the other.' Illingworth's particular strength, and value, lies in his repetition that normal babies cover a wide span, in their physical attributes, personalities, and general progress. Labour at their birth can be short or prolonged. Position at birth can be head-first (95%), face upwards (0.4–0.8%), shoulder first (0.5%) or buttocks first (3.5%). Birth is premature (four to five weeks before expectation) or even earlier, or extremely late. The standard gestation length is thirty-eight weeks, but babies have lived after being expelled a dozen weeks earlier and, legally, it has been presumed that a pregnancy can last for forty-nine weeks. Birth-weight for the prematures can be one-quarter of the standard weight on arrival. Late arrivals have, on rare occasions, been two and a half times that standard. The differential is therefore ten-fold. In Britain the average birth-weight is 8.2 lb. In India it is 6.1 lb.

WEIGHT at birth tends to be higher (says Illingworth) in the 'upper social classes', when the mother 'is above average intelligence', when she is diabetic, and if the infant is destined to be heavier or taller than normal. Very low birthweight is associated with very young or older mothers, and with multiple pregnancy. Maternal smoking, alcohol consumption, high altitude and prolonged labour are all linked to a lower than average birthweight. Pre-term delivery is associated with a variety of (sometimes interacting) maternal factors, such as her youth, height, weight, small weight gain in pregnancy, genes, socio-economic status, illegitimacy (of the offspring), poor pre-natal care, addictive drugs, and several functional conditions like placental insufficiency and uterine abnormality. 'Normally I'm below normal,' said a friend claiming a fever when the thermometer suggested otherwise. Normality is a difficult concept. We speak of it frequently; yet it is only the mid-point in a range of possibilities, not least with human infants when they first see the light of day.

Weight gain following that arrival – girls being lighter than boys – is similarly varied. Average increase during the first three months is

7 ounces per week and 5.3 ounces during each week of the second period of three months. During the second year it falls to 1.5 ounces per week. It then slowly increases, and reaches 2.6 ounces per week by the age of 10 – on average for boys. Girls are by then growing faster, putting on 4.6 ounces per week, again on average, when reaching 10. A few ounces per week can seem a poor gain when, particularly as they age, they can eat tremendously, but 2.6 and 4.6 ounces per week are 8½ and 15 lb respectively per year.

By the second birthday it is possible to assess a child's final height. Foot length, hand length and stature are all about 50% of their final measurements, with weight being about one-eighth. When blowing out (or failing to blow out) two candles on their cake the blowers can seem diminutive, and far smaller than half adult height, but tape-measures do not lie. That 50% rule (with boys, on average, being 49.8% of their adult height and girls 52.8%) is sound, assuming neither disease nor poor nutrition has exerted influence. At ages 4, 6, 8 and 10 the same adult height percentages are 58.2, 65.6, 72.2 and 78.3 (for boys) and 61.9, 69.9, 77.1 and 83.8 (for girls) – all on average. Maximum growth at any age, as Illingworth adds, 'is not necessarily the optimum'. Far more important is the child's 'well-being, abundant energy, freedom from infection and freedom from lassitude'. So too is the fact that average is not the same as normal. A child may be very different from average, and yet be normal. 'Variation from the usual growth is often merely a familial feature, or only reflects the size at birth.' Similarly what is abnormal for one child is not necessarily abnormal for another.

There are ethnic differences in growth, important to remember now that multi-ethnic communities are becoming more frequent. Within the UK, and for children aged 5 to 11, those of African or Afro-Caribbean origin are taller than those of standard British origin while Gujerati and others from the Indian sub-continent are shorter. Those from India tend to have a lower birth-weight of about 7 to 10 ounces, and this smaller start tends to continue as smaller size throughout the pre-pubertal age range. What therefore is below normal for one ethnic grouping may be standard for another.

HEAD SIZE and its maximum circumference can be of concern; so too the fontanelles. Effects of moulding during birth usually disappear in a few days. There can also be head shrinkage at that time, associated with the general loss of weight. The head of a premature baby is larger, relative to its body, than one arriving at term (so much so that thoughts of hydrocephalus may arise). Average head circumference at birth (for boys and girls) is 13.8 inches. By age 1 this is 18.1 inches, by 2 it is 19.1 inches, and by 10 it is 20.7 inches. Many a parturient mother can wish for a smaller head size than 13.8 inches at birth but, although that implies a diameter of 4.4 inches, the newborn head is certainly not circular, being more egg-shaped and ready to undergo moulding when certain bones overlap. The considerable increase in head circumference in the first year post-birth is almost double the increase from age 1 to age 10. If casual talk is permitted it can be argued that Mother Nature, bearing the advantages and disadvantages of big-brained *Homo sapiens* in mind, has been aware that a bigger head at birth would pose even greater problems for the mother than currently exist.

Head size does not demonstrate any significant racial, national or geographical differences, but is helpful in diagnosing malnutrition. Normally (that word again) the head's circumference is bigger than the chest's until the infant is 6 months old. With under-feeding the head is much less affected than the rest of the body. Therefore the head can be bigger relative to the body for much longer. ('How old is this child?' I asked when visiting an Ethiopian hunger camp. 'Six months?' I suggested, having a similar child at home. 'Three years,' replied its mother, most unblinkingly.) Large heads in such camps are rarely hydrocephalus, and may occur if the family has a tendency towards big-headedness. 'Have a look at the parents,' Illingworth frequently recommends.

Oddly, according to work done at the University of California, San Diego, parents do not look enough like their children to allow judges (who do not already know of the relationship) to spot the resemblance. In an experiment, reported in *Nature* in December 1995, judges were shown photographs of offspring. They had to rate the similarity between each of them and three possible mothers or fathers.

This was done with male and female children at the ages of 1, 10 and 20 years. The only reliable similarity to be discovered was between 1-year-olds and their fathers, such toddlers not being matched with their mothers. The result can confound all those who, having peered into pram or high-chair, or having been struck by a resemblance, will immediately exclaim at the likeness to one or other parent, or even both.

The research group, aware of this point, said there is often a desire to please (by the onlookers), that they *only* comment on similarities when these are noticeable and such similarities may be chance events. However, it is conceded that 'resemblance could depend on characteristics that still pictures cannot capture'. The Californians suggest there may be a reason for 1-year-olds to resemble fathers. 'While a mother may be quite sure that the baby is hers, no matter what it looks like, the father cannot. It could then be to a baby's advantage to look like the father, to encourage paternal investment.'

Parents can certainly serve as guidance if the skull is strangely shaped. This too can be a family trait. A baby's head can also be disconcertingly uneven (by the time of the first birthday), with a flattening on one side and a bulge on the other. The cause (almost always) is that baby's preference for lying on one side, a preference so engrained that parental interference will not dismiss it. As for the fontanelles, they also vary greatly in the time of their closure. 'Normally' the front one, small at birth, enlarges for a couple of months and then diminishes. It may close early (at 4–5 months) or late (3–4 years). The rear, or posterior, fontanelle is usually closed by 2 months. There can even, on occasion (6.3% in one study), be a third fontanelle of a different kind between the other two. All fontanelle abnormalities may be a sign of some unwelcome defect, but can be nothing of the sort and no more than one further expression of variability.

DENTITION The arrival of teeth is undoubtedly varied. About one in 2,000 newborns already possess at least one visible tooth, this number of individuals (allegedly) including Julius Caesar, Hannibal,

Louis XIV and Napoleon. Richard III, wrote Shakespeare (ever happy to portray that king unfavourably), could 'gnaw a crust' when two hours old. Lesser mortals may not acquire their first tooth, usually a lower central incisor, for fourteen months, but average time of acquisition is six months. The traditional order of arrival (with their arrival months) is lower central incisors (6), lower lateral incisors (7), upper central incisors (7½), upper lateral incisors (9), lower first molars (12), upper first molars (14), lower canines (cuspids) (16), upper canines (cuspids) (18), lower second pre-molars (20), and upper second pre-molars (24), making 20 in all, the human number of deciduous (milk, baby) teeth.

Only two-thirds of children produce their upper second molars (the pre-molars) within five months of that average date. Only two-thirds of children produce their first tooth between five and ten months after birth, with Napoleon, Richard III and Co. feeling superior. Girls tend to be earlier with their primary teeth but boys are earlier in getting rid of them. The business of teething, much used as excuse for all manner of behaviour, used to be considered a cause of death. In 1842 the Registrar-General reported that 4.8% of all London children dying in their first year had their teeth to blame (which seems harsh, and improbable, with only six teeth customarily erupting during those first twelve months).

Permanent teeth do not mimic the arrival order of the previous set. First to appear are the first molars. These arise before any of the deciduous teeth are lost – at 70–72 months for girls and 73–74 months for boys. First deciduous teeth to go, and then to be replaced, are the incisors – in the seventh or eighth year. Next are the pre-molars – during the ninth or tenth year, but with wide discrepancy. Finally the canines go, and are replaced, during the twelfth year. Whereas the first molars arrived during the sixth year, give or take much variety, the second molars do not appear until the thirteenth year. As for the third molars, allegedly linked with wisdom, they can arise during the seventeenth year, or as late as the twenty-fifth, or even never. The first set (of twenty teeth) therefore spans less than two years in its arrival time. The next lot (of thirty-two teeth) may take three times as long, four times as long, or longer still. While pondering this anomaly it is fruitless, but exciting, to wonder why *H. sapiens* has never created a third set of teeth to parallel this species' length of life. Perhaps, as novel proposition, genetic engineers might bear this undoubted need in mind.

BREASTS AND GENITALIA certainly demonstrate divergence. Four-fifths of newborns of both sexes experience breast enlargement. This starts two to three days after birth and reaches its maximum in the second week. Milk is actually secreted from these infant breasts within seven days of (full-term) birth, but production gradually ceases. By the eighth week, according to one survey, it has stopped completely. As for the breast nodules these can last until the second half of the first year.

Enlargement of genitalia is also a feature of the newborn. In the female the clitoris and the labia are often pronounced. In young males the penis may become erect, and is also frequently grasped. The testes may not have descended into the scrotum at birth, perhaps in 3–10% of newborns, but it is also said that cold (inquiring) hands are a common cause. The testes are able to leap back whence they came, as if resenting over-cool investigation.

VARIATION IN DEVELOPMENT Variation in development is undoubtedly wide-ranging; it is the norm to be abnormal. As Illingworth reported, having confronted the 'range of normality', 'most children pass the various "milestones" of development earlier or later than average'. This seeming tautology must be borne in mind when assessing human alteration from infancy to adulthood. If an average height, weight, competence, reaction-time or other item can be determined, most of the developers will be on one or other side of that average figure. Arguably, if the average figure is sufficiently precise (such as 36.246 inches for a definite age), no child will hit the mark exactly.

For example, most children do not produce tears until they are 3 weeks old but (following one examination of 1,350 babies) 13% were producing them by 5 days and many were not doing so for several months. Squint, as another example, is common in the first few weeks, and only requires investigation if prolonged beyond 6 months. Normal children can have a raised temperature, being 99.8° under the armpits as against the usual 98.4°. Pulse rate on Day 1 is 123 to 126. At 2 months it goes up to 130, and then down until 113 or so at 12 months. Or 127, or even 200, a rare condition that may persist for years.

It is easy to wonder if any aspect of development is not greatly variable. Sea-sickness, which is common, usually begins in the second or third year, but 20% of unacclimatized children never suffer from it. Childish abdomens can be huge, from age 1 to 3, or flat. Almost all children have flat feet at 18 months. Almost all children have lost the fatty pad creating flat-footedness by the age of 10. Knock-knees are common in children aged 3 and the condition has usually gone by the age of 7. Many children aged 1–2 prefer walking on their toes, a normal condition save that most children do not do it. A smile – so eagerly sought by parents – may arise at 2–3 days or not until 8 weeks. Babies may grasp at 3½ months or, more usually, not until 6 months. Normal children can walk as early as 6 months or not until 18 months.

First meaningful words, also pounced on by enthusiastic parents, may be uttered between 9 and 15 months. Thomas Carlyle, having said not a word beforehand, allegedly produced 'What ails thee, Jock?' (to a fellow babe) and thereafter, from 10 months on, only spoke in sentences. Albert Einstein waited until aged 4 before speaking, so it is generally reported. More ordinary mortals are likely to use thirty words by 18 months, 300 by their second birthday (with girls knowing more than boys), 1,000 by their third birthday, and 2,000 by the time they are 5. As for stringing words into sentences, the 3-year-olds are good at putting three together, and the 5-year-olds five together, but Carlyle, Einstein and practically everybody else will be on either side of these generalities. As for sphincter control, so important in child-rearing (and given a host of other names), some offspring can be dry by day from 6 months or not until aged 2 or 3 (and still unreliable at night).

The important point, of concern to all parents and stressed by all investigators, is that no child is mentally retarded if backward in a single field of development and normal in other fields. Mental backwardness may seem difficult to judge with babies who, however splendidly normal, defecate at random, urinate as they please, plaster food around their mouths, dribble, snuffle, snort, cry for no known reason, drop their mugs, drop everything, and are irritatingly incapable in every area, but that is the way with human infants. They are not infant chimpanzees, firmly clutching their mother's fur, finding her nipples, drinking when need be, and looking alert from Day 1. The humans *are* incompetent, and remain so for years. They are not imbeciles, whatever despairing parents may think. They are normal humans, as daft and erratic and half-witted as every other normal human. They each have quirks of behaviour which can be frustrating, particularly when Johnnie from next door seems so advanced, albeit with a different set of peculiarities. A mentally retarded child, by comparison, is backward in *all* fields of development, save perhaps for sitting and walking.

*

A child of above average intelligence is much harder to anticipate. (A statement, frequently quoted, suggests that a child prodigy is one with highly imaginative parents.) Illingworth, in trying to deduce 'advanced maturation' at 6 weeks and 6 months, believed such judgement to be possible, but knew 'of no statistical evidence that prediction in the newborn period is feasible'. One considerable problem involves follow-up. Is an infant, who is allegedly brighter than most at 6 months, still brighter than most at 18 years? Another involves the measurable items. Is early sphincter control, for example, a sign of superior ability? Arnold Gesell, distinguished child psychologist, wrote that 'the scorable end products may not be far in advance, but the manner of [the child's] performance is superior'. The advanced child, in his opinion, 'is more poised, self-contained, discriminating, mature ... The total output of behaviour is more abundant, more complex, more subtle than that of a mediocre child.'

As for solid evidence that is harder to encounter. Some researchers allege that a retarded child responds to pain more slowly than a normal child and, by inference, that a bright child responds more quickly. The bright child is likely to sleep less than the retarded child, and to start reading much earlier. Individuals reputedly able to read at age 3 form a celebrated bunch – Ruskin, Coleridge, Macaulay, Voltaire, Dickens, Samuel Johnson, Dean Swift, Lloyd George, Edith Sitwell. There must, of course, have been very many others who, however literate when aged 3 or even earlier, never achieved such fame.

Many books can be, and have been, written on every aspect of child development, and they frequently contradict each other. Many a parent considers that common sense (however ill- or never-defined) provides sufficient guidance, but three recent long-term studies in the US produced some interesting results (and were not always affirmations of common sense). For example, home-based schemes in which mothers were taught about providing good stimulation for their children had no effect (upon IQ). Big improvements came only from education at special centres. The children of poor mothers with low IQs and also babies born prematurely can greatly improve their own IQs if given high quality education in their first five years of life. ('Give me a child until it is five and I will make it a . . .', runs the adage for a variety of conclusions.) Children most at risk of retardation, according to the surveys, were not from the poorest homes but from those where language abilities and IQ were lowest. The key ingredients, according to a summary of the studies in *New Scientist*, were: encouraging exploration, the teaching of basic skills by a trusted adult, the celebration of achievements, the provision of a rich language environment, and the protection of children from inappropriate disapproval.

Mentally gifted children are less likely to have behaviour problems. On the other hand their divergent thinking makes them less willing to conform. School work (for their age) may bore them, and teachers may in consequence resent them, wishing they could learn by rote 'as normal children do'. Intellectually they may be aged 8 when, physically, they are still 5; therefore school can be a problem. Show me a drug without side effects, as a pharmacist once said, and I'll show you a drug without effects. Show me a child without problems, as could easily be said, and I'll show you a pair of lying parents. All children are abnormal, and therefore problematical, to some degree – for their age, for their sex, for their style of personality.

Girls, usually, walk, speak and acquire sphincter control earlier than boys. First-borns and single children tend to be more intelligent (perhaps because parents spend more time with them). If there are more children the oldest and youngest are, in general, more intelligent. So too if there is greater spacing between children. If siblings are born close in time that can be a disadvantage, and a child's speech may even regress when a new baby arrives. As for stuttering (synonymous with stammering), it affects about 1% of British children, more often boys than girls. Usually (in 70%) it begins before age 5 and almost always (95%) before 7. It has no relationship, according to a major survey, with handedness, change of handedness, or ambidexterity. Many a child passes through a stuttering stage, often no more than a few days. With perhaps 10% of children there is a more prolonged period of rapid, confused and jumbled speech.

Some mixed Illingworth quotations can conclude this section on infant variation. 'All parents take a risk in having children, and do not even know whether they will be mentally subnormal or not.' 'No amount of practice and teaching will enable a child to learn skills unless the nervous system is ready for them.' 'It would be surprising if some children did not show unusual patterns of development.' 'Development assessment is fraught with difficulty, largely because all children are different . . .' 'All children have behaviour problems. All parents have behaviour problems. All teachers have behaviour problems. A child's behaviour problems represent a conflict between its developing personality and that of its parents, teachers, siblings and other children with whom it comes into contact.' 'Behaviour problems have their origin before birth and often before conception.' 'Many mothers seem to be haunted by the fear of "spoiling" their children. It is a paradox that it is to these mothers that most spoilt children belong.' 'Of all parental attitudes favouritism and rejection are probably the most harmful.' Finally, as greatest truism of all: 'The majority of children are born in working order.'

ADOLESCENCE Sex arrives before responsibility, say parents (and many others). It certainly arrives before adult stature is reached. No mammals breed until they have reached a certain size and humans are no exception, being several years short of physical maturity when reproductively capable. Currently, with teenage (or earlier) pregnancies, with sex in full swing long before there is wish for reproduction, with contraception and abortion playing major roles, and with old rules of restraint and family no longer so critical, the former fundamentals of human reproduction become obscure (or even invisible). A comparison between humans and the four nearest primates can demonstrate both our basic differences and all our similarities.

Gorillas, male and female, become far larger than humans, but take less time to do so. Their males reach puberty at age 8–9, and are then potent save that copulation is quite impossible. A further eight years or so must pass before they are large enough to acquire mates, with their strength and size both prerequisites. Female gorillas reach puberty even earlier than males, and then leave their original family. They either join a group of females, a harem associated with one principal male, or behave more independently, attaching themselves to a lone male. They may then forsake him, choose another, abandon that other, and repeat the process before settling down with a permanent partner. Throughout this time they have been sexually active, but solicit copulation only during oestrous (lasting a couple of days in each thirty-five-day cycle). As with humans their first cycles are anovular – unassociated with the production of an egg – but the start of a gorilla's ovulation is quickly succeeded by pregnancy. All mating then stops until the infant is 4–5 years old and successfully weaned.

Orang-utans are similar to gorillas, in that the males acquire harems, but there are many differences. The males leave their mothers when aged 5–7, and then wander through the forest in solitary fashion. At age 18 (or so), having become twice the size of an adult female, each male acquires a territory of his own. He then calls for females with an extremely loud voice. Only those females aged 9–10 who are in oestrous will respond to this invitation, and the sexual contact is extremely brief. If the female fails to become pregnant she will again respond during the brief oestrous period of her next cycle. If pregnancy does follow she will remain on the edge of her mate's territory but have little more to do with him save see him occasionally. She will produce her infant, and care for it as part of his somewhat distant harem. When fully weaned, and having spent at least five years in its mother's company, the infant will depart. Its mother will then, once again, come into oestrous and respond to the booming call. Mating is not entirely simple, being usually performed when both animals are hanging by legs and arms high up a tree.

Gorilla and orang-utan males therefore each experience a time of adolescent infertility. The young gorilla males cannot mate because they are too small – and would easily be pushed aside by a larger individual. Young orang-utans cannot mate because, without territory and the correct adult call, they are unattractive to the females. They may try to mate, and can be bigger than the females they are pursuing, but a successful mating in the treetops needs, at the very least, a willing partner.

As for chimpanzees, often considered nearest relatives to humans, they experience no such adolescent infertility. Mating begins almost the moment it is physically possible. Young chimpanzee males are slightly larger than their female contemporaries, but are otherwise similar in appearance, save for the females's manifest genital swelling. Both males and females are raised in large communities, containing perhaps as many as 100 individuals, and such bands are strongly defended against strangers. (Gorilla groups, by contrast, may even intermingle briefly after an encounter before going their separate ways.) When female chimpanzees reach puberty they leave home (as it were) and attach themselves to another community. They are not attacked by the new group, partly because they are willing (or eager) to permit copulation with a wide range of partners. They are in oestrous for about a third of their thirty-five-day cycle but initially, as with gorillas and orang-utans, do not become pregnant. The earliest cycles, perhaps twenty of them, are therefore presumed to be anovulatory.

When a female chimpanzee becomes pregnant, probably not before age 13, she ceases to be sexually active. She will not come on heat again until four years have passed and her infant has been weaned. Once again there may then be mating by a range of partners, the commonest occurrence, but a couple may unite throughout her period of oestrous, with the male strongly defending his female from other contenders. Even a sub-adult male may acquire a female, but the two of them have to retreat into seclusion if he is not to lose his partner to some larger, older male. Chimpanzee males are therefore able to mate whenever opportunity permits, either with a newly-arrived promiscuous female or with an older female suddenly in oestrous after a span of years.

The fourth ape, the bonobo (sometimes called the pygmy chimpanzee, even though no smaller), is a resident of Zaïre, and has been poorly studied relative to the other three. Its society is matriarchal, with adult males kept on the sidelines. Each female group even engages in homosexual activity, with mutual rubbing of genitals. There is little sexual activity with the males, and nothing comparable to the wanton behaviour of *Pan troglodytes*, the commoner chimpanzee.

As for human sexual behaviour, if compared with these closest relatives, there is little similarity with the gorilla mode of life. Dominant human males may acquire harems (and have done so in the past) but there is not the obligatory period of adolescent infertility before achieving the strength and bulk to acquire a mate. More parallels exist with the orang-utan procedure. First find a home, make certain it is private property, and then invite females to take advantage. As for bonobos it is sometimes argued (by anthropologists) that human matriarchy used to be more prevalent, notably among the higher echelons, for the fundamental reason that only females know which infants come from them, paternity being less clear. It is the promiscuity of *P. troglodytes* that most comes to mind when thinking of human behaviour. The females wish to be accepted in new environments; the males wish to take advantage of all such offerings.

According to C. B. Goodhart, from whom much of this summary has been taken, 'It does look as though harem polygyny (as with

gorillas, in particular) must have been the original human and sub-
human mating system ... We may now be in an unstable transitional
state, abandoning the polygynous harem system of our ancestors, in
favour of one or other of two mutually incompatible alternatives.
These are either monogamy ... [similar to] the lifelong faithful
monogamy found in gibbons, or sexual promiscuity, with irrespon-
sible and indeed anonymous paternity, as seen in chimpanzees.' This
may provide an explanation, he concludes, 'for some of the sexual
stresses and strains to which the human species is now subject.'

PUBERTY can be defined as the latter stages of the process of sexual
maturation. Medically, and individually, it is a series of changes
lasting several years. As with (virtually) every other aspect of development there is wide variation. It is normal for testicles to enlarge –
this being the first sign of male puberty – at age 9½. It is also normal
at age 15. With girls the first event is similar, being a swelling of the
ovaries, but girls are none the wiser about such an internal happening.
Their first external sign is breast enlargement, beginning 'on average'
between 9 and 13. It would be surprising, as already detailed, if some
children did not show unusual patterns of development. It is
somewhat more surprising that this most fundamental development
of them all, the ability to produce another generation, arises in such
a scattered manner. Some children can begin reproducing before they
reach double figures. Others must wait for another half-dozen years,
a 50% increase (more or less) in the time since birth.

There can also be exceptional individuals, extremely different from
the normal range. The *Independent* told recently of one variant:
'Jessica was a four-year-old with a secret. Not only was she as tall as
an eight-year-old, she had already developed breasts and pubic hair.
Then she had a small show of blood – a warning that her periods
were about to begin.' Another child, named Katie, was also evidence
of the condition known as premature sexual maturity. 'She still
believed in Father Christmas and the tooth fairy,' reported her
mother, 'but I'd watch her on the floor playing ... and I'd have to go
upstairs and cry because my baby was wearing a sanitary pad.' The

onset of puberty is associated with diet and body weight 'but we still don't know what actually triggers the process at a certain time', said Tony Price, paediatric endocrinologist, who sees many of the prematures.

The law, which welcomes fixed dates and then abides by them, is more clear cut. It states (in England) that girls reach puberty at age 12 and boys at 14. Similarly, and equally rigidly, neither sex may buy alcohol, tobacco or fireworks until they have lived for 6,574 days. One day earlier, however mature, they are guilty (and so is the person selling these goods to them). One day later, however immature the purchasers, everyone is entirely innocent. Biologically such stricture does not make sense. Many a girl has not even begun to menstruate before, in law, reaching puberty. And many a boy may have been producing sperm for years before reaching this legal form of adulthood. In nations around the world marriages are permitted at age 12 or not until 18, a variance in tune with biological variation but making little sense for individuals within the differing countries.

Puberty is not only a moveable feast; it also spans in its development a discordant set of years. It lasts in general from 24 to 54 months between initiation and completion. Half the children complete the process in three years, and practically all in five years. Those who start their sexual development early, who might therefore be considered precocious, do not mimic this haste in rushing through the developmental stages. They can take as long, or as short a time, as those who begin much later.

*

Boys, beginning with their testicular development, can actually have their development assessed via an orchidometer, named after the Greek for testicle. (Plants of the *Orchidaceae* received their name because some tubers in this family were thought – in 1845 – to resemble testicles.) A testicle volume of one-third of adult testicle size (6 ml as against 20 ml) indicates that puberty has begun. Within the enlarging organ not only has spermatogenesis begun but the testis has started producing the male hormone testosterone. Other masculine organs then start enlargement, such as the penis, the seminal vesicles and the prostate. The larynx also changes, and the voice may abruptly alter its tune or be more leisurely in its shift from boyish treble to a deeper adult sound. Some boys, perhaps at age 14, are still young children in their various male possessions while others are conspicuously adult in their perquisites. (The school shower or changing-room demonstrates differences blatantly, causing almost all present to presume a degree of personal abnormality.)

Growth in height, which can be 6 inches a year (or more but usually less), does not begin its spurt for boys until sexual development has begun. An increase in penis length generally partners the increase in stature. Pubic hair, first downy and slightly pigmented, then curly and positively pigmented, appears before the growth spurt. Armpit hair and facial hair (causing that all-important initial shave) begin to grow a couple of years after the onset of pubic hair. There may be a minor swelling of the breasts, and each areola – the disk around each nipple – will enlarge. As for semen, that will not emerge, either naturally (perhaps at night) or with assistance, until a year or so after the penis has begun to grow. It may not contain mature spermatozoa initially, but will do so within a year. Biologically, a boy then becomes a man, save that he is still short of his final stature and strength.

*

Girls are similar in many aspects of their sexual development. There is that initial swelling of the ovary, an equivalent to the swelling of the testis. There is then a swelling of the breasts. This happens earlier than the maximum rate of growth in height, and begins – on average – between ages 9 and 13. Oestrogen (as against testosterone for boys) is being liberated, and can be measured in urine as one indicator of the ovaries' activity. Uterus and vagina both grow as the breasts grow. As with boys the areolae increase in diameter but, with a major difference, are often elevated, only sinking to breast level in the later stages of development. It is the arrival of menstruation, the menarche, that so distinguishes females from males, partly (of course) because it is a female event but also for being so positive. Either a girl has menstruated or she has not, this fact distinguishing her development absolutely. With boys the process is vaguer, with doubt existing whether or not ejaculation has actually occurred. There is no boy into man certainty as with girls becoming women, and no positive statement of a date.

Menarche does not take place until after the peak of the growth spurt. Many a girl, perhaps worried about excessive stature, can therefore be comforted that growing will soon stop. 'At most', according to one textbook, she will only grow another 3 inches. Just as early male ejaculations may not possess mature sperm so may many early menstrual cycles not be partnered by the liberation of an egg. Sexual competence may therefore not arise in either sex – for a while, despite the outward signs. Average age of menarche is 13.25 years (in Britain). In the US it is just below 13. These ages used to be greater – about eighteen months greater in Britain in 1890 – with bad housing, poor food, and inferior medicine all blamed for the later arrival. It has been suggested that today's average age is more normal, with the squalor of some earlier centuries raising the average age from its traditional level. Menarche is still continuing to arise earlier but cannot continue, for ever, in this fashion. As J. M. Tanner phrased it: 'Common sense dictates a stop in the foreseeable future.'

This average age of menarche should not imply uniformity in age of onset, being the mid-point of ages 9 to 17. Radiological age is a better indicator of menarche's likelihood than chronological age. The number of birthdays a child has achieved is not only crucial in law but in much of human thinking, concerning schools for instance. Radiological age, also known as bone age or skeletal age, is a better indicator of general development than the Earth's passages around the sun. During a body's growth every bone goes through a series of changes, easily identifiable with X-rays. It so happens that the epiphyses – bony bits at the extremities – are uniting with the end phalanges of the fingers at about the date of menstruation. This fusion time is variable – from 12 to 14½ years – but with nothing like so wide a span as the onset of menarche. Therefore, if a girl's phalangeal epiphyses are uniting with those end phalanges, she is likely to have her first menstruation at about that time. Radiological age is useful in informing young ballet enthusiasts, for example, if they are likely to end up too tall for the desired ballet height.

Some other menarche facts are: tall girls reach it earlier than short girls; girls reaching it late are likely to be, on average, taller in the end; it occurs earlier if there are several older siblings; the weight of 103 lb is sometimes said to be critical as a kind of trigger for menarche; malnutrition, anorexia, rigorous training (as in athletics and ballet) and being a twin all delay it. Apart from external influences, such as malnutrition and intensive training, the prime cause of variation in menarche's timing is assumed to be genetic.

Once again the biological basis can make one wonder why there is such variety in the ability to procreate, with women becoming fertile either well before their teens or almost at the end of them. In the modern world the shift downwards of menarche's arrival has caused at least one fundamental change. In 1890, when its average age was 15, the minimum age for British girls leaving school was 12 and therefore well in advance of problems concerning fertility. Currently, when the average age of first menstruation has dropped almost to 13, British girls are not leaving school before 16. What had been a post-school phenomenon is now very much a school event, with many hundreds of pregnancies each year initiated and even completed before school is left behind.

*

Teenagers can and often do live dangerously. As a *Lancet* leading article stated (in 1995): 'Their first sexual encounters may cost them their lives if they choose a partner infected with HIV.' The threat of AIDS is real. Within the USA the largest AIDS increase (in 1994) occurred among the 13- to 24-year-olds. Condoms offer protection, but most teenagers do not use them. There are also non-sexual dangers. School arguments can lead to attacks, often with weapons. Home arguments can also be damaging, more to adolescents than younger children. What the *Lancet* called a 'rite of passage' for many youngsters, the experimentation with illicit drugs, can leave addiction in its wake. Such trials, almost obligatory with tobacco and alcohol, can lead to a life-long enthusiasm that may be damaging. 'I don't mind what something does so long as it is harmful,' is a remark attributed to a teenager who – no doubt – then received a cheer from all contemporaries. Teenagers *do* live dangerously, much to the dismay of parents, physicians, teachers and other elders who, somewhat earlier, went through that phase themselves.

Experiencing sexual intercourse is increasingly a major feature of the rite, and may be earlier than before. In 1996 a 12-year-old girl gave birth in Wolverhampton, England, having become pregnant aged 11. Her consort, attentive in the maternity ward, was aged 14. Their ages thus eclipsed the previous record, achieved by a slightly older 12-year-old girl and an older partner whose baby had been born in Swansea, Wales, a few days beforehand. David Nolan, of the Birth Control Trust, said afterwards that a very few 12-year-old mothers are encountered each year. 'Childbirth is always a risky thing but at that age the risks are even greater.' He believed more effort should be spent on advice and contraception for young people.

It has been feared that sex education classes at school might promote promiscuity. A huge survey, published in August 1995, concluded that the fear was unfounded. More than 19,000 people, aged from 16 to 59, were asked how they learned about sex and when they had their first intercourse. 'Men whose main source was school-based lessons were significantly more likely to have been virgins at the age of 16.' As for girls the age at which their virginity vanished was not influenced by such formal education, but they were certainly losing it earlier than did their predecessors. The average age for first intercourse was 17 in 1990 but 21 for all those aged 55 to 59 at the time of the survey. Nearly 1 in 4 girls had had sex before the age of 16 in 1990 compared with less than 1 in 100 of the older age group. With boys 1 in 4 had had sex before 16 in 1990 as against 1 in 20 for the older group. Some 25% of males were therefore breaking the law (if their partners were also less than 16) when in the middle of their teenage years.

As well as living dangerously many adolescents push this fact to the limit, and die. The suicide rate for youngsters, notably in the US, has been gaining recently. Between 1980 and 1992 it rose by 28%, from 8.5 to 10.9 for every 100,000 individuals between 15 and 19 years old. For the younger group, those aged 10 to 14, the rate rose even faster – by 120% – but with smaller totals (from 0.8 to 1.7 per 100,000 individuals). The act of suicide (about which much more on pages 429–39) is almost always sad, but can seem much worse when its perpetrators are still not fully grown.

INEQUALITY The interval between infancy and adulthood is fraught with contradiction around the world. There are 40 million street children in Latin America (according to the WHO), 25–30 million in Asia, and 10 million in Africa, all living without benefit of family. 'Adolescents are people, not problems,' as was said at the 47th World Health Assembly (of 1994), which focused upon this age group. The young are getting more numerous all the time, with 50% of people now aged below 25, and 33% between 10 and 24. Four-fifths of this total live in developing countries, a proportion expected to rise. Not only is there a gap in schooling between the developed and undeveloped worlds, but male/female educational differences are also wide. Some recent figures affirmed that 85% of adolescent boys and 87% of adolescent girls were at secondary schools in developed communities. The equivalent totals for undeveloped areas were 41% and 28%.

In the developed world there are different problems. Some 250,000 US high-school seniors have used anabolic steroids (and the proportion is increasing). Children in Britain aged between 4 and 15 watch two to three hours of television a day. Their schools allocate one hour a week for physical education, with less than 10% of this time actually spent exercising. Almost 100,000 children in Britain (according to the Children's Society) run away from home each year, some aged 11. 'There is a danger of these children becoming so isolated that they cease to be citizens of our society,' said its chief executive. The commonest reason given is violence, mainly at the hands of parents or step-parents. Such runaways are too young to claim benefit and often resort to crime or prostitution.

There is also inequality of sex as judged by school results. Of the top twenty British schools, designated according to percentages of the highest examination grade, eighteen were girls' schools. It is better – for both sexes – if children are segregated. This is particularly so among private schools, which have a preference for single-sex establishments, but state schools are similar. Out of the top fifty state schools (judged by exam results) only five are mixed. As for boys and girls in general the 1994 figures (for the UK) showed that nearly 46% of girls gained five or more passes at the top GCSE grades (such language may be obscure to outsiders but is well understood by all British schoolchildren) whereas boys achieved less than 37%.

At A level, taken at age 17–18, girls score better than boys even in allegedly male strongholds such as mathematics, physics and technology. More girls than boys, 87% compared with 78%, then continue in full or part-time further education. Nearly half of all newly qualified accountants are now women, as against 10% in the 1970s. But, as Britain's Secretary of State for Education pointed out in 1995, only 2% of big company board members are women, and only 9.5% of British members of parliament.

A lack of segregation may be one reason for the recent shift in exam results at Oxford and Cambridge. The percentage of women formerly achieving 'Firsts' (the highest grade of a university degree) was higher than for men. Their A levels were also better (from exams taken at age 17–18). Then came desegregation, with all-male colleges accepting women for the first time and many all-female colleges also accepting men. Instead of women choosing women for places at university the still male-dominated and formerly male colleges employ mainly men in their selection process. This change may have been influential along with the desegregation. At all events women began to fare less well. They were still equal with A levels but got only half of the university 'firsts'. Their worst subjects were physics and mathematics. In law and classics the men were only just ahead, and chemistry became the sole subject for which women were superior. When these facts were publicized it transpired that Oxford and Cambridge were not unique. The gender gap, as it has been named, exists throughout UK higher education.

ACNE Puberty also equals acne, or rather it does for 70% of the teenage population to some degree (and over many years for an unhappy proportion). Just when youngsters are becoming most concerned about appearance, along comes a form of skin eruption caused – somehow – by the surge in sexual hormones. (Eunuchs do not get it, but will if given such hormones. It will then vanish when the hormones are withdrawn.) Acne can appear at age 8, with males tending to get it worse than females. Those with oily skin, as opposed to dry or normal, are more likely to be sufferers, and the contraceptive pill can sometimes make it worse.

Not to blame, despite entrenched (parental) opinion, is diet, or lack of exercise, or lack of hygiene, or unkempt hair hanging over the face. The approach of menstruation can make it worse. So too a hot, humid environment, and unemployment, and stress in general (with the very fact of acne adding to that strain). For some reason, or set of reasons, sebaceous (oil-producing) glands become blocked with the sebum they create. Secondary infection causes the unsightly acne, as blackheads and whiteheads create bacterial breeding grounds. The complaint is so common it is said to be physiologic rather than pathologic, normality rather than disease. It can be treated with topical antibiotics. Benzoyl peroxide is also widely used.

The intriguing aspect of acne – if any adolescent can accept the notion of this complaint arousing intellectual interest – is its prevalence. Why should it appear, and so frequently and lengthily, just when the sexes are being mutually attracted? Sebaceous glands are ubiquitous on skin, but acne only develops on the face, chest and back, with the face most susceptible. A theory has therefore been expressed that this scattered mat of pustules acts as a form of repulsion. Humans are sexually capable when acne occurs (probably) but not yet sufficiently mature (probably) to reproduce. The apes (see the earlier pages) experience a variety of adolescent infertility, with gorillas not big enough to mate despite sexual capability, and with sexually mature orang-utans incapable of holding territory until several more growing years have passed. Human beings are also short of adult stature and (let it be said) of adult intelligence when first sexually mature. Therefore acne, according to this theory, prevents more premature pregnancies than might otherwise occur.

All theories have a right to existence, before proven to be fallacious, and this one may be daft. On the other hand the fact is bizarre that such an event seems so entrenched, as it plagues adolescents, destroys their esteem, and undoubtedly reduces their appeal. There is a cure. Acne does vanish with age, a fact as poorly understood as the complaint's arrival months or years beforehand.

OVERWEIGHT is one more feature to disturb adolescents (and others). It is becoming increasingly prevalent, notably in the United States. A major survey (of 10,039 American youngsters), which examined them before and after a seven-year interval, indicated that 'overweight adolescents and young adults marry less often and have lower household incomes in early adult life than their non-overweight counterparts, regardless of their socioeconomic origins and aptitude-test scores'. A British survey, more concerned with energy expenditure than weight, concluded that children are getting fatter not because they are eating more but because they take less exercise. Indeed it discovered that children of the 1990s are eating 20% less than their equivalents of the 1960s.

Most energy is consumed in keeping the body ticking over, the resting expenditure. Thermogenesis, the maintenance of body temperature, uses another 2–3% – or less if warm clothing and central heating reduce demand. Growth, however much used as justification, is not a major consumer of energy, being only 1–2% of the total after the first birthday. All blood cells, for example, are renewed every month. A youngster, weighing 56 lb (and possessing 4 pints or so of blood), puts on less than 1 lb of weight a month. Therefore blood manufacture goes up each month from 4 pints to 4.06 pints, an increment making little difference to the energy spent in creating blood. Overall, and in an adult, about 100,000 cells divide every second and a near-identical number is destroyed. Hence, in part, the major expenditure of energy, some 60–70%, in maintaining a body's equilibrium.

Modern children in developed communities have become markedly less active, according to the Dunn Nutrition Laboratory, Cambridge. They watch television, play computer games, get driven more often, and roam less in the neighbourhood (owing to parental apprehension). Fat children are less active than lean children, but cause and effect are less straightforward. The fat individuals, by reason of their extra bulk, require more fuel simply for maintenance. They may become fat because of a slower metabolism. Different research (at the Energy Metabolism Research Unit, Alabama) has concluded that no difference exists in resting energy expenditure between lean and obese children.

It would seem that fatness is one more human variable during growth, along with sexual development, final stature, teeth eruption, acne, intelligence, sphincter control, stuttering, head size, birthweight, fontanelle closure, skull shape, tear production, speech, behaviour and all the rest. It would be strange, particularly for a species that likes to measure everything, if there were not concern about virtually every aspect of this variety. 'Half of our schoolchildren are below average,' thundered a British politician when wishing to berate the system. It must also be repeated that no one is precisely average and all normal individuals develop within a broad range of possibilities. Abnormal individuals are those who exist beyond the extremities of this range.

The strangeness is that sexual development occurs so much earlier than physical development. This is the case in other primate species but is a considerable confusion for human society. Biologically it must be presumed there is reason. Sociologically it can frequently be regretted when children, still growing apace, enter the sexual arena to acquire this single form of adulthood many years before other aspects of maturity.

WHO MAKES WHOM? Child mortality, infant mortality, perinatal mortality, stillbirth and miscarriage all amount to the same conclusion: every individual who dies early, by whatever means, does not help to create another generation. Biological success entails reaching maturity and then fathering or mothering a child which, itself, must do likewise. Approximately half of all implanted human pregnancies do survive the various forms of attrition from implantation to adulthood. If 50% success is taken as truth (a figure which varies widely up and down the planet) this also means that only 25% of their grandparents' generation were responsible for the generation now reaching maturity. Going back another six generations to the year 1800 (and assuming similar mortality) means that 0.4% of the offspring then conceived were responsible for all of today's young adults. A further century back in time, and to 1700, means that 0.03% were responsible. Therefore, even in human society where death is kept at bay more than in most species, there is considerable loss as well as considerable opportunity for natural selection to operate.

Putting such truisms another way, it is intriguing (and somewhat flattering) that each one of us alive today is the end product of steadfast success since life began. All of our fathers and mothers not only survived to adulthood but succeeded in producing a child. So did their fathers and mothers, and so on back in time. For any male there is a further fact for such introspection. His Y chromosome could only have come from his father, and from his father's father, and his father's father's father. That small fragment of genetic material was (in all probability) identical in every one of his direct line forefathers, and alive and well when Christ was born, when agriculture was beginning, when cave painters were at work, and even further back. (J. B. S. Haldane bemoaned the ending of his Y chromosome – he had no children – but was comforted that some of the genetic material he shared with his sisters was surviving through their children.)

For each female there is similar but more confusing scope for reflection. Her two X chromosomes come from her father and her mother, but she only received one of her mother's two X chromosomes. The X she received from her father came from his mother, and there is not the same form of continuity as with males. A further point is that each X chromosome, longer and possessing more genetic material than each Y, is more likely to have experienced a mutation and therefore not be identical to the Xs borne by female ancestors of the distant past. The major point, pleasing in its way to each and every one of us, is that we and we alone are the survivors of our kind.

PERSONAL GENETICS

THE PENDULUM   GENETIC TESTING   WHO SHOULD KNOW?

It was hereditary in his family to have no children.
Quoted in *Webster's Dictionary* (exemplifying Irish bull)

THE PENDULUM has been active during the twentieth century, having started substantially in favour of the power of inheritance. This was thought to be critical, with good breeding perpetuating good stock and bad leading only to bad, when Gregor Mendel's work was first properly encountered in 1900. First published from 1865 to 1870, it lay dormant, and largely unread (in the *Proceedings of the Brünn Society for the Study of Natural Sciences*), until that opening year of the new century. It seemed to confirm all the old prejudices. His 'particulate inheritance' proved that flower colour, pea shape, and other properties were passed on according to rigid laws. Therefore, as was then argued (pushing the pendulum further in its old direction), human characteristics – drunkenness, sloth, incompetence, brilliance – were passed from one generation to the next in similar style. The good created good, and the bad increased the bad – unfortunately in greater numbers. 'It is clear that the general quality of the world's population is not very high, is beginning to deteriorate, and should and could be improved,' wrote Julian Huxley, and so did a good many others with similar certainty. The pendulum was swinging towards eugenics in headlong fashion.

A dictionary defines that concept as the production of fine offspring. Every potential or actual parent is therefore, in essence, a eugenicist. No one wants lamentable offspring, deficient in some major degree, and the science of eugenics (about which much more on pages 359–72) was dedicated to that end. Two aspects embraced this seemingly laudable intent: either discourage less-satisfactory breeding or encourage the more satisfactory. Discouragement was simpler to effect. California performed almost 10,000 sterilizations by

1935. The contrary aspect of encouragement was also vigorously pursued. Theodore Roosevelt, after referring to the yellow peril and race suicide, urged citizens to breed more 'native Americans' (not meaning Amer-Indians). The Immigration Restriction Act (of 1924) was introduced against 'genuine human weeds', such as those 'biologically inferior' from around the Mediterranean.

The revelations of Germany's death-camp conduct, surfacing mainly as the war was ending, kicked the genetic pendulum smartly in the opposite direction. Anything smacking of eugenics, either by promoting or demoting various kinds of human being, was then vigorously opposed. Nature, via its genes, had to yield, and nurture, via the environment, took its place. All men were equal, provided they were given opportunity to be so. Nations should acquire self-government. Race should be disregarded. Gender should not be divisive. Society in general, and not prejudice or bogus science, should determine a better world. Genetics, with its stress on differences, should never be allowed to rule.

At the same time it was being realized that Mendel's laws, although precise and true, could not easily be applied to major facets of human behaviour, such as ability or general worth. There was also a reaction to the notion that intelligence, the most formidable aspect of *Homo sapiens*, was largely a matter of inheritance. Dark skin, hair type, eye colour – yes, but not intellectual skill. Some promoters of the idea that mental prowess was a matter largely of inheritance even argued that 80% of this trait arose from genes, with only the remainder influenced by family, by opportunity, by education and environment. Such a thought was anathema to others in the opposing camp who, on occasion, even suggested that nothing mental was genetic, much like the acquisition of a particular language (as distinct from the ability to speak). No child gains English, French, Chinese or Mongolian save for its environment. Similarly intelligence was acquired by external influence, the newborn brain being clay for appropriate moulding.

Gradually this extreme view became eroded, and equally gradually the pendulum began to swing back towards genetics. Various diseases helped it to gather pace, these plainly being inherited within afflicted families. The genetic mechanism also became better understood, first by an awareness of DNA, then by an unravelling of its structure, and next by the coding system it employed in building proteins. Comprehension therefore accumulated how genes could be inherited, duplicated and divided in the business of reproduction. Environment might influence progeny but genetics held the key. It lay at the base, of shape and form, of proneness to disease, of ability in general and intelligence. Genetics was also more complex than originally imagined. Genes were multifactorial, with dozens or hundreds influencing some character, but it was no less critical. The environment could influence but was not the prime determinant. With each fresh discovery the pendulum swung more and more towards genetics. It was not always the main concern but at least it was being viewed more sympathetically.

GENETIC TESTING If the first half of the twentieth century took many wrong genetic turnings, the second half provided knowledge which could have prevented many of those errors. However it is the twenty-first century which will witness a very different – some say better, some say worse – response to all this knowledge. At first glance genetic testing appears entirely beneficial. At second, or more prolonged sight, this form of progress is not without some disturbing drawbacks.

Tay–Sachs disease is an immediate example of benefit. Only very rarely are the genes which cause this most unwelcome form of death found in the general population. Jews from Eastern Europe are the exception because their ratio of the abnormal gene is 1 in 60. Therefore, as every individual possesses one gene from each parent, the ratio of carriers is 1 in 30. The chance of marrying someone who is also a carrier is therefore 29 to 1 against, but this can happen (arithmetically) in 1 out of every 900 potential pairings. The incidence is therefore small, and smaller than the incidence of Down's syndrome, for example, in the general population, but Tay–Sachs is particularly foul. Victims appear normal at birth, but tend to die within a few years, miserably, expensively. Hence enthusiasm for testing, for finding (at modest cost) those who are carriers, and for circumventing the lethal possibilities. It is very easy – for any outsider – to recommend that pairs of carriers should not mate, but the pairs themselves may not welcome the thought of alternative partners. Worse still, should they choose to go ahead with pairing, they also may not welcome the thought of abortion, should they subsequently be informed that the odds of three to one against have proved too short and a Tay–Sachs foetus is being incubated.

Fortunately there is now an alternative course of action: pre-implantation genetic testing. Renee and David Abshire, as detailed in *Scientific American* (June 1994), chose this new technique. They were both Tay–Sachs carriers (and knew it) but had proceeded with their partnership before the fates then turned against them. Their first daughter succumbed to the disease in 1989 and both her parents vowed not to have another child – unless it could be proved free of the disease. A pre-implantation checking procedure was therefore attractive. The couple donated sperm and eggs, learned that four embryos had been successfully created and then learned that one of the four was victim to the disease while the other three were not even carriers. As a result Brittany Nicole Abshire was born in January 1994, the first child ever to be certified free of Tay–Sachs even before implantation within its mother's womb.

Any feeling of smugness – among those who are not Ashkenazi Jews and are therefore even less likely to be a carrier of Tay–Sachs – must immediately be dispelled. 'We're all mutants,' as Michael M. Kaback, of the University of California, has proclaimed: 'Everybody is genetically defective.' Various estimates agree that each of us carries at least five to ten genes that could make either us ill or our children ill. Genes do not necessarily cause disease, but can cause a predisposition to disease, a proneness in some suitable circumstance. There may, for example, be a proneness to alcoholism which cannot – of course – manifest itself unless alcohol is available. There may be a resistance to malaria but this is of little worth unless mosquitoes and their parasitic trypanosomes are in the area. (Many a nineteenth-century visitor to West Africa died within days of stepping off the boat whereas many others lived full and normal lives.)

Genetic testing is still a fringe activity, with us mostly unaware of any deleterious genes we may be harbouring. Nevertheless it is the fastest growing area in medical diagnostics, according to the (US) Office of Technology Assessment. 'Potential new genetic tests roll off the conveyor belt ... almost once a week,' stated Norman Fost of the University of Wisconsin-Madison Medical School. The amniocentesis and chorionic villus sampling now conducted, almost as routine if thought advisable, is a form of genetic testing. Are there extra chromosomes (as with Down's)? What is the child's sex, if sex-linked disease is a possibility? And what other errors are detectable by having a look at the offspring's chromosomal possession instead of waiting for the offspring to be born? Children and adults can also be examined. Genes have already been discovered with links to Alzheimer's, to Huntington's disease (chorea), to colon and breast cancer. People can therefore be warned of their above-average susceptibility to particular afflictions, or even of the certainty.

Such testing can be either simple, as with counting the chromosome number, or extremely complex, as with searching for a distinct genetic sequence within someone's DNA. Consequently the tests are either cheap (perhaps $50) or expensive (twenty times as much). The convenient notion of one gene causing one disease, or a predisposition to that disease, is becoming increasingly confusing as more and more is learned. Perhaps only 3 per cent of all genetic disease is caused by a single gene defect. In any case the relevant genes may only add up to susceptibility (and therefore far less certainty). As an article in *Scientific American* (of June 1994) phrased it: if researchers find a gene 'that confers a 60 percent predisposition for gross obesity, is that a genetic defect? What about a gene that gives a 25 percent predisposition for cardiovascular disease at age 55?'

The situation has even become more complex in cases believed to have been clarified. Cystic fibrosis, one of the most common hereditary disorders, became more famous (and slightly less worrying) in 1989 when researchers in Canada found a responsible gene on chromosome No. 7. This was shown to affect cell membranes, and therefore tied in with the disease's disruption of, for example, the mucus glands of the lung's bronchi. Unfortunately (for simplicity) further work revealed that chromosome 7's error applied more distinctly to a particular kind of cystic fibrosis, often with only minor symptoms like asthma or bronchitis. Traditional cystic fibrosis kills its victims, some as infants, most as children or adolescents (although modern medication can prolong life to forty years or so). Therefore the single name of cystic fibrosis is now known to relate to several kinds of this disease. In fact almost 400 mutations (changes in the genetic material) have now been linked to cystic fibrosis, and only one is associated with most CF cases. Despite a single name it is no longer a single disease and is not caused by a single genetic aberration. Almost inevitably the more work that is done the less will it resemble the beautiful simplicity of Mendel's elementary laws. He never mentioned susceptibility, or susceptibility only in certain circumstances.

Therefore, to generalize, cause and effect are not necessarily straightforward. A might not always cause B, but will in certain circumstances, that is if it does not then cause C or D instead. Or, as Thomas H. Murray (of a centre for biomedical ethics) phrased it: 'Because cystic fibrosis, heart disease, cancer, autoimmune disorders, multiple sclerosis and other conditions arise from an unfortunate confluence of genetic and environmental factors, genetic tests for those illnesses can never by themselves predict an individual's future with perfect clarity.' This lack of simplicity, and the considerable possibility of diagnostic error, may even have advantages. A future of increased predictability is not to everyone's liking. It is like the disarming nightmare of learning one's actual date of death.

To exemplify this point take sickle-cell disease, and also take some unforeseen consequences of its diagnosis. The disease is particularly unpleasant, every symptom being caused by a defective form of haemoglobin. This leads to stiffer blood and therefore to clogged capillaries which, in their turn, cause long-term damage to various organs and then (in general) premature death, notably from pneumonia. Some forty years ago it was called a disease of childhood, mainly for the poor survival of those afflicted. Thanks to better treatment, principally with stronger antibiotics, the median age at death is now (according to a recent survey from Boston, Mass.) 42

years for males and 48 for females. Such relative longevity is likely to be improved still further, partly for a greater ability to detect when blood viscosity is deteriorating dangerously and also via drugs (such as hydroxyurea, first reported in 1995) which assist in the production of healthier red cells. However there is (as yet) no cure for this disease, it affecting all those individuals who receive the faulty genetic material of sickle-cell anaemia from both their parents.

There is no health problem in being a carrier of this genetically ordained disease (and in receiving the faulty gene from only one parent) but, during the early 1970s, a major screening programme was initiated so that carriers could learn of their possession. They could also learn if their possible partner was a carrier, and therefore whether any future children would stand a 1 in 4 chance of developing sickle-cell anaemia. So far, in a sense, so good; but then came consequences.

Informed carriers started to believe they were ill. Massachusetts ordered all children to be screened before entering school, a piece of legislation aggravating black–white relations because those principally at risk were Afro-Americans. Some insurance companies refused coverage to carriers. The US Air Force Academy rejected them. It was even proposed, on television, that black carriers should not breed, a suggestion smacking of genocide. The original screening policy for all likely to be at risk, actively promoted by church ministers and even members of the Black Panthers, suddenly seemed much less satisfactory.

Thalassaemia, another inherited disorder also relevant to malaria, is similar in its genetics but has not been involved with such worrying consequences. It is often thought of as a Mediterranean disease (being named after the Greek for sea) but it is also frequently encountered in a belt extending through the Middle East and the Indian sub-continent to Malaysia and Indonesia. Worldwide, according to the WHO recently, there are 'at least' 70 million carriers (largely in Pakistan), and 'at least 42,000 patients with thalassaemia major are born each year'. As with sickle-cell, it is a thoroughly unpleasant disease, its treatment demanding regular blood transfusions, spleen removal, and expensive drugs for withdrawing iron from the body. It is therefore far better if carriers of this defective gene do not mate with other carriers. In Sardinia, for example, the disorder has a carrier frequency of 12.6% among the general population (of 1.5 million). Consequently, if random mating were to proceed as in former times, one couple in every sixty would risk having a child with thalassaemia, this risk also being 25% (as with sickle-cell).

In 1977 a 'preventive genetic programme' was initiated by the WHO, its intent being to cut down on the number of new cases by detecting carriers and then by informing these prospective 'carrier' parents of the risks. The result has been a 'massive', a 'considerable', a 'major' reduction in thalassaemia, particularly in Greece, Italy and various Mediterranean islands. In Cyprus it has been virtually eliminated. Unfortunately, because of those 42,000 patients being born worldwide each year, the health burden is still enormous and this will increase with the rise in life expectancy, particularly – as is such a refrain – in developing countries.

*

Tay–Sachs work is being continued, and is generally considered to be advantageous, save that the harmful genes are spreading through the community. (Each death from this disease does at least take with it two of the harmful genes.) As for cystic fibrosis, its dilemmas are considerable. In families with a history of the disease the individuals are well worth screening to discover who is and who is not a carrier, and therefore to discourage potentially disturbing matings. Unfortunately four-fifths of all CF victims occur in families where there is no such history (or, rather, none that is known, many a family being ignorant of its less than immediate history). Therefore, as John Rennie wrote in *Scientific American*, 'nearly all couples in the US would need to be screened to prevent most cases', a formidable task in a country with some 4 million pregnant women a year.

There would also be the labour of counselling, of explaining about CF, and certainly of explaining the consequences of defective genes. 'Given that the number of professional genetic counsellors in the US is barely more than 1,000,' added Rennie, 'cystic fibrosis screening alone would swamp the country's counselling resources.' All the relevant information would have to be sought from personal physicians, not necessarily well informed themselves about the niceties of genetics (according to surveys of the medical profession). In an attempt to discover the benefit (or otherwise) of CF screening one major Wisconsin investigation found no evidence that 'identifying the children at birth was better than waiting for symptoms to emerge', a dampening (or edifying) conclusion. This revelation should discourage other states from making the screening mandatory, as has happened in Wyoming and Colorado.

*

Mothers (and fathers) inevitably worry about any abnormalities in their recently delivered offspring. They always (according to obstetricians) count the fingers and toes of the new arrivals, and certainly pay attention as doctors go through the more intensive examination of each newcomer. Genetic testing gives further opportunity for anxiety. Worse still, there can be error. One survey examined families in which children had been wrongly diagnosed as CF victims. Even after the errors had been corrected, one-fifth of the parents continued to believe their children were afflicted. It is often so easy to make genetic tests but generally so difficult to be aware of all the outcomes. Scientifically there is a love of gathering information for its own sake, and with little concern for consequences. As Norman Fost said: 'Currently there is very little to stop someone from implementing genetic tests on a population basis without the sort of institutional review and informed consent required for other new technologies.'

WHO SHOULD KNOW the results of genetic testing? A major poll in 1992 learned that most Americans thought test results should not be kept private. Almost all (98%) considered spouses should learn, and 70% that other close family members should be told. Information about some diseases should be disclosed (by physicians) even against a patient's wishes, thus breaking the normal doctor–patient relationship. A majority of those polled (58%) considered that insurance companies should be informed, with healthy individuals perhaps realizing their premiums might rise if there were not such disclosure (by the less healthy). As for employers, 33 per cent of the polled Americans considered these should also be informed, which presumably means that employers could be permitted to discriminate against bad-risk employees or individuals who might – otherwise – have been accepted as employees.

Currently DNA tests are not required by employers or insurance companies, partly because they are not yet good enough, not cheap enough, and not sufficiently all-embracing to be of interest. As for the future, it is certain, so say insurance associations, that companies will want to know as much as possible about policy-holders' prospects. The old simplicity of 'How old are you?' and 'What diseases do you have or have had?' will become (and, in some instances, has already become) 'How healthy are your parents? How old are they? If dead what did they die from? And what of your siblings, your grandparents, your great-grand-parents?' In general people have become accustomed to providing facts about relatives, but genetic information is in a different league. (There is more on this problem in the subsequent chapter on the genome.)

New Scientist called it a 'moral minefield'. A book reviewer within the *British Medical Journal* preferred that 'Genetic screening tends to serve as a flypaper on which our hovering fears of a Brave New World alight and stick.' Most certainly there is current dilemma, but there is lots more to come in the years ahead. Already genetic testing can show who is likely to develop Huntington's disease or (to some extent) cystic fibrosis. A particular form of breast cancer is next in line. Then will come a genetic susceptibility to colon cancer. And then, probing even further into the future, will come a predisposition to heart trouble, or other cancers, or – eventually – a certain kind of personality coupled with a particular level of intelligence. The minefield (or flypaper) is fraught with hazard, and seemingly endless in its possibilities.

If genetic testing becomes mandatory, permitting governments to know what problems are in store, do each nation's individuals *have* to learn of their particular misfortunes? And, if they do learn, can they then keep the facts to themselves (like keeping quiet about a

prison sentence despite the state's awareness of that fact)? Employers can employ whom they choose (provided unwarranted discrimination is kept at bay), but what about an employee's susceptibility to some relevant industrial chemical which may lead to a shorter life-span? And what about the life-span itself, either killing an employee early or causing an exceptionally long retirement with annual pension payments going on and on and on? Warranted discrimination is quite another can of worms.

As ever, human beings can behave differently from the fashion in which lawyers, ethicists, politicians and sociologists expect them to behave. A woman visited relevant authorities at Columbia University to ask if her two young children could be tested for Huntington's disease. She knew of the debilitating effects of this genetic inheritance, and knew these did not strike until middle age; so why did she require such advance warning for her two small offspring? Did she wish to apprise them early of a terrible fate or happily clear them of any such concern, with the tests giving bad news or good to one or other child? No, she only had enough money to send one of them to Harvard and wished to lavish all possible educational assistance upon the one more fortunate. (This tale does not relate if both were shown to be at fault – or neither.)

In October 1994 the governor of California signed into law a bill prohibiting health insurance companies from discrimination against anyone who has no symptoms of genetic disease. For this piece of legislation it defined the onset of disease as the onset of its symptoms. In consequence a genetic susceptibility is not a disease; nor is a proven forecast after genetic testing has demonstrated some unwelcome inheritance. Insurers and employers would not have access to the results of genetic tests, a piece of legislation thought to be the strongest such law in the United States, but such leadership has not been greatly followed. Other bodies, notably within the insurance industry, consider it fair to ask about greater risks. Are you young? Do you parachute, hang-glide, fly in balloons, go scuba diving? Are you overweight and do you smoke? Do you leave your car out at night, are your windows secure, do you wear glasses, do you have dizzy spells, what operations have you had? Such questions have long been tolerated by policy holders, with truthful answers being demanded by insurers. So what is new (or wrong) with the general query: Do you know of any reason why you might die early and be a greater-than-average risk?

In January 1996 the UK Health Departments set up a committee, said to be 'the first of its kind in the world', to ensure that genetic testing is not only provided safely but used ethically. This Advisory Committee on Genetic Testing is to investigate arrangements 'for the protection of personal genetic data' even though, as was immediately pointed out in numerous articles, all sorts of organizations already protect personal data. The police keep note of villainy, the banks keep note of wealth or poverty and doctors keep quiet about their patients – all in theory, but facts have a way of slipping past locked doors. On occasion they should slip past, if known bankrupts, psychopaths or fraudsters, for example, are likely to endanger others.

The committee does not foresee an easy time for itself. What if someone refuses to be tested? Is that a form of guilt? And might the famous Fifth Amendment to the US constitution become standard fare around the world: 'I refuse to answer that question on the grounds that my answer might incriminate me.' Both US houses of Congress passed health-insurance reform bills in the spring of 1996 which would explicitly bar insurance companies from basing coverage on genetic information, but the proposed alterations do not stop there (and therefore the bills may not become law). Also at issue is whether the DNA registry held by the military, helping it to identify (bits of) those killed in combat, should be suitably deleted whenever anyone leaves the service. The possibility of such a registry for all of us therefore comes to mind. It would create an undoubted loss of liberty but, in days when bombs are often aimed at civilian targets, it could be useful for authorities (and others) to know precisely who had become unrecognizable, save for their DNA.

It is easy to believe that future questioning will become more invasive. 'Has your DNA been tested? What did it reveal – about disease, longevity, health in general? What about your parents' DNA?' Insurance companies are in the business of making money. They do not like high-risk individuals (save by charging them higher fees) and do welcome all relevant information. But, as the biomedical ethicist Thomas Murray, already quoted for his concern about the less fortunate, chose to add: 'No one can control his or her genetic make-up, [therefore] it seems wrong to penalize individuals for it.'

The *Scientific American*, also already quoted on this topic, received replies from readers (following its article) which emphasized a strong dichotomy of opinion on genetic testing. In favour, for example, was C. Owen Paepke (of Arizona) objecting to Rennie's statement that 'ethical problems surrounding testing (were) as ominous as the diseases themselves'. This, wrote Paepke, equates potential discrimination by insurers or employers with the slow, painful death of Tay–Sachs. 'Parents hope for children who are healthy and smart ... The emerging ability of genetic testing to fulfill this hope does not render it malign ... Within 10–12 years, couples will test embryos before implantation with as few qualms as they now vaccinate their children. Articles that agonize over whether we should or shouldn't serve no purpose when we so obviously will.'

Less enthusiastic, and near the spectrum's other end, was Hans S. Goerl (of Maryland). 'Unless society responds swiftly to the current and future misuse of genetic information by insurers, employers, governments and other institutions, millions of healthy Americans will be deprived of access to insurance, credit (especially mortgages) and career-track employment. They will become members of a new genetic underclass ... Progress in genetics will eventually lead to wonderful treatments; in the meantime it threatens everyone's civil rights. Our response must not be to slow the science but to strengthen our commitment to true equality.'

*

The discovery in 1987 of the gene for Duchenne muscular dystrophy led to similarly divergent opinion. Various health authorities in the US chose to test every newborn boy for the DMD gene, the causing agent of a muscle wasting disease which kills most affected males in their early adulthood. Testing was largely carried out without parental consent. (It is easy to comprehend that the counselling of parents and the request for their acceptance might be a time-consuming irritant, particularly in a country producing over 500 babies an hour.) The Institute of Medicine, part of the National Academy of Sciences, gathered experts in 1993 to review the matter. In essence they considered the mandatory screening for DMD misguided. A professor of paediatrics went much further, arguing that such testing should not be carried out even *with* parental consent. Parents forewarned of a child's certain fate 'might develop a different outlook' on that child. Indeed they might. Would they bother about his education? Would they withhold love, if possible, or provide even greater warmth? And what of the child himself; how would he respond if made aware of his condition?

The fact that genetic testing is arriving piecemeal – first for chromosomes, then for single genes, and later for more complex causation – is at least enabling humanity to appreciate the issues involved. Satisfactory conclusions have not been reached – far from it, but at least the various issues are being confronted. In time, in very little time, the possibilities of genetic testing will become a deluge rather than the current trickle. Investigation of the human genome (see the next chapter on this subject) will unleash a torrent, of potentiality, of ethical upset, of consequences for a brave new world unlike anything already seen. It is therefore good, like individuals downstream making preparations in case the dam should break, that virtues and drawbacks of genetic manipulation are actively debated. The onslaught of possibilities might then be less traumatic. It might even be of benefit to the great majority.

*

British babies are all screened for two genetic disorders, phenylketon-uria (PKU) and congenital hypothyroidism. Mothers tend not to be informed as medical staff take a drop of blood or a urine sample, but controversy has not arisen because the testing is wholly beneficial. If PKU, for example, is not detected soon after birth, a form of mental deficiency is inevitable. If detected such a consequence can be prevented by special feeding (with protein foods from which the amino-acid phenylalanine is absent). The choice is therefore between an idiotic child and a normal one, with dietary inconvenience (frequently described as 'hellish') to effect the change. Babies with a likelihood of sickle-cell anaemia and thalassaemia, both already mentioned, are also screened, notably if their parents are, respectively, of Afro-Caribbean or Greek-Cypriot origin. The disorders of haemo-globin which are caused, if defective genes are inherited from both parents, can be better treated if detected early. Once again the consequences of such screening should be entirely beneficial, provided insurers, employers and others (of an inquisitive kind) are kept at bay.

The gene for cystic fibrosis is not routinely sought, even though large numbers of the British population (of European stock) carry it. One in 2,000 children born to white parents has the disease, but it is much less prevalent in ethnic communities. Some screening pro-grammes have been begun, such as 'couple screening' which investi-gates both partners. The twosomes are only informed of their possession if each of them carries the gene – as they then suffer a 1 in 4 chance of their children succumbing to CF. Relations of CF victims are sometimes sought and tested, but the procedure (involving saliva) only picks up 85–90% of carriers. Therefore a number of people allegedly free of the gene (10–15%) are not so fortunate, and are then perhaps worse off than those never tested.

Motivation behind testing is not necessarily as humane as might at first appear. With any health service the relatively small cost of screening may be set against the huge cost of illness. If the act is financially beneficial, say the proponents, it should be performed. On the other hand 'we are in danger of slipping into a eugenic system by default, through the impersonal amoral operation of a money-pinching bureaucracy', argued Angus Clarke, clinical geneticist at Cardiff. A hugh DMD screening programme in Wales, which tested 95% of all Welsh male births (5% of parents refused), was not prompted – it is claimed – on the basis of any cost-benefit analysis. The testing was followed by considerable support for all those parents whose children were found to be affected. In assessing the Welsh endeavour an American pathology professor stressed (in his variant of a common tongue) that doctors should 'pay particular attention to the careful, extensive community outreach organization, which undoubtedly led to the minimum emotional impact of the programme'.

Follow-up is crucial. Not everyone understands the implications of being a carrier. And not everyone is versed in risk, probability, statistical inference. 'Only a scientific people can survive in a scientific future,' wrote Aldous Huxley, and no one believes we are yet a scientific people. H. G. Wells made a similar point: 'Statistical thinking will one day be as necessary for efficient citizenship as the ability to read and write.' For most of us there has to be interpretation, a time-consuming procedure out of proportion to the few minutes necessary for taking blood and checking its credentials.

People do not behave in straightforward or expected fashion, as when told: 'You do not have the disease, and are not even a carrier,' leading to 'Thank goodness for that!' A Dutch team encountered 'long-term emotional numbing' and 'survivor guilt' among a group at risk from Huntington's after being informed they did not have the gene. On occasion relatives with the gene 'banned' the gene-free and 'privileged' individuals from the family because the 'tie' which formerly united them 'had been severed'. Sometimes the relief at being gene-free was transformed into obligation 'to be continuously available in order to support affected or at-risk relatives'.

Huntington's is a rare disorder, but there are hundreds of rare disorders. The number of distinguishable diseases known to be genetically determined was reckoned to be 1,500 even twenty years ago. In one North American paediatric hospital at that time 30% of its admissions 'were due to genetic disorder'. In Britain 40% of all child deaths are related to genetic disease. Some 70% of childhood deafness and 80% of blindness (in non-tropical countries) is caused genetically. According to the National Institutes of Health 15 million Americans are thought to be affected by genetic disease. Or, as Robert J. Pokorski wrote in *Nature* (July 1995), 'A list of common diseases with a genetic component now includes virtually every condition encountered in an average medical practice.'

The analogy of being downstream of a dam, and worrying if it might break, is not too far off the mark. Genetic testing could become a deluge, very easily. Conversely it might remain a trickle, with people reluctant (and resistant) to know what fate has handed them. To repeat Michael Kaback's words: 'Everybody is genetically defective,' not necessarily on the surface but in the genes which form part of all inheritance. We do not wish to know when or how we will die, as many surveys testify, and we may also not wish to know some details, such as our proneness to a certain kind of cancer, our likelihood of heart malfunction at age 55, or what we have unwittingly donated to our children, particularly if our chosen mate made similar donation, thus guaranteeing some gross error in their inheritance. We might refuse testing for our sake, but what about their sake? And will each nation tolerate personal wilfulness, knowing – as it might – that testing is a lesser drain upon its wealth than the full flowering of some genetic disorder.

*

The DNA Mystique, written by Dorothy Nelkin and Susan Lindee (published in 1995), addressed the personal impact of genetic testing. Clinicians tend to view this form of advance in medical terms, thinking of Parkinson's, of DMD, cystic fibrosis and the like, but their patients take a broader view. They are more interested in behavioural traits, a topic which has surfaced repeatedly in the last decade. Nelkin and Lindee have compiled a list of traits allegedly attributed to heredity: 'Mental illness, aggression, homosexuality, exhibitionism, dyslexia, addiction, job and educational success, arson, tendency to tease, propensity for risk taking, timidity, social potency, tendency to giggle or to use hurtful words, traditionalism, and zest for life.' Such ideas have already, as the authors point out, entered popular culture in novels, movies, soap operas, and advertisements.

To some degree this overt reaction to genetic testing is similar to the misguided belief in genetics during the earliest years of the twentieth century. No sooner had Mendel's laws been discovered than they were, in many minds, applied wholesale to drunkenness, immorality, bad breeding and general degradation. Eugenics lay along that path. Now, with genetic testing already in the news (and much more of it fluttering in the wings), it is also being applied to the wide range of human activity. Eugenics will not follow from this enthusiasm, but a form of opting out may well do so. As Nelkin and Lindee explained, people 'can relieve personal guilt by implying compulsion, an inborn inability to resist specific behaviour'. 'It wasn't my fault,' as every child says (so repeatedly), 'it was an accident.' Or, as Byron opined in *To a Youthful Friend*, 'The fault was Nature's fault not thine, Which made thee fickle as thou art.' There is comfort to this thesis, and it may well prove popular, however impossible it may also be to disentangle the genetic involvement in such features as risk taking, timidity or educational success.

*

Most certainly there has been newspaper coverage of the new genetics. A flick through current cuttings shows the range of interest, of future possibilities, of present understanding. A bishop has blamed God for giving us 'our promiscuous genes'. A gene causing 'werewolf syndrome' in a large Mexican family has been pinpointed to a short section of the X-chromosome. On 7 October 1994 'a beautiful article by Mark Skolnick and 44 colleagues [described] the long-awaited discovery of the gene for hereditary breast and ovarian cancer, BRCA1'. 'A single "letter" out of place in the genetic code of chromosome 4 causes the most common form of dwarfism in humans.' 'Researchers have identified a gene that causes the commoner form of diabetes, suffered by at least 500,000 people in Britain.' A large study of twins 'suggests that genes contribute about half the risk that women have of becoming alcoholic'. 'That a group of errant genes plays a part in determining the susceptibility of large numbers of people to developing psychiatric disorders is not in doubt...' 'Identical twins show a correlation of about 0.88 in intelligence tests when they grow up together, and only slightly less when they grow up apart.' 'Genes and hormones, and not how much you eat, make you fat', according to scientists at New York's Rockefeller University. 'The conventional wisdom that myopia is purely hereditary is presently being challenged...' The results of another twin study 'suggest that about 38 per cent of irritability could be put down to genetic make-up'.

It is easy to wonder, following such a mish-mash, if there is any attribute of humanity which has not come under the genetic spotlight. It is also hard to realize that, until the 1970s, there was virtually nowhere in Britain that could offer training in medical genetics. An implication of all this publicity, and of the scientific effort promoting it, is that our lives are to be radically altered, if not overnight, very speedily. There will be change, but not necessarily in the areas of greatest interest.

As a dampener on so much effervescence some quotations – on mental ability – may serve to quell expectation. 'Ransacking the human genome in search of genes that determine intelligence is a fool's errand,' said a geneticist at America's greatest scientific gathering in 1995. 'The influence of any single gene is most certainly small,' said another. One more delegate, adding a further douche, considered 'people don't know what intelligence is any more than they did in the 1930s; so how can you talk about how it is inherited?' With so many problems to confront, said the 'fool's errand' individual, proving how genes affect intelligence 'would be a squandering of taxpayers' money, possibly as much effort as putting a man on the moon'. The question of intelligence is assuredly of major interest, far more than some rare disease or errors like criminality, but an inconvenient law is taking hold: the more exciting/intriguing/fascinating a topic the harder it is to unravel. The convenient and much-used term – multifactorial – means many genes, many environmental influences, many complex interactions, with all of them leading to some complex end – such as intelligence or personality.

*

There are two sides, as someone said, to every triangle. There is value to any community in testing for unwelcome genes if this prevents costly treatment. There is less value to the individual if, being forewarned, this means knowing of unpleasantness in store. Insurance companies are being informed they should not penalize individuals with genetic defect. Some individuals are already purchasing more profitable insurance having learned of their genetic errors. US employers who discriminate against the genetically unfortunate are now breaking the law (following pressure from the Equal Employment Opportunity Commission). As employers have to reject some applicants they may be forced (in order to escape the label of discrimination) to refuse only those without any known genetic misfortune. Some genes undoubtedly influence behaviour (with Down's patients generally and abnormally good-natured). Lawyers for Stephen Mobley, condemned for murder (of a pizza parlour cashier), argued in Georgia that he and his family possessed genes making them abnormally aggressive.

Such a plethora of embranglement can make us relish former simplicity, as with early death certificates: 'He got sick and died'; 'Died without the aid of a physician'; 'Had never been fatally ill before'. The Georgia supreme court, in the case of Stephen Mobley, ruled that tests to prove he shared genes with his family for abnormal aggression could not be allowed. Sociologists argue that environment is often to blame for unwelcome behaviour, such as the influence of drunken, neglectful, violent parents, such as a local community steeped in crime, such as poverty, poor housing, ill health. The sociologists may be right and the geneticists may be right, but where does that leave the guilty individual? With both aspects influential, and responsible for development, each human product is no more than the result of their power. They have ordained. They have made all of us, our form and stature, our proneness to disease, our susceptibility. If we are a natural born killer (this term surfacing in 1995) so be it, along with natural born haemophilia, natural born ability and also disability. Aldous Huxley's brave new products, from the alphas to the epsilons, were pre-conditioned (and his readers were appalled by such a circumstance). Today we are also, and increasingly, disturbed by our origin, whether inherited or imposed. There are genes and there is the world in which they operate. There is nature and there is nurture. Together they lead to us.

What they will lead to in the next century is difficult to contemplate. As Neils Bohr said, comfortingly: 'Prediction is very difficult, especially about the future.' On therapy, and future practice, one recent cartoon showed a room full of the geriatric, surrounded by ear-trumpets, Zimmers and wheelchairs, in which one ancient was yelling to another: 'If we hadn't listened to that doctor, we wouldn't be seeing all of this.' Without doubt many new forms of therapy prolonging life will be based on genetic principles.

GENOME

DNA   GENETIC CODE   GENES   CLONING DNA   GENOME

> We hold these truths to be self-evident, that all men
> are created equal.
>
> Thomas Jefferson

DNA It was 130 years ago when Gregor Mendel, cultivating peas and roses by the abbey where he worked in Brno, also sowed the seeds of genetics. Inheritance was thereby shown not to be a blending, like mixing two colours which cannot then be separated, but a more precise arrangement whereby distinct units remained distinct. His revolutionary work was neglected until the opening year of the twentieth century when three scientists chanced upon it, separately and in different countries but more or less simultaneously. (Charles Darwin, major exponent of evolution via natural selection, was ignorant of Mendel, having died eighteen years before that three-pronged rediscovery.) Genes and genetics were suddenly all-important, this study not even having a name until William Bateson coined the term some ninety years ago. But what precisely was the gene, thus 'unit of particulate inheritance' (as it was also known)?

Mendel was still alive when the other crucial ingredient of genetics had been discovered in 1871, the substance named deoxyribonucleic acid, but this too had to linger before being properly appreciated. Friedrich Miescher, a Swiss living in Germany, was the discoverer, and knew only that it was a major component of cell nuclei. The lengthy name, conveniently compressed into the three initials of DNA, can be broken into deoxy (short of an oxygen atom), ribo (short for ribose, a sugar, although that oxygen lack means the sugar is no longer ribose) and nucleic acid. (For readers whose eyes are glazing at this paragraph a leap over the next one may be advantageous.)

There are two forms of nucleic acid in cells: DNA, most of which exists in the chromosomes, and RNA (ribonucleic acid) which exists primarily in the cytoplasm surrounding the nucleus. DNA and RNA differ only in their sugars, but considerably in their functions. As for the nucleic acids, these consist of long chains of nucleotides. And as for nucleotides they each possess one molecule of phosphoric acid, one of a five-carbon sugar (pentose) and one of either a pyrimidine or a purine. These pyrimidines are either thymine or cytosine, and the purines are either adenine or guanine. Once again (as with DNA itself) there have been convenient abbreviations, these four bases being shortened to T, C, A, and G. Of course the actual chemistry is critical but, for those of us enthusiastic about short cuts, it is sufficient to say that DNA and RNA each contain T, C, A and G (and these four will surface most emphatically in the section on coding).

Nothing much immediately happened about DNA after Miescher made his discovery. Not everyone believed it might be the carrier of genetic information, and all sorts of proteins were being proposed for that role, but the big leap forward occurred in 1943 – therefore seventy-two years later. Oswald Avery, a Canadian, took some DNA from dead bacteria and injected it into living bacteria of a different kind. To everyone's amazement, including Avery's, the living bacteria acquired some of the properties of those bacteria from which the DNA had been taken.

The next huge leap occurred in 1953. By then it was known that the complex molecule of DNA not only carried the genetic information to make a new individual, whether worm, snail, fish, bird or man, but could also divide itself into two identical parts (when each cell divided) and then recreate itself as it had been before the division. Linus Pauling, brilliant chemist (and eventual winner of *two* Nobel prizes), postulated that DNA was a spiral structure, much like a rope with three entwining strands. He was right about the spiral but wrong about the strand number. Had he been a biologist, it has since been argued, he might have been more immured in the doubleness of everything – two arms, legs, kidneys, breasts, testes, cerebral hemispheres, nostrils, lungs, etc. – and he might therefore have thought of two strands rather than three. Biologists live in a world of doubles (and are hard pushed to think of threesomes in their experience).

Reading more like a detective story (with love interest) than a piece of science, James Watson's *The Double Helix* (1963) is a thrilling account of the race between Pauling (in the United States) and Watson and Crick, then working at Cambridge, England. James Watson, 24, and Francis Crick, 36, together dreamed up a double arrangement for the structure of DNA that satisfied all requirements, how the four bases, the sugars and phosphates could not only be united (according to chemical laws) but fit the picture of DNA obtained through the novel technique of crystallography. The two men published a 900-word article in *Nature* which will be read and re-read for all time, it describing the most significant milepost of them all in molecular biology.

The leisurely progress which had occurred between DNA's discovery in 1871, and the postulation of its structure after a gap of eighty-two years, then dramatically changed gear. In 1956, just three years later, certain experiments (by Arthur Kornberg, American biochemist) supported the hypothesis that DNA conveyed its information via those four bases. Then, two years later, came proof (as foreseen by Watson and Crick) that replication involved the separation of both strands of their double helix. (Their *Nature* article had ended with a splendid, and even cocky forecast: 'It has not escaped our notice that the specific pairing we have postulated suggests a possible copying mechanism for the genetic material.')

GENETIC CODE   Then came the problem of information transference: how could that DNA and those bases – adenine, guanine, thymine, cytosine – organize the proteins which create a living organism? It was all very well knowing there was a blueprint, and that it could replicate itself, but how could its instructions build the mass of a living being? Proteins are composed of amino-acids, often called the bricks of the building material, and protein manufacture needs the correct assembly – somehow – of those amino-acids. There are twenty of them which, when suitably arranged, make up the countless thousands of proteins. Therefore DNA has to order them precisely to create each particular protein; but how? How could the four bases possibly give such exact instructions, bearing in mind that an average protein has 150 amino-acids joined together and some proteins are far, far bigger?

If there were only four kinds of amino-acid it would be simplicity itself for the bases to be arranged in the appropriate order on the DNA stairway, but there are twenty rather than four. So what about pairs of bases arranging the twenty amino-acids? Unfortunately there are only sixteen ways in which four such bases can be organized, namely (if thought of as A, B, C, and D): AA, AB, AC, AD, BA, BB, BC, BD, CA, CB, CC, CD, DA, DB, DC, and DD. This range of possibilities is therefore too small to select properly and correctly from that assortment of twenty amino-acids. However, if the bases are arranged in threesomes, there are sixty-four ways in which they can be ordered, namely AAA, AAB, AAC, AAD, ABA, ACA, ADA and so on to DDD. In short, more than enough to select from the twenty amino-acids, and to have some left over to make for punctuation, controlling when one protein is completed, another is starting, and so forth.

The triplet idea was first proposed at the start of the 1960s. In 1961 the first association was made between one particular triplet and one particular amino-acid, and by 1965 the code for every amino-acid had been learned. (So, also, had the fact that the same code is used in the creation of all living things, be they man, mouse, or buttercup. Nature is nothing if not economical with good ideas, apparently able to appreciate that 'If they work, don't fix them' which wise humans also learn.) Within a very short time it was even being postulated that 'about' 100,000 genes existed within each human cell. Individuals with telescopic vision peering into the future were proclaiming that science might, one day, know the location of each one. Others, with feet more squarely on the ground, scoffed at this idea.

The scoffers did seem to have some wisdom on their side, with all new information merely adding to the vastness of the task. It was, for instance, being estimated that 3,000 million base-pairs – every base upon the stairway being arranged in partnership with another – existed within the twenty-four different human chromosomes. (Humans *do* have forty-six chromosomes, having received twenty-three from each parent, but the two sex chromosomes, the X and the Y, are markedly different from each other whereas the twenty-two other components of each pair are not. Hence that total of twenty-four.) It was also being learned, which did not make for immediate simplicity, that the disease of cystic fibrosis could be caused by any one of very many mutations. Similarly, leading to more bewilderment, the error of sickle-cell anaemia resulted from the switching of one thymine molecule for one adenine molecule (much like altering a garden catastrophically by removing one grain of its earth). For each advance in genetics more and more problems seemed to be revealed, and no one – back in the 1960s when the coding was unravelled – realized (out loud) how much more would be achieved by the 1990s; but first a word about comparisons.

GENES Since Mendel's day, as theories of evolution have progressed, it has steadily been appreciated that humans and other primates have common ancestors and, if regarded further back, *a* common ancestor. Even without this acceptance of evolution it is obvious that the apes, in particular, have much in common with *H. sapiens*, not just in four limbs, four eyes, two ears and so forth but in their faces, in their teeth and certainly in their brains. Nevertheless many individuals expressed surprise when told that chimpanzees and humans have 98% (or so) of their genes in common.

Perhaps they should be more surprised that the chromosomes of any human become increasingly different as that person ages, possibly (it is said) by some 500 mutations per chromosome, or one base in every half million. If two unrelated human beings are examined (unrelated in that they know of no common relation) they are found to be different in about one in every 1,000 bases. As for chimpanzees, related but far more distantly, they differ by about one in every 100 bases, or ten times that average difference between two human beings. One result of such estimates is that better guesses can be made about the time, perhaps 5 million years ago, when humans and chimpanzees parted company to go their separate ways.

The actual chromosomes, such as the largest (No. 1) in both apes and humans, look extremely similar even though, as Christopher Wills has written (in *Exons, Introns and Talking Genes*), about 3 million differences have accumulated on this pair of chromosomes alone. 'Most of these differences are not visible until one gets down to the level of DNA sequence.' That is why so much of genetics is so disarming to those of us who find it difficult switching from something obvious, like eye-colour, a nose's shape or even (with a microscope) chromosomes to the molecular level which lies at the root of everything. (I was once told – J. B. S. Haldane having done the calculation – that a human being is about half-sized, i.e. halfway between an atom and the universe. Both ends of this spectrum can leave most of us incapable of comprehension, but I found the statement helpful in making me understand – a little – about the astonishing smallness of things, as well as the astonishing bigness.)

It *is* difficult absorbing that the DNA in one human cell weighs 6 millionths of a millionth of a gram, or – if this is preferable – 0.000000000006 grams. As DNA is the entire system on which all those 3,000 million base-pairs are arranged this means adding another nine noughts to the quantity in the previous sentence in order to contemplate the weight of one molecule pair of adenine or guanine, thymine or cytosine. The fact, already mentioned, of one thymine being switched for one adenine to cause sickle-cell anaemia (and, quite possibly, death) can make us not only reach for the nearest hand-rail for support but wonder how on earth the system ever works as correctly as it does.

With 5 billion human inhabitants on this planet their combined weight of DNA, which organized every one of them (and made them all unique), only totals 0.03 grams in weight. As perhaps 20 billion people have lived since the time of Christ that means the controlling DNA for all of them still weighed far less than one gram. Hand-rails are crucial equipment, from time to time, when contemplating much of science, such as the minutiae of genetics.

CLONING DNA   In 1970 the first enzyme was isolated that could cut DNA molecules at specific sites. Two years later, not so much contrary as complementary, a different enzyme was used to join DNA fragments together (the result being known as recombinant DNA). One year later some 'foreign' DNA (from quite a different kind of organism) was inserted into bacteria, and the possibility of cloning therefore achieved prominence. Thenceforth any number of copies could be made of that foreign gene, the inserted piece of DNA. With bacteria reproducing every twenty minutes or so a swift production line is assured, with two becoming four becoming eight and so forth very rapidly.

What this (and later work) added up to, as scientists (and others) began to realize, was that molecular biologists had acquired the power to alter life on a scale, as Watson phrased it, 'never before thought possible by serious scientists'. The thought caused tremendous joy on one hand, and some anxiety on the other.

As for the joy, it was realized that chromosomes could be altered. Entire segments could be removed, transferred, adjusted. The genes within those segments could therefore be amended, destroyed or improved. With so many genes known to be harmful there suddenly seemed to be scope for unlimited good. Haemophilia is caused by a most unwelcome gene on the X chromosome; therefore amend it (and let males, who all inherit only one X chromosome and who suffer grievously should their single X be faulty, be rid of the disease). Hundreds of other diseases are known to be caused by erroneous genes. Adjustments of such genes could lead, in consequence, to a general improvement in human welfare.

Anxiety followed hard upon the heels of joy. Unlimited good, to quote again from Watson (and his co-author of *The DNA Story*, John Tooze), 'might simultaneously be setting the stage for discovering the power of unlimited bad'. There were frightening possibilities. 'Might some of the new genetic combinations that we would create in the test-tube rise up like the genie from Aladdin's lamp and multiply without control, eventually replacing pre-existing plants and animals, if not man himself?' As one example of a natural disaster there was the very lethal flu epidemic of 1918. Humanity's tinkering with DNA 'might have consequences orders of magnitude worse' than nature had ever been able to contrive. Scientists were preparing to alter

DNA, but none had concrete facts 'by which to gauge', as Watson and Tooze asserted, 'these scenarios of possible doom'.

If the scientists could not foresee possible outcomes how were other people to assess what might be waiting in the wings? Science for the People, a protest organization, considered that everyone should join in the debate and have a say; but molecular biologists did not agree. Watson and Tooze concluded: 'We did not want our experiments to be blocked by overconfident lawyers, much less by self-appointed bioethicists with no inherent knowledge of, or interest in, our work. Their decisions could only be arbitrary. Given that there were no definite facts on which to base "danger" signals, we might find ourselves at the mercy of Luddites who do not want to take the chance of any form of change . . .'

It was the scientists themselves who proposed, in July 1974, a moratorium on their recombinant work. Inevitably this stoppage raised curiosity among others. What, precisely, was going on? There was some talk in laboratories of advancement within guidelines but there was rather more talk outside them concerning the need, within a democratic society, for matters of such potential consequence to be more generally discussed. This, in fact, did happen and, as Watson and Tooze declared, 'with such a vengeance that it has left many of us temporarily in the state of shock that accompanies not knowing whether it is oneself who is stark staring mad or whether the problem is only that all others have gone out of their minds ... The recombinant DNA debate has taken its place as a major event in defining the never cosy relationship between the practitioners of science and the populace around them.' In stating their own position the two authors 'personally believe that recombinant DNA research is best left virtually unregulated'. Nevertheless they accept 'that others saw – and may still see – the matter as less clearcut'.

Science is not good at any form of moratorium. Various individuals, troubled about nuclear fission, have suggested that scientists tear up (or somehow bury) the information which led to the splitting of the atom. Scientists (and others) have made a mockery of such sentiments, however well intentioned, but there was a difference with molecular biology. A moratorium was imposed, most effectively, and

DNA cloning work was stopped. Then, after 'crippled' bacteria were produced that could be used for cloning without the risk of infection, the moratorium was lifted. Research then proceeded faster than ever before, the temporary hiccup having been left behind. With all the wisdom inspired by 20/20 hindsight it is hard to see what else could possibly have occurred. Of course human beings would be fascinated by themselves in particular, and by their genetic make-up, but few could have foreseen the swiftness with which our entire genetic blueprint was to be laid bare.

GENOME The Human Genome Project, a most audacious endeavour from any observer's point of view, was initiated in 1990 by the US government (via the National Institutes of Health and the Department of Energy). It was subsequently joined by the UK, France, Germany and Japan. As for the cost, when the (possibly) fifteen years of work are finally ended, that may be $3 billion. The incredible intent is to sequence the entire 3 billion base-pairs of DNA which exist, as mentioned earlier, on the twenty-four human chromosomes. (At $1.00 per base-pair the price can suddenly seem cheap, particularly when an alteration to one base-pair can cause some dread disease.)

The first five years were concentrated on genetic and physical mapping, with the genetic map a representation of the order of genes along each chromosome and the physical map (as those in charge of the project have explained) 'a collection of identifiable overlapping fragments of DNA'. All sorts of other work has shown which bits of DNA tend to stay as bits and which genes are on which chromosomes. Therefore, as with someone knowing certain streets and certain people who live in them, map making might as well begin with what is known.

The second phase has now begun (well ahead of schedule, trumpet the enthusiasts). This will create the ultimate map, the ordered sequence of all bases which describe the genes and the proteins that they make. Remember that each protein may contain hundreds of amino-acids. Remember also the tremendous quantity of proteins performing their separate roles. The task of detailing this entire blueprint is therefore mind-numbing but will be completed before very long.

On 28 September 1995 *Nature* published some 350 pages (as a massive supplement to its regular weekly issue) entitled *The Genome Directory*. This contained a description of 88,000 'expressed sequence tags'. For those of us not in the know the directory does not make exciting reading, being mainly numbers and letters, and those sequence tags are not immediately revealing about their importance or interest. Similarly the proteins they create – adrenodoxin reductase, and the like – are also for a specialist audience, but the breadth of information is fascinating and well on the way to being completed. It *is* extraordinary how we were each manufactured from the sperm and egg that gave us birth. It is almost as remarkable that we, the end product, have managed to divine the blueprint which causes everything to happen.

(Christopher Wills, also attempting to come to terms with the quantity of information being unearthed by the genome teams (genome meaning the set of chromosomes within a cell), examined his *Webster's Third New International Dictionary*. This has sixty characters to a line, 150 lines to a column, three columns to a page, and 2,600 pages. Therefore it contains 70 million characters. As the genome contains those 3 *billion* bases this suggests that forty-three volumes of Webster would be necessary to include its data. Or, instead of all those tomes, just 6 millionths of a millionth of a gram of DNA.)

*

The UK's Advisory Committee on Human Genome Research has predicted that 'the increased knowledge of basic biology will eventually transform medical practice'. John Timson, writing in the *Galton Institute Newsletter*, was a touch more cautious: 'Optimists believe ... the data will lead to successful treatments for at least some genetic disorders. If this happens there would be a revolution in medical care perhaps equal to the introduction of antibiotics ...' Considerably more cautious has been Steven Rose, professor of biology at the UK's

Open University, who believes the advocates of human genome therapy have been 'guilty of extraordinary hype ... The research is of course valuable, but it is not the answer to the whole human condition. Let's face it, most deaths are not tied up to specific genetically related conditions.' As for doctors in general, and their understanding and acceptance of new possibilities, even the latest generation coming out of medical schools have been called 'genetically illiterate'. Pat Jacobs, of the Wessex Regional Genetics Laboratory, added that there was an urgent need to boost the training of all medical students.

*

During the late 1990s, with the approach of the next millennium, it is not so much the therapy possibilities which are exercising most minds in the business but the ownership of all this information. Is it 'the common heritage of mankind', as UNESCO phrased it in its World Science Report of 1996? Or, as Human Genome Sciences (of Rockville, Maryland) has stated, is it to be sold if there is money to be made from it? This organization, HGS, possessing details of DNA sequences that could identify 35,000 human genes, has spent a lot of money (mainly from SmithKline Beecham) in discovering this information. Craig Venter, key man at HGS, has said that any researcher who wants to use the information for academic reasons can do so – without cost – but if a researcher then learns something which could be commercialized, such as a disease treatment, HGS wants to be able to negotiate a contract. This would be like the patenting of a new drug. A pharmaceutical company invests its millions and then, after marketing the novel remedy, re-acquires its millions (or even, say some, its many more millions).

In its leading article (of 1 February 1996) *Nature* stressed that 'few implications of modern genetics generate as much heated debate as its consequences for medical and life insurance'. This is a separate issue but also involves ownership. Who owns the most fundamental information about any individual? The insurance industry has already been mentioned, but the genome's unravelling will involve it even more. Without doubt someone's genetic profile can provide important clues about that person's future health and life expectancy. It has already done so (a little), but will do so increasingly as that profile becomes better understood. Even without any genetic involvement certain people have been refused insurance, after their lifestyle, financial acumen and general reliability has proved faulty. Will others, who are genetically at fault, also be refused insurance, their prospects being poor, their genes erroneous? Smoking, a form of life-style not welcomed by insurers, certainly shortens life (in general, if not for everyone), but is that tendency to smoke a predisposition, a genetic fault and therefore as blameless in its way as cystic fibrosis, sickle-cell or other genetic error? To what extent is each of us a product of our genes? And will such a question ever be answerable in full about every aspect of our lives, about all our illnesses and the reasons why we die?

Insurance companies, putting forward their point of view, can sound most honourable. They want all relevant information, partly to be equitable. If defrauded by high-risk people (who know of their higher risk) the low-risk individuals will have to pay for that deceit by having their premiums raised. Paul Brett, an actuary based in Cologne, has said that, if applicants know their risks and insurers do not, 'this can lead to the breakdown of the entire insurance market'. Genetic testing (say the insurers soothingly) could even benefit those who, at present, have extreme difficulty in acquiring insurance cover, such as the unfortunates known to be at risk from Huntington's disease because a parent has died from it. Testing would either clear such individuals completely (permitting them to buy insurance in standard fashion) or, less satisfactorily, make absolutely certain of their difficulty in obtaining cover. The argument has a double-or-quits flavour to it, but is wholly beneficial – either way – to the insurance companies.

Current legislation on this matter is varied. In the UK, Germany, Japan, Italy, Spain and Portugal there are (as of early 1996) no laws governing the use of genetic tests by insurance companies. Currently the Association of British Insurers does not make genetic testing a precondition of issuing policies but 'reserves the right' to have access to tests. In Germany health insurance companies (not 'life' companies) can ask to see any genetic test results for policies exceeding

DM 250,000. France and the Netherlands have each imposed a moratorium, a sort of 'don't know' until the subect is better understood. (The Netherlands is expected to compromise, to permit insurance companies to ask for test results but not to be able to demand them.) Belgium, Norway and Austria have banned both notions.

In the US the states have voted differently. Ten of them 'prohibit' genetic testing and the use of genetic information for health insurance purposes, but impose no restrictions on 'life insurers' requesting genetic tests. Members of the American Council of Life Insurance do not demand tests – yet. As life insurance is based on predictions of survival, and as genetic tests should be able to improve such estimates, they will provide 'more accurate calculations' of insurance premiums, said a spokesman for ACLI. It is difficult expecting insurance companies to refrain – for long – from wishing to know more (and more), particularly when genetic testing becomes more commonplace.

In part genetic testing is already here. In another and greater part much of it may never be simple, will always be costly and may never be satisfactory. Huntington's disease is at the simple end. This incurable assault affects 1 in 20,000 Europeans (and European-Americans) in mid-life. It has been identified with a single mutation on a single gene on a particular chromosome, and individuals certain to develop the disease can be accurately pinpointed. Cystic fibrosis is also caused genetically (and affects 1 in 2,500 of European stock) but its cause is far from simple. As very many mutations can give rise to it there is no such certainty as with Huntington's. Such rare diseases, however lethal and sometimes expensive, are of lesser concern to insurance companies. The organizations are much more curious about commoner errors, such as cancers and cardiovascular disease, but the genetic involvement of these is far from simple.

According to a *Nature* 'Briefing' on this subject mutations in five separate genes are known to increase susceptibility to colon cancer, and two genes are also known to predispose to breast cancer. An absence of these mutations does not mean that either disease will not develop and, with breast cancer, only about 5% of its cases are now known to be hereditary. With one-quarter of men and women under 65 (in the US) and one-half of those over 65 dying from diseases of the cardiovascular system any genetic information about these ailments would be far more interesting to insurers (and lots of others) than the disease rarities. A few mutations have been found guilty but most kinds of heart and blood vessel disease are a combination of genetic susceptibility and life-style. Alcohol, tobacco, exercise, diet, stress – is there anything which does not modulate the risk of developing a circulatory disease? Besides, most of our afflictions are polygenic and multifactorial; many genes are involved (out of our total of 100,000) and many environmental factors (out of the infinite range of possibilities). The discovery of what causes what in whom is never going to be plain sailing, even when the genome has been thoroughly dissected.

As for the patenting of genetic information, and who owns what and why and how, that form of sailing is already on stormy water. Several laboratories are keeping their genetic sequences as trade secrets. That is not unexpected; it is what they do with them that is the issue. Moreover the discovery of a sequence does not immediately (if ever) lead to commercial gain. Much more work will always be necessary – just how does the genome do its job in critical areas? – before there will be benefit, and that work is most likely to progress if existing secrecy is not maintained. *Nature* stated that 'better medicines must come from the identification of genes whose products are involved in metabolic disturbances linked with disease'; but, as two scientists from Merck and Co. responded to that 'Briefing' article, 'the route from discovered gene to developed drug is long, expensive and has a high failure rate'. Science works best in an open society, and science needs 'the intellectual and experimental resources of the world's best biologists [as the Merck men phrased it] for ultimate success'.

*

In short, and as conclusion to this chapter, the completion of the Human Genome Project is only an intermediate step in the widespread application of its knowledge for the benefit of all. From today's standpoint the next step, involving gain for humanity in general, appears no less formidable as an undertaking. If all goes well it will help to eradicate disease, to improve lives and even to improve people by making them less prone to genetic error. That last phrase begins to sound much like eugenics, the subject of the pages which now follow.

EUGENICS

GALTON   IMMIGRATION   GERMANY

THE EUGENICS MOVEMENT   THE ENVIRONMENT

CHINA AND *YOUSHENG*

> The progress of biology in the next century will lead to a
> recognition of the innate inequality of man.
>
> J. B. S. Haldane

GALTON This chapter has to start with Francis Galton, genius, prodigy, statistician, medical doctor, mathematician, and a man as well connected as endowed. He coined the term eugenics in 1883, and was its principal promoter for many of his eighty-nine years. The word has since fallen out of favour, and Galton is in consequence often vilified. The American Eugenics Society voted (in 1972) to change its name. The Eugenics Society (of Britain) did so later, but more reluctantly (and perversely selected the Galton Society as replacement title).

In essence eugenics means the science of improvement of the human race by better breeding. It is therefore opposite to dysgenics, the deliberate deterioration of the human race. For a species that has acquired much current eminence by the steadfast selection of better traits of both domestic animals and plants, choosing the tastier, the bigger, the longer lasting, it is odd – at first glance – that anything smacking of improvement should not also have been dominant in our personal history. We did choose the more rewarding grasses, and turned them into staple cereals. We did likewise with tubers, fruits and vegetables. As for the wolf, we fashioned it into mastiffs, greyhounds, pugs, spaniels, poodles and countless more, affecting not only their size and shape but personalities. 'You can get beasts to weigh where you want them to weigh,' said Robert Bakewell, the eighteenth-century Leicestershire livestock-breeder who, more than anyone, turned farm animals into the plump, woolly, beefy, docile, milk-producing stock we recognize today. This was all eugenics, however much the word had not arrived. It was improvement – on behalf of humans – brought about by better breeding of plants and animals.

On therefore to Galton and human eugenics. 'It would be quite practicable to produce a highly gifted race of men by judicious marriages,' he wrote. 'Each generation has enormous power over the natural gifts of those that follow.' 'An enthusiasm to improve the race is so noble in its aim that it might well give rise to the sense of a religious obligation.' Galton, an agnostic, was not alone in such pronouncements. Charles Darwin, his cousin, so much an inspiration concerning fitness, wrote that 'civilized men should do their utmost to check the process of elimination [of civilized men] ... No one who has attended to the breeding of domestic animals will doubt that this must be highly injurious to the race of man.' George Bernard Shaw, publicly (and dramatically) on the side of prostitutes, pacifists, socialists and most minorities, wrote that 'there is now no reasonable excuse for refusing to face the fact that nothing but a eugenic religion can save our civilization from the fate that has overtaken all previous civilizations'. H. G. Wells, Sidney Webb, Havelock Ellis, Dean Inge, and Leonard Darwin (son of Charles), for example, were all active members of the Eugenics Society.

They were worried not only at the lack of breeding by the eminently civilized, but by the procreation of those at society's other end, such as (to quote Galton again) 'paupers, criminals, and the feeble-minded'. Every eugenicist knew of the Jukes family, of its 709 members (over a lengthy span) which included 128 prostitutes, eighteen keepers of brothels, seventy-six convicted criminals, and 200 who had, at some time, been on relief. So too of Kallikak, this man a one-time member of the US Revolutionary Army. During the war, and at a tavern, he and a feeble-minded girl had created a son. Over the post-war years this union had led to 480 descendants, of whom 143 were also feeble-minded, forty-six were normal, and the rest 'doubtful or unknown'. The same soldier, on returning home, had married 'a good Quaker girl'. Their descendants became doctors, lawyers, judges, educators. Game, set and match, it would seem, to eugenicists and their new religion.

Somerset Maugham, in *A Writer's Notebook*, summed up much

contemporary thinking when he wrote: 'The only means of improving the race is by natural selection; and this can only be done by elimination of the unfit. All methods which tend to their preservation – education of the blind, and of deaf-mutes, care of the organically diseased, of the criminal and of the alcoholic – can only cause degeneration.' Winston Churchill, no less, personally drafted the act which proposed 'shutting up people of weak intellect and so prevent their breeding'. F. C.S. Schiller, outspoken Oxford don, stated that 'race-suicide' among the best human stocks was 'a far more urgent evil' than overpopulation. Rudyard Kipling thought that assistance for the needy was 'making milksops of the democracy'. The President of the British Medical Association considered that 'the National Health Insurance Act [of 1911]' was 'a long step in the downwards path towards Socialism ... [It would] increase that spirit of dependency which is ever found in degenerate races.' A zoologist informed the Fabian Society: 'Nothing kills character so much as the shifting of responsibility from the individual to the State.'

Galton was aware of opposition to such thinking, but dismissed it: 'There exists a sentiment, for the most part quite unreasonable, against the gradual extinction of an inferior race.' Maugham had added: 'Failure or success in the struggle for existence is the sole moral standard. Good is what survives.' There was not much talk of an unfavourable environment affecting the unfortunate. 'Take care to get born well,' Bernard Shaw had said. Without doubt many millions had taken insufficient care.

The Boer War in Africa and then the outbreak of the First World War gave a sense of urgency to eugenic thinking. Benjamin Rowntree, a caring employer, estimated that half of England's workforce was unfit for military service. Three out of four men attempting to enlist in Manchester for the fight in South Africa were rejected as medically unsound. David Starr Jordan, president of (the new) Stanford University in California, said that war eliminated the brightest and best. It was therefore a eugenic disaster, particularly when nations were subsequently forced to breed from inferior stock, from those left at home. Major Leonard Darwin approved of conscription because it would be more nearly representative of the nation, but suggested that older men should be recruited first so that younger men would survive to build up the postwar stock. The fact that casualties were greater among junior officers, perhaps three times so, than common soldiers caused many to appreciate that wars could be dysgenic. Therefore eugenics would be necessary to restore the earlier situation.

IMMIGRATION was one area where the principles of eugenics could most speedily be applied, with Galton happy to concede that much of Britain's greatness had come from Europe. It is 'remarkable how large a proportion of eminent men ... bear foreign names ... We cannot fail to reflect on the glorious destiny of a country that should maintain the policy of attracting eminently desirable refugees...' Spain, in Galton's opinion, had suffered by being 'deprived of free-thinkers at 1,000 a year from 1471 to 1781, with 32,000 burned, 17,000 destroyed in effigy (having died or escaped), and 291,000 imprisoned or otherwise penalized'.

The United States, prime recipient of immigrants in Galton's day, certainly reflected on its destiny immediately after the First World War. It initiated a Restriction Act designed to reduce intermixing, such miscegenation being thought deleterious. Annual immigration from each European country, according to the new law, would not be greater than 2% of the number of US residents who, in the 1890 census, had given that nation as their place of birth. Those in charge of eugenics (and behind the Act) were therefore urging that all newcomers would be from the same background as those in charge of eugenics. (Bill Bryson sums up the situation neatly in his always readable *Made in America*: 'If one attitude can be said to characterize America's regard for immigration over the past two hundred years it is the belief that while immigration was a wise and prescient thing in the case of one's parents or grandparents, it really ought to stop now.')

In the years leading up to 1890 many immigrants had arrived from 'Nordic nations', this favoured term (of the time) being almost synonymous with 'non-Mediterranean', another much-used grouping. Under the new laws Great Britain was permitted an annual quota of 65,000 migrants a year, Italy 5,666, and Greece 308. This rigid restriction was only repealed (with the passing of the Celler Act) in 1965, the annual maximum then being fixed at 20,000 per country, irrespective of that country's earlier migrations.

(Today's situation is made more complex by the number of illegal migrants, thought to be 300,000 in 1993 – as against 900,000 for the 'legals'. The size of the foreign-born population is now tremendous and, at 20 million, bigger than it has ever been – and bigger than the entire US population in 1850. In the earliest days of the United States almost every arrival came from Europe. By the 1950s that proportion was down to 68%, even including those from Canada. Currently it is 13%.)

GERMANY   Eugenics can be sub-divided: negatively it prevents breeding, positively it encourages. To castrate is negative; to promote, by whatever inducement, is positive. The farmer is positive when permitting only certain stallions, bulls, rams, boars to cover, mount, etc. selected females, but is negative when preventing access by unwanted males. The American immigration laws of the 1920s were both positive and negative, welcoming arrivals from chosen areas, discouraging those from less favoured locations. Germany's laws of the Nazi period (detailed in this volume under Contraception) were largely negative, preventing (via sterilization or death) procreation, but there was also a positive side to Germany's eugenics.

Lebensborn, spring of life, was founded in 1935. At the Nuremberg trials (of 1946) this organization was described by its defendants glowingly, as if it had been some beneficent welfare group, taking care of expectant mothers (of good stock), supporting large families (of good stock), and helping children. It also permitted, and encouraged, favoured young men to father offspring who would be cared for by the state. Heinrich Himmler, active promoter of Germany's racial edicts, had said that no members of his *Schutz-Staffel* should lay down their lives without first creating replacements 'of the same high quality' (*Edelprodukte*). It was up to *Lebensborn* to provide homes where such unselfishness could be achieved.

Suitable girls, also happy to donate offspring to the state, were selected, given a private room, permitted to choose from the available SS consorts, and then made pregnant by them. 'As both the father of my child and I believed completely in the importance of what we were doing, we had no shame or inhibitions of any kind,' reported Fräulein Koch, one such girl. In due course the child was handed over. Her man went back to his duties, and she became engaged to a Gestapo NCO. These two, as was mandatory, were then investigated concerning their suitability for marriage. When both had been awarded their *Ahnenpass*, the certificate of 'pure breeding', they could go ahead with the ceremony. Her husband was later said to have been enraged when he learned of the earlier pregnancy, but apparently stayed loyal.

Many of its citizens expressed amazement, as Germany's war of conquest proceeded, that innumerable individuals in other lands were no less Aryan, if not more so. Poland was rich with the blue-eyed and fair-haired. So too Scandinavia and also, to a lesser extent, the Soviet Union. Many children from these lands, if suitably endowed, were then transported to Germany – as orphans, whatever their actual status.

These numbers swamped those emerging, also as orphans of a kind, from *Lebensborn*'s breeding homes. The war inevitably interrupted this organization's carefree matings; but, shortly before the conflict began, its authorities announced 'that 832 valuable unmarried German women had decided, in spite of the greatest hardship, to donate a child to the nation'. By then many tens of thousands of other individuals were paying the price of negative eugenics. The contrary number who had arrived, more positively, stood at 832 – and that, more or less, was that. At the war's end not one of the *Lebensborn* offspring was more than 10 years old. They had therefore not had time to serve the state which had so thoughtfully arranged for their existence.

*

Germany's actions showed, devastatingly, that the pursuit of eugenics could go wrong. So did the study of genetics in general. Mendel, with his sweet-peas (coloured or plain, wrinkled or smooth), had indicated simplicity. Genetics *is* simple when single genes are involved, when one individual either does or does not possess some characteristic, a proneness to disease perhaps, a defect that either is or is not detectable, but as the science of genetics advanced, and as the interplay between nurture and nature became more fully understood, that simplicity was steadily eroded. Whether someone was criminal or alcoholic had seemed – at first – as straightforward as Mendel's laws. His red flowers always produced red flowers. Therefore alcoholics and degenerates were doing likewise. Stop them from breeding, and their problem would vanish, along with their erroneous genes.

Geneticists increasingly thought otherwise. So did sociologists, who wondered at the complex causes of unwelcome behaviour. Mere parents, not often asked for opinion by scientists about their progeny, also knew that a similar environment and similar genetic backgrounds could produce wildly different children within a single family. Sibling personalities could be chalk and cheese; so too their intelligence, their physique, their attitude to life. Parents of a single child can nourish solid beliefs about cause and effect in rearing offspring. Parents of two can be less certain, and parents of more can be bewildered. Even physical traits, like hair-colour, feet size, ear-shape, can be quite dissimilar. As for behaviour – adventurous, parsimonious, moody, belligerent, unassuming – the children might be from different families for all their similarity. Yet eugenicists were daring to assert that criminality, alcoholism, immorality, infirmity and other shortcomings could leap, unaltered and unhindered, from one generation to its successor, much as dark skin or pale skin will breed true if no earlier mixing has occurred.

THE EUGENICS MOVEMENT advanced through three phases. The first was largely a matter of prejudice and observation. Environments possessing either worthy or unworthy individuals tended to produce more of the same. (Galton never questioned why lawyers, doctors, writers, financiers traditionally failed to surface from an impoverished working class.) Oliver Twist was a perfect example of genetic wholesomeness. He had been thrust into the workhouse and then into crime, but emerged untainted when opportunity permitted. Colonialists thought it right to govern; peoples deemed inferior needed quantities of care. Some races, some classes, some kinds of individual were lower down the scale. Why therefore did they have to propagate and manufacture more?

It *was* prejudice but then, as second phase, came science. Mendel's laws were laws. Inheritance did obey rules. Dominant characters did overshadow the recessive, and a double dose of recessive inheritance was necessary for its existence to be shown. The old rivalry between Lamarckism and Darwinism was being settled, with the inheritance of acquired characters (and the blacksmith's son possessing a stronger arm) yielding to a fixed inheritance, with parental genes the prime determinants. What had been done blindly and pragmatically by animal breeders could therefore leave the barn and join with science. So too the human race. Improve the environment by all means, but improve the stock as well. Do not do one without the other. Make a better house, and also make better people to inhabit it, with Shaw, Churchill, Inge, Wells, Darwin, Maugham, Kipling, Webb and countless others nodding eagerly. They did not necessarily expect improvement of mankind; they wished to halt its headlong degradation.

By no means were Germany, the United States and the United Kingdom alone in their concern for a 'downgrading' (always a popular term) of the human stock. Alexis Carrel, of France, received a Nobel prize in 1912, being then – at 39 – the youngest ever recipient. Later he wrote a book, *L'Homme cet inconnu* (Man, that unknown), in which he proposed replacing democracy with a 'biocracy' because medical progress was in danger of degenerating 'the great white race'. This form of advance was promoting the survival of the weak and, in consequence, an increase in mental illness and criminality. His book was translated into nineteen languages. (In February 1996 the University of Claude Bernard, Lille, France, decided that the Carrel Faculty of Medicine should be re-named. It did not like to be reminded of the man's association with eugenics, nor his alleged links with the Vichy regime during the Second World

War. The faculty is now named after René Théophile Hyacinthe Laënnec (1781–1826), inventor of the stethoscope.)

Phase three, starting in the 1930s, involved a gradual erosion of certainty. Genetics advanced on a broad front. Even the seeming simplicity of hair-colour or eye-colour became more complex. The experimenters preferred to work with fruit-flies or mice, where events were controllable, but their results were applicable to life in general. With humans, breeding so casually (from any scientist's viewpoint), the facts that could be gleaned came, in the main, from rare diseases – haemophilia, cystic fibrosis, thalassaemia, amaurotic idiocy – whose genetics could be partly or even wholly understood. As for personality – that would have to wait a while, perhaps for ever. Meanwhile, as news arrived from Germany, eugenics began to acquire its bitter taste. The science of genetics was being used as a basis for sterilization, and then for euthanasia. Like a sudden landslide, which gives only seconds for anyone to reach security, many individuals abruptly realized how near they had been to advocating something similar. Eugenics soon died – along with millions of human beings.

THE ENVIRONMENT then took over (some say excessively) but people could do something about the environment. In a world recently rid of its most devastating war there could be better housing, better health facilities, more care for the unfortunate and for the underprivileged. There should be more equality, less bigotry, and should even be concern about the planet. Human genetics, far from being a panacea, took a back seat. Genes, it was argued, were largely irrelevant in determining, for example, intelligence, as if the human brain, unlike human stature, human anatomy, and the human form, was (almost) independent of its genetic origins. Therefore forget about genes. Intelligence could come from outside, from experience, from opportunity. A few scientists of the time were stating that mental ability was 80% genetic, 20% induced. It was not so much this precision that others so resented, but the affirmation that so large a share could be inherited.

As the 1960s became the seventies and eighties the pendulum began to settle down. In the century's earliest decades it had swung, stead-fastly, pro-gene. It then, post Second World War, swung the other way. Finally it achieved a more central stance, as the list of genetic diseases grew longer, as information was collected widely, as the science of genetics solved a few old problems (and put further riddles in their place). Genes could *predispose* an individual towards some out-come, neither causing it nor being irrelevant. The simpler days had gone for good (and see the sections on Genetic Disease, the Genome and the Genes themselves for much more in this general area).

CHINA AND *YOUSHENG* As for improvement of the human species by better breeding, this concept has not entirely vanished. In 1994 the Chinese government approved a law forbidding those with mental disabilities or contagious diseases to marry, thus – according to the nation's official news agency – 'helping to reduce births of physically or mentally abnormal babies'. There had been an earlier, and stronger, proposal (in 1993), and it had been hoped by the International Genetics Federation (and others) that subsequent silence – following widespread protest around the world – meant the idea had been dropped. But not so.

The full text of the new legislation, which arrived after the news agency's report, stated its aim: 'To improve the quality of the newborn population'. Not only would pregnancies be terminated when the foetus 'has a serious defect or suffers from a severe genetic disease', but an adult 'with certain genetic disease of a serious nature which is considered to be inappropriate for childbearing' may be married only if the couple agrees to long-term contraceptive measures or steriliza-tion. The Chinese Health Minister, aware of foreign criticism, blamed problems in translation. *Yousheng*, interpreted as eugenics, did not refer to eugenics 'as practised by Hitler'. It indicated better birth, better rearing and better education.

Yousheng did not banish the ethical issues. Before marriage, under the new law (designed to start in June 1995), both sexes had to obtain a 'pre-marital physical check-up certificate'. This would identify whether they suffered from any disease that may have 'an effect upon childbearing'. The legislation affected three areas: infectious disease 'such as AIDS, gonorrhoea, syphilis, leprosy...'; genetic disease 'that may totally or partially deprive the victim of the ability to live independently (which is highly possible to recur in generations to come...)'; and mental disease, defined as 'schizophrenia, manic-depressive psychosis, and other mental diseases of a serious nature'. Marriage must be postponed if either parent is thought to be incubating any of the defined diseases or is suffering from mental disease. As for genetic disease, the couple must be prevented (by contraception or sterilization) from breeding, even though no list had been drawn up of what constituted a 'serious' genetic disorder.

It is illegal in China for the sex of a foetus to be identified, but ultrasound technique is being used increasingly with this aim in view. *China Information* recently reported that at Zhangye (north-western Gansu province) the sex-ratio at birth was 131:100. More boys than girls are born everywhere in the world, but not to that degree without intervention. China does have a formidable population problem, but anything smacking of sex-determined abortion, female infanticide or eugenics has the rest of the world eager to condemn. One-quarter of the world's people already live in that single nation. Neither it nor anyone else wishes that quantity to increase, but a desirable end does not justify all means to reach that end – and certainly not eugenics.

The European Society of Human Genetics, and Britain's Genetic Interest Group (among others), have stressed that China's new legislation infringes the right to have a family. Worse still 'serious genetic disability' could mean 'just being Tibetan'. The International Genetics Federation, worried by the new laws to reduce 'inferior births', threatened (in 1995) to abandon its choice of the Chinese capital for its next five-yearly congress scheduled for 1998. That threat was later withdrawn when the Chinese agreed to include a symposium on eugenics during the meeting (that symposium likely to witness the hottest debate of all). China does have massive problems, not least in its numbers, but the remaining three-quarters of the world is not indifferent to the fashion in which such problems are attacked.

*

This chapter began with Sir Francis Galton, and can end with him. He considered that statistics were always 'full of beauty and interest'. He pursued them wherever possible, as in counting the number of fidgeters in an audience, or in totalling 'attractive, indifferent or repellent' women he encountered in the street. Statistics can be interesting, appealing, and even beautiful. It is what is done in their name which can be so grievously at fault.

FAMILY

PROFILE   CHURCH v. STATE   PRESENT-DAY PARENTING

CHILDHOOD MORTALITY   CONFLICT

If it were possible decently to dissolve marriage during the
first year not one in fifty couples would remain united.
Somerset Maugham

Italians used to find their mate at an average distance
of 600 yards. When the bicycle was invented this average
leaped – to 1,600 yards.
news item

Human beings are the only creatures on Earth that allow
their children to come back home.
Bill Cosby

PROFILE More often than not, whether mammal or bird, there is
a form of family. The male may not assist in rearing the young, and
may even be entirely absent, but sometimes there is pairing for life,
with both parents collecting food, or sharing labour, often with the
male as protector and female as carer. The human family is not
greatly dissimilar, it being equally fundamental, equally vital for the
next generation, and a basis of our lives. The WHO has called it 'the
social unit upon which societies are built and maintained'. A family
provides its members with 'health, nutrition, shelter, physical and
emotional caring, and personal development'. And yet, fundamental
as it is, there was not an 'International Year of the Family' until 1994.
May 15 was then the first ever 'International Day of Families', with
its theme: 'Building the smallest democracy at the heart of society'.

Within the context of external fertilization, assisted birth, genetic
manipulation, embryonic selection and every other aspect of a
changing reproductive world, there is probably no aspect altering
faster than that of the family, the basic human unit, the initial

framework. Reduction in family size has been tremendous (in many places to 20% of its former level and, on average, by one-third in the developing world since the 1960s). The extended family, with cousins marrying cousins, used to be commonplace (in Loire-et-Cher, France, 6% in 1918, and 1% in much of America before this century), but not these days with so much people movement. Even so, one in 200 of all British marriages is still between first cousins (according to *Nature Genetics*), and one in 20 in some parts of Asia and Africa. Israeli Arabs seem to hold the record for consanguineous marriages, 'hovering over 50%' according to a *Lancet* report by Rachelle H. B. Fishman. Marriages are frequently between first cousins or even double first cousins (as with a father's sister's offspring getting together with a mother's brother's offspring). Major congenital malformations are up to 16% in the progeny of such close pairings.

Marriages are now frequently ending in divorce or breakdown (with 3 million British experiencing divorce in the decade before 1995, and 1 million of their children cut off from all contact with one parent in that time). Single-parent families are becoming remorselessly commoner: 1 in 7 in 1986 for the UK, 1 in 5 in 1992, and 2.5 million children in lone-parent 'families' by 1993. Worldwide, 20–30% of all families in Africa, Latin America and the Caribbean are single parent, 15–35% in European nations, and 15% in Asia and the Pacific. Women head these families in 90% of cases. More women are entering the workforce. More people are living together in casual relationships. There is more domestic violence (although this may result from an increased willingness to report poor treatment). And AIDS is not irrelevant to family well-being. It is therefore a wonder, when confronted by such change, that the WHO did not initiate its international family year until 1994.

The family does deserve a higher profile. Britain, for example, does not have an explicit family policy. There are no formal instruments of such a policy, no explicit programmes with clearly defined intentions towards the family. In March 1994 the government announced that the Minister for Health would also be known as Minister for the Family, but this change was not partnered by any transfer of responsibilities from other ministries or any additional resources. In *Farewell to the Family* Patricia Morgan states that tax alterations in Britain during the 1980s mainly benefited single people and double-income couples. The most recent tax increases have fallen disproportionately on couples with children and, in particular, on couples with one main income. 'Soon,' she writes, 'the family man will pay all national taxes at the same rate as a single man, while paying 25% more council tax as part of a couple.' Couples, whether married or not, rear over 80% of the nation's children, and yet have been bearing more and more of the tax burden. The Minister for the Family could well introduce financial measures that disproportionately benefit couples raising the next generation. It is not as if child-rearing is in any sense a financially profitable occupation.

The rise of the single-parent family has earned more attention than the traditional kind, particularly the never-married one-parent family. In 1986 divorced lone mothers made up 5.6% of all families with children, separated mothers 2.6%, and never-marrieds 3.2%. By 1991 the equivalent figures were 6.5%, 4% and 6.6%. One year later these proportions had changed to 6.4%, 4.5% and 7.3%. Therefore, in the few years between 1986 and 1992, the number of single never-married mothers rose from 3.2% to 7.3%, an increase of 128%. There is no evidence, according to the National Council for One-Parent Families, that this trend will soon be halted. Women are becoming more and more economically independent and, almost always, lone parents are female. In 1992 this ratio was 19.1% single women (out of all the families in the land) to 1.8% single men, a ten-fold difference.

There are advantages for a child born to a single parent, one being that the parent may gain a partner, which might add to the child's well-being, but – short of accident, or death – will not lose one. Children of divorced parents, on the other hand, do suffer economic and social consequences, according to the Centre for Family Research, Cambridge. Its director has stated that they are not only likely to suffer increased poverty but do far less well at school, marry earlier, and suffer higher divorce rates themselves. Effects of divorce are particularly powerful for daughters of middle-class families. They have less chance of going to a university than girls from stable families, and a 45% chance of marrying by the age of 20 (as against a

15% chance for those whose parents are together). Martin Richards, the centre's director, added that the effects of divorce were greater than when a parent died or when the children had been raised by a single parent. 'There's something specific and special about divorce. The children's esteem tends to fall ... [They] have to face the fact that their father appears not to want to see them. That is a very heavy message for the children ...' Work in the United States has shown that damaging effects are as apparent today as in the 1960s or 1970s when divorce was relatively rare, and Richards envisaged a child sueing its parents who, by their divorce, had damaged its future.

CHURCH v. STATE Religions have, in their various ways, wished to appropriate marriage, it being (according to the Church of England) 'an honourable estate, instituted of God ... reverently, discreetly, advisedly, soberly...' Inevitably the Church is concerned by most current changes – more divorce, more single-parenthood, less marriage. Marriages in Britain fell below 300,000 in 1993, the first time for forty years. Births outside marriage increased from 4 per cent to 9 per cent during the Second World War, and have since reached 30 per cent. In February 1995 Britain's Archbishop of York (second in the hierarchy) described the record divorce rate as 'sad and traumatic'. As for the drop in weddings 'those [individuals] who are looking at the possibilities of different life-styles see no particular advantage for themselves in actually getting married, and society seems to be saying through the tax system that it doesn't matter'. He therefore urged politicians to give marriage a boost with tax incentives for those who commit themselves to it.

The United States has affirmed a family policy more than Britain. Walter Mondale, when vice-president, said in 1978: 'Our task must be to do more than lead a government that tolerates family life or pretends to be neutral about it. We must help shape a society which nourishes families ... There is no more important task.' The relevant committee of the US Department of Health, Education and Welfare has outlined the elements of child care services, these including a 'guaranteed minimum income system to ensure that all families have sufficient income for one or the only parent to choose not to work outside the home, without sacrifice to family or children ... Parents should have the option, first and foremost, of raising their children at home, without sacrificing a reasonable standard of living.' The fine words may intrigue many a parent who knows, full well, the degree of sacrifice incurred in raising even a single offspring.

In the numerous forms of marriage around the world there are – almost always – taboos. The Church of England used to forbid marriage with thirty different kinds of relation, ten of which were consanguineous (such as a man marrying his sister or mother) and twenty of which were considered socially unacceptable (such as a man marrying his son's wife or his wife's sister). With the Marriage Act of 1949 ten of the forbidden thirty were struck from the list (such as a man marrying his father's brother's wife or his wife's mother's sister). Not one of those struck off was of a genetic relationship.

It is sometimes thought that cousins should not be able to pair, as they have one-eighth of their genes in common, but British law (and the English Church) permits such partnering although Roman Catholics need special dispensation from Rome before proceeding. Without doubt the offspring of such cousin–cousin pairings do suffer a greater chance of congenital abnormality. Research in the United States concluded that actual mortality was more probable, being 8.1% in the first ten years (for offspring of cousins) and 2.4% (for the offspring of unrelated marriages). Equivalent figures from France for deaths of neonates (less than 1 month) were 9.3% as against 3.9%. In Japan, for children aged 1 to 8, the figures were 4.6% and 1.5%.

It is therefore strange that so many blind eyes – even from the cousin couples themselves when informed of the hazards – should disregard the extra risk. It is also strange, from this viewpoint of genetic similarity, that few people – even in these well-documented times – know of their great-grandparents. It is therefore perfectly possible for pairings to take place, allegedly unrelated, that are closer than the participants realize. As for forbidden pairings still considered socially unacceptable, such as a man marrying his step-mother or mother-in-law, the future may think otherwise about this ban. Strange as it might seem to music-hall comedians, a man's mother-in-law

could provide the perfect match following the death of her daughter/
his wife. (When Bernhard Grzimek, the famous zoo-man of Germany,
lost his son in an African flying accident one result was the departure
of Bernhard's wife. Thereafter he lived with his son's widow, the two
of them united and equally supportive in their grief. Such a pairing is
still forbidden by the Table of Kindred and Affinity as outlined in the
Book of Common Prayer published in 1662.)

Part of the confusion, of who can pair with whom, rests in the
relationship between Church and state. In Britain either the Church
or the state (as from 1836) may perform a marriage. In most
European nations there has to be a civil ceremony, whether or not
there has been a religious ceremony. No marriage is considered valid
in the Western world (however pregnant the girl may be) if either of
the partners is below the legal age – 16 for both sexes in the UK
(which is before they can vote, buy alcohol, etc.). The various states
of the United States are disunited on marriage age, with New
Hampshire, for example, permitting girls to marry at 13.

It is argued – in various nations – that the current blurring between
Church and state, a hangover from earlier centuries and not simplified when the head of state is also head of the Church (as in Britain),
should be rectified. 'It is no business of the State to define marriage,'
proclaimed Matthew Parris in *The Times* (of London). 'We should
return matrimony to Christianity, Judaism and Islam, to consecrate
and interpret ... as each thinks fit among its own flock ... This
would leave the State free to do what it is the State's business to
do ...' The State does have much to do. It has, for example, to make
certain that the vulnerable and weak do not suffer unduly, such as
children. Whether a marriage has taken place should not affect a
child's rights to proper care. Children in the future will not live with
the earlier unfair simplicity of legitimate or illegitimate. They may
perhaps be raised by homosexual couples or single homosexuals.
They may have been born in surrogate style, or re-adopted midway
through their childhood, or fashioned from donated embryos, or –
well, who knows what the future might also bring? Wedlock can seem
most archaic as a concept in such a changing world, but responsibility
is quite another matter. It will not go away.

PRESENT-DAY PARENTING Various patterns have been classified, and six have been identified:

1) One partner as full-time parent until end of compulsory schooling.
2) One partner as full-time parent until child first goes to school aged 5.
3) Truly shared parenting between both partners.
4) Both parents working (either full-time or part-time) and delegating some child-care to others.
5) Single-handed parenthood by choice.
6) Dispensing with any attempt at 'normal' parenting as in, for example, communes.

There is also the problem of nomenclature in today's divergent world. Africans often say (almost as if the phrase is single-syllabled) as an introduction: 'This is my brother, same mother, same father.' In other areas such fraternity can be called full-brother, as against half-brother or step-brother. Science refers to 'uterine brother', indicating one mother but different fathers. Conventional humans, in introducing a modern family, can say: 'This one's mine. That one's hers. And this one's ours.' Adopted children can resent being called 'adopted', as if their status – rather than their origin – is somehow different. So what of the AIDs or AIHs, the offspring of donated sperm or paternal sperm? Or the children of surrogate mothers? Or of donated embryos? In future years, with increasing frankness about reproduction, there may be a similarly forthright attitude concerning derivation, and a whole pack of new words to define the varied forms of procreation.

Or there may be nothing of the sort. Dependency will always be important – who is looking after whom – and the actual method of initiation may be considered of little concern, particularly when artificial reproduction in all its guises becomes more commonplace. Former fascination with illegitimacy, bastardy, out of wedlock, and the wrong side of the sheets, has already receded. Perhaps all forms of origin will follow suit, and new phrases will achieve more prominence. 'This is my dependant'; 'This is our dependant'; 'She is my carer'; 'They are my carers'. Animals cannot use words, but know precisely who is dependent upon whom, who is the carer and who receives such care. One day humans may also consider such knowledge to be quite sufficient.

CHILDHOOD MORTALITY Sir George Newman wrote in 1916: The 'death rate of infants is the most sensitive index we possess of physical welfare and the effect of sanitary government'. Were he alive now he could still pen those words but be appalled that welfare and good government are still so unequal around the world. Childhood mortality does indicate a nation's, a region's, a family's well-being. Most such deaths are preventable, even in the wealthier nations. In Africa and Asia, where malnutrition and infection remove half the children before their fifth birthdays, there is no need for medicine or science to provide new answers. More than sufficient is already known to combat these two killers. Economic factors are certainly to blame, but only in part. Local culture is also guilty, as in abruptly transferring babies from sterile, nutritious, near-perfect maternal milk to the awesome, spiced, non-sterile diet much beloved by adults. Poor education (about what to do) and poor health services (often at a distance) are also at fault. So too burgeoning numbers which, from a government's viewpoint, are already sufficiently troublesome without the threat of extra mouths to feed, to educate, to find work, to vote unsympathetically.

Sir George Newman might be comforted that the death rate of infants is diminishing (virtually everywhere), with each earlier decade seeming archaic. Britain can make this point most clearly, and many of its older citizens ought to be amazed that they personally triumphed over such adverse circumstances, much like soldiers realizing how great had been the slaughter from which they have escaped. Before Sir George's time, during the middle of the nineteenth century, infant mortality in English towns was 200 per 1,000. By 1910, when he was active, the national rate (for infants dying in their first year) was 116 per 1,000. By the time of his remark it was 108 per 1,000, a figure not tremendously different from the 200 deaths per 1,000 of British military engaged in the First World War.

By 1918 infant mortality was down to 90 per 1,000, much of the drop attributed to a greater awareness by the authorities of public health, so many of the wartime conscripts having been found unfit (and not fit enough to die in battle). By 1923 the official figure was 72 per 1,000, and by 1928 60 per 1,000. In 1935, after a recession period of extreme deprivation, the number rose to 62 per 1,000. At the outbreak of the Second World War it was 57 per 1,000, that war then improving, rather than worsening, the figures. Food rationing (meaning adequate food for everyone) and full employment are credited as dominant causes for lowering infant mortality to 45 per 1,000.

In 100 years the death rate had therefore dropped to 25% of its former level, but it was to drop still further. By 1950, after the arrival of the National Health Service, the figure reached 38 per 1,000, by 1975 it was 27 per 1,000, and by 1980 12.5 per 1,000. That number, although indicating the death of 8,000 babies in their first year, was one-eighth of the 1850 level and even a third of the level existing when the mothers of those 1980 babies had themselves been born. By 1991–2, according to the Office of Population Censuses and Surveys, the figures (in England and Wales) had dropped from 7.4 to 6.4 for boys and from 6.4 to 5.8 for girls. Some recent decades, such as the 1920s and 30s, can seem most medieval by comparison, with mortality rates ten times as high.

Even current years may – in time – be regarded as archaic, with the lowest recorded figure still adding up to some 4,000 infants dying within the relatively prosperous United Kingdom during their first twelve months. In Europe, the discrepancies between nations, and between regions of those nations, show that much death should not occur. Deaths under 1 year of age per 1,000 live births in a recent survey of six European nations were: Sweden 11.1, Holland 11.4, France 16.0, England and Wales 17.2, Belgium 20.5 and West Germany 23.5. For children aged 1–4 (and per 1,000 live births) the mortalities were Sweden 42.0, England and Wales 70.5, France 79.0, Holland 83.0, Belgium 87.0 and West Germany 95.0. The best country, from this standpoint, was therefore doing twice as well as the worst one. This disparity continues, although less markedly, into the children's later years. Deaths among those aged 5–14 (and per 100,000) were: Sweden 32.3, England and Wales 33.7, Holland 37.9, France 38.6, Belgium 42.1 and West Germany 45.7.

Such figures become more revealing when divided into 'Mainly preventable' and 'Mainly non-preventable'. There is plainly a difference between fatal disease – for which there is no known cure – and some lesser ailment, or even an accident, which should not have been responsible for death. Sweden again achieves prominence, with 40.4 mainly non-preventable deaths (per 100,000 1–4 year olds) as against 26.4 for (mainly preventable) accidents and 7.2 for (mainly preventable) infections. Comparable figures for England and Wales were 28.6, 27.5 and 18.1, therefore down on non-preventables but up on both preventable causes. West Germany's figures were 23.7, 33.0 and 12.6, while France's were 17.5, 32.6 and 10.0. Of Sweden's deaths the ratio of non-preventable to preventable was 1 to 0.83 whereas that of England and Wales was 1 to 1.59, of Germany 1 to 1.92, and of France 1 to 2.43. It may be simplistic to conclude that Sweden is doing three times better for its children than France (as other factors must be involved) but there is an undoubted differential.

Dissimilarities certainly exist even within the UK, according both to region and to social class. It is with classes, ranging from I through II, III, IV and V (with I being most prosperous), that differences are most pronounced. In Scotland, for example, and with neonatal (first month) death, the five groups (in descending class order per 1,000 live born infants) were 7.5, 9.4, 12.3, 12.6 and 15.5. Those born less well were twice as likely to die. From one month to one year of age the differences were even greater, being 2.7, 4.0, 5.8, 7.5 and 12.0, a 4.4-fold variation. For all infants under one year, combining these two sets of figures, the numbers were 10.2, 13.4, 18.1, 20.1 and 27.5. As some religions/philosophies/individuals believe all children choose their parents there should be curiosity why so many of them choose Class V.

Considerable variation also exists within each area of the United Kingdom. Low birth-weight is commoner in the north of England than the south, and within major towns and cities. For England as a whole the perinatal mortality rate (stillbirths plus first-week deaths) is 17.6 per 1,000, a figure covering a range of differences, such as 23.7 for Dudley, 22.9 for Rochdale, 22.4 for Barnsley (all three being northern industrial towns) and Suffolk 12.0, Camden and Islington 11.0, Oxfordshire 10.4 (all three southern and, in general, wealthier). These trends are similar if first-year deaths are totalled. Rochdale has 20.7 deaths per 1,000 1-year-olds; Oxfordshire 10.1. In short, if prospective infants do have a choice (and prefer to live rather than die), they should not only choose their nation and their social class but which portion of that chosen nation they should best inhabit.

*

In theory all accidents are preventable. In practice they will always occur, however protective the society. They figure proportionately less in the statistics of developing nations, partly for being swamped by infection or malnutrition but also because simpler housing can be less lethal. (The number of ingredients in richer homes able to cause death range from cleaning fluids to mechanical contrivances and even to flights of stairs, let alone the traffic just outside the door.) In developed countries accidental injuries are the most common cause of death in children aged over 1 year. In England and Wales there are 700 accident deaths to children annually, with 120,000 admissions to hospital and 2 million attendances at accident and emergency departments. For children under 5 the most lethal place is the home. For those between 5 and 15 the road takes precedence, with 60 per cent of all traffic accident fatalities in that age group being of pedestrians. (Don't walk; get in a car.)

Accidents seem to be random events, with no rhyme or reason to them, save that generalities can be made. Children aged 1–4 are the most vulnerable. Boys are one and a half times more prone than girls. ('So what else is new?' say all parents of male offspring.) There are more accidents to children in economically deprived areas, in large families, in single-parent families, or with teenage or old mothers. Being accident-prone is actually a valid statement. Denise Kendrick, of Nottingham University, has reported that 'children who have already had an accidental injury requiring medical attention are at greater risk of future injuries than those children who have not'. Parents, aghast at the antics of their children when not under immediate supervision, can be astounded that only 1 in 10,000 actually die from mishap in any year.

*

Teenagers are also not averse to bringing misfortune upon themselves. Research by the University of Wales discovered that large numbers had unhealthy life-styles, involving smoking (20%), alcohol (54%), illicit drug use (6%). Some 10–15% have behaviour problems – school engendered difficulties, relationship difficulties. Another survey concluded that teenagers felt more unwell than could be inferred from their low consultation rate with general practitioners. It is unfortunate that teenagers do not conveniently fit into any of the current medical specialities. There are paediatricians caring for children, and there are geriatricians caring for the elderly. There is argument that teenagers, so affected by weight and skin problems, so confused by burgeoning sexuality, and neither wholly adult nor wholly child, deserve specialists of their own. The United States is most advanced in such specialist services, these allegedly having resulted in improved teenage health.

Providing information is not necessarily rewarding. Some school-based smoking education was effective in advancing knowledge, but did nothing to cut down the smoking level (of 12% among 13-year-olds). One teenage survey learned that 98% already knew the dangers of smoking. Teenage health appears as an inverse care law: those most at risk from pregnancy, smoking, alcohol and drug abuse are less likely to seek advice on their life-style. Young teenagers are rarely seen in the average surgery. Those above 15 consult more often but rarely if they are male. Researchers in the US asked teenagers not so much how but why they had become pregnant. Over 50% thought they could not become pregnant, 25% believed it was the wrong time of the month, 20% had not anticipated intercourse, 5% thought they were too young, and 3% did not know about contraception. Another survey learned that 23% of pregnant teenagers had intended to become pregnant.

The over 50% who thought they could not become pregnant highlight some specific teenager attitudes. These have been listed (in the *American Family Physician* of 1992) as:

1) Adolescents' thought processes seldom include a realistic view of the future.
2) Adolescents who feel unworthy and unimportant may view conception as an acceptable outcome.
3) The 'myth of immunity' sometimes supervenes (as in 'it won't happen to me').
4) Pregnancy demonstrates independence, but also allows continuing dependency on parents.

When does a child become an adult, medically, socially? At age 6 (or so) children are unable to consent to any medical undertaking. At age 16 (or so) they can have responsibility for full consent. The Medical Research Council (of the UK) has defined 12 years as the cut-off point. Below it the child is judged to be incompetent in all circumstances. Above it he or she should be assessed whether or not they have sufficient comprehension (of the problem, the circumstances) to give consent. Minors who have left home – runaways, students, or early spouses – do not need parental consent; they are doing everything else themselves, and might as well give consent as well. US states have, as ever, reacted individually to this problem. In Oregon a minor must be 15 before agreeing to medical treatment (and that treatment must be of direct benefit to that minor). In Alabama the age is 14. In Mississippi there is no age boundary, merely an assessment of the patient's competence to give assent.

A child's ability is highlighted by baby-sitting. From many a parent's viewpoint there is no time interval between a child having to be baby-sat (causing repeated donations to various sitters from the neighbourhood) and that same child earning good money as a sitter (and generally milking the neighbourhood). Both the United States and Canada have initiated after-hour school classes enabling 'young people to develop the self-assurance, knowledge and skills required to carry out their baby-sitting duties effectively'. Such sitting is, or can be, more hazardous than sitters realize, and much more than sitting, watching TV, raiding the fridge, and putting the child back to bed. It can involve asthma or sudden infant death.

A Canadian has done some calculations. Every year 400,000 children are born in Canada. About 5% of them have asthma attacks when between the ages of 2 and 6. Therefore 80,000 children, all between the ages of 2 and 6, are subject to asthma. Assuming two asthmatic attacks for each child per year, with half of these beginning at night, and further assuming that Canadian parents use a sitter once a month, this means that 2,700 asthmatic attacks take place when a youthful sitter is in charge. Even if such sitters stay – on average – only until midnight this means they will be subjected, terrifyingly, perhaps disastrously, to the sight of 900 attacks a year. For Sudden Infant Death Syndrome, occurring primarily during the first six months of life, the possibilities can also be alarming. If the same assumptions are made, it is probable that about twenty-six babies a year will suffer SIDS when in the care of a sitter. Not only will that older individual subsequently suffer the trauma of guilt and remorse should the baby die, but be bound to suffer a coroner's questioning so that foul play is ruled out.

CONFLICT As a final family point there is the near-inevitable disruption when dependants become independent. Offspring not only learn to dispense with parental care, but wish to have nothing more to do with it. Parent–offspring conflict can then take the place of parent–offspring care. As further cause for family strife there can be sibling rivalry, with each offspring wanting 100% of the parental love and resentful of what it feels is less than it deserves. Biologically this all makes sense. Each child of a particular parent possesses 50% of that parent's genes; therefore the parent maximizes reproductive success (as biologists phrase it) by investing the same degree of care and attention in each child. Children can dislike such unanimity, even seeing it as unfair. They share 50% of their genes with their siblings, but are primarily involved in the 100% which they possess. In the animal world this often leads, as with various raptors, to the expulsion of other chicks from the nest. In the human world jealousies can also be paramount, if stopping short of murder.

Conflict of a different kind begins even earlier. Mother and foetus are fighting (after a fashion) for hormonal control of the pregnancy as from its seventh week. According to Louise Barrett and Robin Dunbar, of University College London, 'morning sickness and the high blood pressure experienced by many expectant mothers can be interpreted as signs of a disagreement between mother and offspring ... The dispute is over what the mother should eat and how much oxygen the fetus should receive.' Each embryo has, as they described it, evolved the ability to fight back. It does this by masking (and actually neutralizing) the molecules – known as histocompatibility antigens – which trigger the mother's immune system. 'In other words, the embryo effectively "disguises" itself as maternal tissue so that the mother's body has no reason to reject it.'

A different form of competition then begins at birth. Pelicans can act in extreme fashion to gain more food, throwing themselves at their parents in a frenzy, beating and even biting their own wings, and attacking every chick in the area. Vervet monkeys adopt a strategy of deceit, exaggerating their actual needs to extract more care. Such a subterfuge can backfire when the mother has become used to excessive screaming, particularly if she then disregards a genuine call for help. Human parents may find many parallels in their own experiences by also watching animals, notably the primates. Each side – parent and child – learns tricks to outwit the other and then, when these tricks no longer work, devises alternate strategy. The only difference with humans, Barrett and Dunbar concluded after considerable field-work with monkeys, is that human children learn faster and are more skilled at social manipulation. 'They soon learn that most parents will do anything rather than let their children scream in public ... So, if you find yourself being emotionally blackmailed ... just remember that it's all part of the game of life.'

Perhaps wedlock, however it is interpreted, will never go away, despite the rise in singleton mothers, and all those absent fathers, and the divorces (in which Britain leads Europe), and the donated embryos and so much artificial reproduction. Couples, who live together, breed together and share mortgages together, often get up one day and, much to the surprise of all their friends, go through a marriage ceremony. Is marriage more than obeisance to a church or obedience to the state?

Hollywood is not known for bowing to convention, to spouse loyalty, to lengthy bonding, but even surprised itself in 1994. Richard Gere and Cindy Crawford, superstar and top model, the 'golden couple', paid £20,000 to advertise their continuing marriage in the London *Times*. 'Crude, ignorant and libellous' speculation had caused them to use an entire page 'to alleviate the concerns of their friends'. 'We got married because we love each other and we decided to make a life together. We are heterosexual and monogamous and take our commitment to each other very seriously...' (Unfortunately the crude and ignorant speculators were proved right the following year as yet another marriage bust.)

*

Marriage, divorce, separation, reunion, cohabitation, infidelity – there are fewer rules than ever before, and the game of life is more bewildering than it has ever been. As for families they too are more varied than ever in their history. It is even comprehensible why many a government has no explicit family policy, however fundamental this unit in which (almost) all of us spend at least the early portion of our lives.

Besides, what is a family nowadays? How can it best be protected so that children, the all-important generation to come, lead lives that make them better citizens in the years ahead? They are (often) happy to state that their forebears got things wrong, made a mess of everything, made the world a poorer place. Families in former times were more prominent than now, with quantities of children, with cousins galore and abounding relatives of every kind. Those days have now gone in the developed world, and are even receding elsewhere. So is that good or bad for the offspring? It would be nice to know but time (and all manner of further research studies) will indicate whether the reduction in the size and scope of family actually boded good or ill. And whether the current crop of youngsters – indulged, transported, studied, financed, heated (all more than ever before) – will do better than those of us who, lovingly, caringly, were the ones who gave them birth.

AGEING

I don't believe in dying . . . It's been done.
George Burns (who did eventually die aged 100)

I don't want to achieve immortality through my work . . .
I want to achieve it through not dying.
Woody Allen

Everyone seems to have had something to say about ageing, with more statements about it than for any other aspect of the human condition. On the positive side are Groucho Marx – 'A man is only as old as the woman he feels' and Shirley MacLaine – 'I don't understand the idea of slowing down' (a sentiment expressed when she was turning 60) and James Thurber – 'I'm 63, and I guess that puts me in with the geriatrics, but if there were 15 months in every year, I'd only be 43.' Less positive, but not entirely pessimistic, is Bob Hope – 'You know you're getting old when the candles cost more than the cake' or Joseph Heller (in *Catch 22*) – 'He had decided to live forever, or die in the attempt.'

Neither positive nor pessimistic but overtly realistic were John Lennon and Paul McCartney – 'Will you still need me, will you still feed me, when I'm 64?' – or George Bernard Shaw – 'Life is a disease, and the only difference between one man and another is the stage of the disease at which he lives.' Happily contemplative was Eubie Blake, jazz pianist, when interviewed on his 100th birthday (and one week before his death): 'If I'd known I was going to live this long, I'd have taken better care of myself.' Miserably blunt is David Lovibond – 'Old people are not only costly but horrid'. This individual, writing in the *Sunday Times*, referred to 'whingeing pensioners . . . usually characterized by lewd reminiscences, fatuous opinions and illimitable tedium'. It is possible to hazard a guess at

Lovibond's age while longing for some form of accelerated ageing to blow his way.

LONGEVITY A man of 104 once complained of pain in his right leg. 'What do you expect at your age?' countered an unsympathetic friend. The old man answered that his left leg, of equal age, was still in good condition, giving no cause for complaint, and he therefore had reason for objection to the one-sided, retrograde performance of his unfortunate right leg, a point of view cogently cited at an ageing conference to stress the inequality of ageing. Many individuals do die, in effect of old age, when in their 60s. Others continue blithely, healthily and lengthily, with a few even becoming centenarians. About 85% of individuals over 85 (in the UK) 'remain mentally intact' and 82% look after themselves. The smaller fraction do not do so.

At no time has old age inevitably meant incapacity. Leonardo da Vinci's extraordinary chalk renderings of physiological anatomy were drawn when he was 61. Winston Churchill was 65 when the Second World War began (and, one year later, he was relishing leadership). His ancestor, the first Duke of Marlborough, was 61 when achieving his most admired military success (remaining in the saddle that day for seventeen hours). Gladstone was 83 when he formed his last government. Goethe was over 80 when completing *Faust*. Francis Chichester was sailing, solo, around the world when aged 66. Anton van Leeuwenhoek (first man to see sperm through a microscope) lived actively to the age of 91. Michelangelo was 89 when working on the Rondanini Pietà – before plague called a halt to it and him. Heraclitus of Ephesus lived to be 96, and the Athenian orator Isocrates 98. Pythagoras was in his eighties when he died, Sophocles in his nineties and so too Hippocrates, each having lived at a time when, or so we assume, conditions for survival were rougher than now.

Conditions for survival have not necessarily been better for those at the top. No English monarch ever achieved three score years and ten between AD 871 (when age of death began to be recorded) and 1760 (when George II died aged 77). His forty-seven predecessors in the previous 889 years did nothing like so well, with only nine of them reaching 60. Since his death UK sovereigns have fared much better, all but one of them (George VI) reaching 60, and all but two more (George IV and Edward VII) reaching 70. Their average age at death, including the still living Elizabeth II, is over 71. US presidents, since the first inauguration in 1789, have also lived a long time – partly because they first have to achieve their greatness instead of, as often with royalty, having (via inheritance) it thrust upon them. Of the thirty-six who died before Lyndon Johnson joined their ranks in 1973 their average age at death was almost 69. Youngest to die was John Kennedy (at 46) and the two oldest (both reaching 90) were John Adams and Herbert Hoover.

Without doubt there is also frailty to counterbalance our span of years (longer than every other mammal, including elephants). Psalm 90 tells of three score years and ten, but biblical mention of plagues, pestilence, leprosy and affliction, seemingly on every page, affirms that longevity was not always partnered by good health. Nor is it today. The famous Kennedy family of the United States, so assaulted by premature death (assassinations, aircraft crashes, a drug overdose), also experienced prolonged dying. Joe Kennedy, patriarch, was so damaged by a stroke when aged 73 that he was unable to move or speak for his final eight years. Rose Kennedy, his wife, lived to 103, but suffered five strokes and required continuous nursing.

It is likely that late death, relative to three score and ten, will become increasingly the norm, with labour and sorrow the certain partners (as also promised by the psalmist for anyone reaching four score years). Perhaps the writer/composer felt aware that ageing was a novel phenomenon which we humans were not intended to see. Remains of *old* prehistoric people have rarely been found. Wild animals do not normally live long enough to age, and wild humans – if they can be so described – were probably similar. Ageing, as Leonard Hayflick (of California University) has described it, is an aberration of civilization.

THE GREYING   In 1900 there were between 10 and 17 million
people over 65 alive on earth. Global population numbers were less
well known than now, but the elderly proportion certainly formed
less than 1% of the total. When old age pensions began (for men over
65 early in the twentieth century), 5% of the British population was
eligible. Today's proportion is nearer 25%. Worldwide the number
over 65 is now about 200 million for women alone, and is expected
to be 326 million by 2015. In 40–50 years, according to the WHO,
the aged total – men and women – will be 678 million. By 2025 Japan
is expected to have 29% of its population over 60. Sweden, already
with 20% over 60 and believed to possess the world's oldest
population, will have 30% by that year. China, with figures blanketing
those of every other nation, will have 80 million over 65 even by
2000.

Speed of change has also been different. A total of 115 years passed
before the proportion of elderly in France rose from 7% to 14%, but
only thirty years in China. With stable communities, such as England
and Wales, forecasts are still disturbing. The total population of those
two areas is expected to rise by 8% between 1991 and 2031, with the
60–74 age group rising by 43%, the 75–84 group by 48%, and the
over-85s by 138%. In British society the elderly are the only group
whose absolute numbers and whose proportion of the population are
both increasing.

A child born now (in Britain) can expect to live ten years longer
than one born in 1950 and twenty-five years longer than one born in
1900. Improved length of life seems satisfactory, but is not a total
blessing. One Canadian survey, after noting an increase in life-span
between 1951 and 1978 of 4.5 years for males and 7.5 years for
females, also pointed out that disability-free life only increased by 1.3
years (for males) and 1.4 years (for females) in between those dates.
'For each active functional year gained, we have added about 3.5
compromised years,' the report concluded. Much of that compromise
concerns cash. In Britain the elderly cost the Health Service seven
times more per head than adults aged 16–64. Old people are now
12–20% of the population of Western countries, but they receive

30–40% of prescribed drugs (and are particularly at risk from adverse reactions to those drugs).

A somewhat contrary report, published for Britain by the General Household Survey 1994, stated that Britons are not only living longer but staying healthier in old age. Bob Barnes, head of the survey, said: 'Despite the ageing population, the proportion who cannot get out and look after themselves hasn't changed.' Nearly 40% of all elderly people said their health had been good in the previous year compared with less than 25% who said it had not been good, an improvement over 1980. Less of an improvement, from the viewpoint of home care, was the reduced proportion of elderly people living with their children or close relatives, a drop from 21% to 15% since 1980. Of the 39% living on their own, a tenth needed help to bath and a third could not cut their own toenails.

Care for the elderly varies widely, not least in the provision of home helps. These are 33 per 1,000 people over 65 in Denmark and 2.4 per 1,000 in Germany. In Denmark home help, financed by local government, is available to all old people. In Germany it is free only to those receiving basic social welfare. (Germany is also different in obliging offspring to look after their parents.)

Within the UK there are 7 million people with care responsibilities for people other than children, and 1 million of the carers are over 65. They save the government (stated a recent estimate) £34 billion a year in care costs. Most of those cared for are over 65 (80%) and 20% are over 85. Of the carers 1.5 million spend at least twenty hours a week on this work, and some of this army of volunteers – as they have been called – spend fifty hours a week. According to the British Medical Association only 250,000 of the 7 million receive any state benefit for their caring responsibilities, and the formidable army provides more care than the combined efforts of the National Health Service. A team from Manchester University has estimated that 40,000 children, aged between 11 and 18, look after a dependent relative. The need for carers, whether paid or unpaid, will also not diminish in an ageing society.

There are many reasons for higher health expenditure among the old (as any old person will be quick to affirm). In one British practice (at Exeter) its patients aged over 75 (average age 85) suffered from cataract (20%), depression (15%), diabetes (10%), cancer (8%), hip replacement (8%), stroke (7%), and dementia (5%), all of which are diseases of the elderly. In that same practice the average old patient was suffering from 2.5 to 3 conditions at the same time. As an article in the *Journal of the Royal Society of Medicine* stated: 'Medical care of the elderly represents one of the greatest, if not *the* greatest, organizational challenge to health services all over the world.' The increase in life expectancy during the first half of the twentieth century is generally attributed to public health measures rather than to medical intervention; but during the second half (and from now on) medical science and health care are shouldering the burden. Neither science nor care comes cheap.

In the US the life-time medical costs for a single individual are approximately $100,000. Of that total $18,000 is spent in the last year of life, with two-thirds of this sum being consumed in the final month. As for the Medicare budget, half is spent on the last few months of life. Therefore should the elderly receive this disproportionate quantity of health care funds? The British Medical Association has pronounced 'that no patient should be denied medical diagnosis and treatment just because of advanced age'. But health care *is* a limited resource. If the greatest good is for the greatest number a form of ageism will have to operate. In America the question of 'intergenerational equity' is already being addressed, with children's advocates claiming that the elderly now receive too many of society's resources and children too few.

According to a 1994 article in the *Journal of Medical Ethics* ageism 'already flourishes in British hospitals'. A. B. Shaw quotes his own infirmary (at Bradford) 'where there is a useful coronary care facility'. Patients aged less than 65 with suspected myocardial infarction (blockage of blood supply to parts of the heart) are 'routinely' admitted to it. Those over 65 are admitted to other wards, but transferred to the care unit only 'if a clinical condition arises'. The relatively young are therefore favoured, with the relatively old on the sidelines until specialist care becomes more critical. As for the public, says Shaw, they 'have never been consulted'. With the number of British people over 75 increasing by 25% even between 1980 and 1990, with this trend continuing, and with the length of stay in hospital varying from ten to twenty-five days for those who are, respectively, aged 60–65 and 80–85, the problem is set to increase, to become more expensive, and a matter for more general discussion. Ageism is 'widespread but erratic', says Shaw, 'and often covert. It should be open and agreed.'

The current practice is 'insidious', added John Grimley Evans (professor of gerontology, Oxford), with 'bad ethics' creeping into the health service and people 'valued according to their life expectancy'. A 1991 study showed one in five coronary care units to be operating an upper age limit, usually 70 or 75, for admission, and in the US, according to Evans, 'older people were much less likely to get good and effective cancer treatment'.

*

More is involved than chronological age, as Evans has also pointed out. 'The lower social classes have lower life expectancy than the upper, blacks have lower life expectancy than whites, and men have six years lower life expectancy than women.' Therefore a higher social class white female should have a dozen more years ahead of her than a lower class black male of the same chronological age. Equal treatment for both would, in consequence and in probability, be more rewarding for her than for him. Similarly, if she leads an unhealthy life-style, being a fat, heavy-drinking smoker (when he isn't and doesn't), she could almost certainly be erasing her sex, class and race differential. Should her doctor take this information into account? Should society be creating guidelines? Should the fact be openly proclaimed that limited resources ought to be used effectively, with the young more deserving than the old, the healthier life-style more favoured, and should expectation be taken into consideration?

This particular can of worms, exposing horrendous decisions, is swelling in size even faster than population growth. And, as with population, it is growing fastest in countries still developing. The world population over 55 is increasing by one million a month, with

80% of this growth occurring in developing countries whose aged population is expected to reach 1 billion by the year 2020. No one predicts equality among nations at that time, only two decades from now. The developing world will still be developing, and be behind-hand, and relatively poor. The problem of its elderly will therefore be a massive extra problem added to its already lengthy list.

APPROACHING DEATH   The deadliest disease of all is life: it forever leads to death. (At least it does when a sexual system of breeding is in place.) Life also leads, with increasing age, to a greater risk of degenerative disease, like Parkinson's and Alzheimer's (about which more shortly), like cancer (with 50% of all cancers occurring among people over 65), like influenza (which accounts for 10,000 to 40,000 deaths annually in the US, 80% of these among the elderly), like heart disease and stroke (the greatest killers in developed communities), like hearing loss (affecting 29% of those over 65 or, in another statistic, 28 million Americans), like osteoporosis (also more later) and broken bones, such as hip fracture.

Each year in England 300,000 older people are seen by their local accident departments, with some 59% of their accidents having occurred at home. About 4,600 people over 65 die annually as a direct result of falls. Hip fractures are the commonest consequence, twice as common with women as with men, and – at any age – are associated with high mortality and morbidity. In 1994 there were 54,028 cases of hip fracture in England, a total likely to increase with an increasingly aged population. Some 30% of older people sustain a fall each year.

A report from New Zealand, where hip fractures 'occur in 15% of post-menopausal women', considered that the increased rate of hip fracture could be explained 'by an increase in the length of the hip axis'. There were 317 hip fractures in 1950 among women over 65, and 2,153 among such women in 1987, almost a seven-fold increase despite the number of women over 65 increasing only two-fold. Increase in female height, brought about – it is believed – by better nutrition, has been partnered by an increase in hip axis length, and therefore – it is also believed – by a greater susceptibility to falls.

*

Better life expectancy has not reduced one extremely fundamental ageing fact – after the age of 30 the risk of death doubles every eight years. It did a century ago (when mortality statistics began to be reliable). It does so now, despite all the changes, and despite a declining death rate throughout the century. A 38-year-old man of today has a greater life expectation than his father, and even more so than his grandfather, and more so yet again than his great-grand-father, but he is still twice as likely to die as is a 30-year-old of today, just as his ancestors were when they were eight years older than their 30-year-old contemporaries.

To what extent is ageing partly in the mind? Becoming forgetful is said, by many, to be an inevitable accompaniment of later years. In China, where the elderly are held in high esteem, older citizens taking memory tests do not differ significantly from younger compatriots. Certain capacities can actually improve with age. Freshmen, aged 19 at the University of Iowa in 1919, then took an intelligence test. When located forty years later they scored as well or better with the same tests. But wherever 'timed' tests are involved, fewer individuals over 65 will answer correctly. There is frequent mention of neuron loss in the cerebral cortex with increasing age, but Robert Terry (in 1987) found no loss between age 20 and 80. What he did find was atrophy, a reduction in brain weight.

Physically there is undoubted deterioration, even when attempts are made to deny this fact. Clarence de Mar, American enthusiast, ran the Boston marathon for fifty consecutive years, and won it several times, but his speeds declined at an average of two minutes a year. At his death (aged 68 from cancer) he had large-calibre cerebral arteries but quite a bit of atherosclerosis. 'Fitness for Older People', the title of an article in the British Medical Journal of July 1994, reported that healthy elderly individuals lose strength (the ability to exert force) at 1–2% a year, and power (force × speed) at 3–4% a

year. Old people do not necessarily take exercise, but can gain by
doing so. The *BMJ* referred to a study where women aged 75 to 93,
after 'training' three times a week for twelve weeks, increased their
strength by 24–30% (precisely the same increment observed in young
people after similar training).

✦

Exercise, nourishment, environment, genes, welfare, habits – all may
affect the ageing process, but they only delay or advance the
inevitability. A team from Chicago's Center on Aging stated, in
Scientific American (April 1993), that average life expectancy is
unlikely to exceed 85 years 'in the absence of scientific breakthroughs
that modify the basic rate of aging'. Fifteen years earlier Jean
Bourgeois-Pichat, demographer of Paris, calculated that 'average
human life expectancy' would not exceed 77 years. To reach this
figure he theoretically eliminated all deaths (such as accidents and
suicides) unrelated to senescence. He then estimated the lowest death
rates possible for diseases definitely associated with ageing (cardiovascular, cancer, etc.) before concluding that 77 was the best we could
expect from birth; but several nations are already proving him wrong.
The Japanese, world leaders in this field, can now hope to achieve 76
years (if male) and 82 (if female), thus averaging 79. (The Japanese
have also grown in physical stature by an average of 3–4 inches since
the Second World War, and are currently only 3% short of average
British height.)

The doctor/patient ratio is not always the proportional blessing
one might expect. Japan, the leader, has only 1 doctor for 610 people,
as against 1 in 575 for the UK, 1 in 440 for the US, and 1 in 296 for
Spain where life expectancy is lower than for the US, the UK, and
much less than for triumphant Japan.

CENTENARIANS For those who welcome the notion of longer
life there is no indication that humans will be able to mimic tortoises
(thus contemplating two centuries rather than one). Churchill,
however warlike when aged 65, did eventually die – at 89. His

victorious Marlborough ancestor may have done well on the battle-field in his sixties, but he died at 72. Leonardo da Vinci succumbed at 67, Chichester at 71, and not one of those long-lived individuals already mentioned – such as Plato, Hippocrates, Michelangelo, Sophocles, van Leeuwenhoek – achieved a century.

In theory the British monarch sends congratulatory telegrams to every Briton who achieves 100 years. In practice this only happens when an acquaintance (or even the centenarian) sends Buckingham Palace a note, plus the necessary evidence of a birth certificate. Certainly the palace must be busier these days. In 1951 there were 100 individuals within the UK who had been born in 1851 (or earlier). By 1991 there were 4,390 who had started life in 1891 (or earlier). In Japan there were 4,100 (at the last count) who had scored 100. (Japanese life expectancy was 44 when those centenarians were born.)

Despite these growing numbers the chances (for any individual) of reaching 100 are not good. Of the 650,000 or so who die in Britain annually only about 400 have achieved three figures (roughly 1 in 1,500). As with winning lotteries, those who do reach 100 (or more) certainly achieve renown – along with their telegram. Everyone in Britain knew when Emmanuel Shinwell, member of parliament from 1922 (when aged 38) scored 100. By then he had become Lord Shinwell, and was thus – as one of the British peerage – a member of the most extended cohort in the world for reliable pedigrees. It has witnessed only three centenarians in the last ten centuries.

As for the less well-documented British population, a number of births were entered in parish registers (after 1538), but official birth certification – for all – only started in 1837. Therefore not until 1937 could an 'official' centenarian be recorded. In fact not until 1943 did the first individual (a woman) die who was over 100 *and* in possession of a proper birth certificate, this having been issued in 1837. The first British man to follow suit died in 1945 when he was 105. The former Soviet Union's allegedly long-lived citizens, reputedly still siring offspring when deep in their second century, usually lived in Georgia, Stalin's home base. Since his death the number of Georgian centenarians, as one man phrased it, has 'declined from a torrent to a drip'.

They are not necessarily centenarians but the people of

Campodimele, a community seventy-five miles south of Rome, rarely
die before reaching 85. More than ninety of its 280 inhabitants are
aged between 75 and 99, according to a report published in 1995.
They have consistently low blood pressure and cholesterol levels, with
some 80-year-olds mimicking the levels of newborn babies (of whom,
it would seem, there are very few in Campodimele). Some of the
villagers emigrated to Toronto in the 1960s, and they too are living
long and healthy lives. Pietro Cugini, who studied the villagers for
five years, reports that they eat at the same time every day, rise at
dawn and go to bed at sunset. 'Early to rise and early to bed,' wrote
James Thurber, 'makes a man healthy and wealthy and dead.'

A European centenarian, who hit the world's headlines in February
1995, then celebrated her 120th birthday. Jeanne Calment, of Arles in
southern France, had already outlived her husband, her daughter and
her grand-daughter. It was not so much her extreme age that amazed
the world, but her personal involvement with much of history (and
her remembrance of the facts). She had had a meal on the Eiffel
Tower when it was still being built. She could remember Vincent van
Gogh (when he entered her father's shop to buy canvases) as 'ugly as
sin, a bad-tempered "nut" smelling strongly of drink'. She was already
a pensioner when she acquired the right to vote. In time she earned a
place in the *Guinness Book of Records*. Towards the end of 1995 she
became the oldest person ever to have lived with verifiable proof of
that achievement. On 18 October she overtook a Japanese man who
had died in 1986 at the age of 120 years and 237 days. Mme Calment
was then supreme. (In France her longevity has been cited as a boost
for the French life-style and the low-cholesterol diet, common around
the Mediterranean, of olive oil, fruit, vegetables and red wine. Mme
Calment's comment on achieving the world record was: 'The good
Lord seems to have forgotten me.')

This extreme centenarian was examined by a leading dementia
epidemiologist, Karen Ritchie, over a period of six months. In that
time no evidence was found of 'senile dementia, delirium or Parkin-
sonian features'. The widow, who had lived alone after her bereave-
ment until aged 110, and independently until 115, was both alert and
curious. Her memory was within normal limits for people over 80,

and her IQ above average, being normal/superior. She did have memory difficulties, and also dysarthria (impaired speech), partly for refusing to use her artificial teeth. The technique of neuro-imaging discovered, as the *Lancet* recorded, that 'she had pronounced atrophy of the temporal, parietal, and occipital lobes, widening of the sulci, and enlarged lateral ventricles'. This editorial concluded: 'Bright people can sustain ... neuropathological morbid changes and yet retain considerable intellect.'

The brain does indeed shrink with increasing senility, but proper function can survive. Less than half of all centenarians are demented. (Senility is often used as a synonym for dementia. It should only mean the changes due to old age, with particular reference to the brain. Dementia is a mental disorder. To be senile is not necessarily to be demented. To be demented is not necessarily to be senile.) In Ritchie's opinion 'high premorbid intelligence may be a prophylactic'. In other words, to be intelligent beforehand can diminish or offset 'neuropathological morbid changes', such as that inevitable shrinkage of the brain with increasing age. Van Gogh's paintings sell today for many millions (beyond belief in his lifetime) and it *is* intriguing that one contemporary of ours can still remember, clearly and vigorously, the ugly, bad-tempered nut who made them (and who died in 1890, more than a century ago).

Madame Calment will follow in his wake, some time, along with everyone else. Intriguing questions which follow from that most basic statement are How and Why. One curious fact, reported in April 1996 (by David Smith, of the Northwestern University Medical School, Chicago), is that those who do achieve a century may owe their success to abnormally strong cancer defences. He had found that, of those who had died in a recent year aged between 50 and 69, some 40% were victims of cancer. Of the centenarians who had died that same year only 4% were similarly afflicted.

THE PROCESS OF AGEING is a conundrum, with the end result obvious but its mechanism obstinately obscure. A recent attempt at classification, in *Aging, Sex and DNA Repair* (by Carol and Harris

Bernstein) lists about 300 hypotheses on the causes of ageing. As T. B. L. Kirkwood phrased this dilemma: 'The phenomenology of senescence is rich in the abundance of model systems that it offers for the experimental study of ageing.' Alex Comfort, noted writer on both sex and death, has worried about this excess of hypotheses – 'Throughout its history the study of ageing ... has been ruinously obscured by theory, and particularly theory of a type which begets no experimental work.' Or, to quote Kirkwood again, 'the multiplicity of ageing theories, together with the multiplicity of experimental model systems, has generated a body of knowledge which is conspicuous for the looseness of its connections'.

An engaging book, *Reversing Human Aging* by Michael Fossel (published in 1996), suggests that medical science in the next ten years will 'extend your life dramatically, make you younger and keep you that way'. His argument starts from the fact that cancer cells are immortal, forever dividing. It then leans upon the work – by Leonard Hayflick in 1964 – which showed that ordinary human cells divide only a set number of times, this number depending on the cell type. Therefore, says Fossel (who gained his doctorate in neurobiology from Stanford University), normal cells should be induced to behave – in part – like cancer cells. This could/might be done by interfering with telomeres, the bits of DNA at the end of each chromosome which play a key role in cell division. Pieces of telomere are apparently lost each time the cell divides. Therefore, so this argument runs (headlong), use the enzyme called telomerase to rebuild the telomeres. But, as Joanne Silberner wrote (when warmly but cautiously reviewing this book), science does not necessarily find it possible to achieve what is apparently achievable, particularly in a timespan like ten years. She reminds us that the US Secretary of Health and Human Services announced in 1985, following the discovery of the virus which causes AIDS, that a vaccine would be available in a year and the disease would then be vanquished. Well, the virus known as HIV is still wreaking havoc and not many people would put money on the ageing process being seriously delayed in the decade now ahead.

In short, there is not yet a general agreement how the ageing process works, however much it is so emphatically known that the system never fails. Every kind of biological error seems to occur in other areas – infants are born with all manner of deficiencies, they grow on occasion as dwarfs or giants, they are extraordinarily intelligent or quite the converse – but the error of immortality has never occurred. Nor is there is an equivalent to progeria, the accelerated or premature form of ageing which may show its first symptoms during the first year of life and kill off its victims – often from coronary thrombosis – long before the normal span is reached. There is not even a word (antigeria?) for a major delay in the manner of growing old.

Chernobyl, site of the world's worst nuclear accident, has caused its own brand of progeria. People who cleaned up the reactor, and were therefore greatly exposed to ionizing radiation, are showing an increased incidence of cardiovascular and gastrointestinal disease, along with more cancer of lung and stomach. 'It is as if they are suffering from accelerated ageing,' said Alexei Nikiforov, head of the Russian Centre of Ecological Medicine.

Progeria itself, also known as HGS (Hutchinson–Gilford syndrome), is extremely rare, only about eighty cases having been reported since it was first described (by Johnathan Hutchinson in 1886, with Hastings Gilford reporting the second case in 1904). Its incidence has been estimated as one in 8 million births. It is commoner in males than females (1.5 to 1) and in Caucasian rather than Afro-Americans (97 to 3). It is characterized, acording to one definition, by 'growth retardation and accelerated degenerative changes of the cutaneous, musculoskeletal and cardiovascular systems'. The cause is unknown, there is no effective treatment, and average life expectancy is thirteen years, with a range of seven to twenty-seven years.

Normal ageing starts to operate the moment life begins. (Or even before birth itself. Foetal surgery, currently feasible – even if rare – is said to leave no scar in its wake.) Skin damage at birth is more likely to be properly (and invisibly) repaired than at any later age. A human aged 18 is, to some degree, at the threshold of life, but is also 18 years *old*, having suffered illness, much natural radiation, and general wear and tear during those 18 years. It is difficult learning a language perfectly at that threshold (and is easiest of all when very much younger). All wounds heal less well as the years progress. Hearing becomes less acute with the passing of time (babies often reacting to modest sounds as if bombs are detonating). Most contrarily ageing seems to happen fastest when life is at its newest – and, at the other end, can be most leisurely for those about to die.

Some say that DNA, the basis of our genetic inheritance, is also at the base of ageing, with accumulated damage to DNA causing increasing inefficiency. DNA is likely to be relevant, but accumulation of damage is thought too easy as a concept. It does not explain why the ageing process exhibits so many different forms and rates. Numerous animals have no opportunity to age, being predated so remorselessly that youthful death is almost certain; but they still grow old and die, as in a zoo, when predation has been quite removed. ('A robin redbreast in a cage lives to a tremendous age,' states the skit on William Blake who preferred that such captives 'put all heaven in a rage'.) There are creatures who do not noticeably age, such as salmon (known as semelparous species), which expire the moment they have given their all in reproduction. They have therefore compressed their senescence into minutes more than years (much as Dorian Gray's portrait altered so rapidly at the awful conclusion of Wilde's tale).

There are also creatures that do not reach a fixed size at maturity and show no obvious ageing changes, such as the Galapagos tortoise and sharks. A 55-year-old American lobster weighing 44 lb could close its claw as fast as any juvenile. Such animals are not immortal. They do eventually succumb, with predation, disease and accident taking a steady toll. The odds against prolonged survival get them in the end.

Neat definitions of senescence exist, such as 'the decrease in reproductive success with increasing age', but none explain the process. In a major book Caleb E. Finch (of Northern Arizona University) stated: 'It is safe to say that there is no single explanation of senescence at the genomic, cellular or physiological level that is applicable to all organisms.' He considers senescence to be a trait 'that has repeatedly undergone evolutionary modification'. It is certainly under genetic control, but the form of that control has been modified, over and over again, throughout evolution.

Perhaps no precise cause, or set of causes, will ever be elucidated. It just so happens, for human beings, that their hotch-potch of inheritance leads them to die with increasing frequency after the onset of reproduction. The trend does not wane, as with exponential curves that never reach the base, but always hits that bottom line. This happens, more or less, at three score years and ten for most individuals, just as the psalmist first sang over 2,000 years ago. It also happens at five score years for a very, very few, and at six score years for extraordinary exceptions like Madame Calment.

THE CHANGES that make up ageing are considerable. Height decreases. Hair thins (save where it flourishes, as in the nose and ears). Skin wrinkles, becomes dry, and has dark patches. Limbs shake. Walking becomes more of a shuffle. The extremities get cold. So does the body as a whole (and a 'normal' youthful temperature may suggest fever in the elderly). The brain uses less oxygen, and becomes lighter. Its grooves become wider and deeper (much as a walnut may open up and shrink with age). The senses deteriorate, with eyesight lengthening (usually), with presbyopia probable (as the lens becomes less flexible), with smell and taste diminishing, and with hearing (as already mentioned) suffering, notably for the higher tones – but not in everyone.

Lungs shrink. The liver shrinks. Both uterus and ovary atrophy. In females there is the menopause. In males, much to their night-time annoyance, the prostate gland enlarges, causing problems with urination, and both sexes can become incontinent (but much less so with faeces). Cramp becomes commoner. Pulse rate lessens, but can also increase in the very old. Hearts vary, either shrinking or enlarging. Blood pressure goes up. Reaction times slow down (but UK motorists over 50 have fewer accidents than younger drivers, despite using their cars more often). Speech alters, becoming – in some – little more than a tinny tremble. The body's water content diminishes. So do cell solids and bone mineral, but fat goes up. Intake of food and drink goes down, accompanying a drop in basal metabolic rate. Teeth are often lost. Joints stiffen. Both jaws shrink, and the chin may jut. Reflexes, such as the ankle jerk, can nearly vanish (or can remain virtually unchanged). 'Old age does not come alone,' said Plato. It certainly does not.

Such changes have always occurred (even though not all, such as hair thinning, are linked to increased decrepitude). Jaques, in describing the seven human ages (within *As You Like It*), defines the 'last scene of all [as] second childishness, and mere oblivion, sans teeth, sans eyes, sans taste, sans everything'. Currently there is a plethora of 'avec' to counteract the 'sans'. The elderly have sticks, one, two, or four (with Zimmer frames). They have hearing-aids and spectacles. Their hip-joints may be replaced; so too portions of their knees. Defective blood-vessels can be circumvented, as in bypass surgery. Whole organs – liver, kidney, heart – can be transplanted. Prostheses (of various kinds) can be supplied. Corneas can be changed, and so – with difficulty – can nerves. Bone marrow, as in leukaemia therapy, can be killed and then replaced. Pace-makers maintain a beating heart. Teeth are certainly altered, first with fillings, then by crowns and dentures.

Jaques's speech, already substantially rewritten, will be altered increasingly as the years progress, as the brand-new world arrives. There is – for some – cryogenesis, keeping the body in cold storage (in the hope that later generations might restore its life). There will assuredly be further transplantation, as well as physical substitutes for the heart, the lungs, the kidneys. Currently no one foresees immortality, but further prolongation of life, using methods as yet unknown, is a certainty – for those individuals (and nations) able to afford the benefits on offer.

Jaques did not mention urinary incontinence, a particularly unwelcome attribute of old age. In the US, according to a report in the *Lancet* (of January 1996), it 'afflicts 15–30% of elderly people living at home ... and half of those in long-term care institutions'. Medically it can lead to rashes, pressure ulcers, infections, and even falls and fractures. Psychologically it can involve embarrassment, stigmatization, isolation and depression. The annual US figure for 'managing incontinence' is $10 billion, or more than the amount spent on dialysis and coronary-artery-bypass surgery combined. Numerous happenings predispose the elderly to incontinence – bladder capacity and contractility decline, night-time excretion increases along with sleep disorders – but none *cause* incontinence. The (faint) good news is that conditions which promote a loss of continence are often factors outside the urinary tract, and are amenable, as the *Lancet* phrased it, to medical intervention. Suicide (about which much more in its own chapter) has also been rising with an ageing population. Within the US, between 1980 and 1992, the suicide rates fell in the 65–69 and 70–74 categories, but increased in the 75–79 (11%), 80–84 (35%) and over 85 (15%) age groups. Older people make fewer attempts per completed suicide, a chilling fact indicating greater determination to have done with an elderly life.

*

Some old age changes have received more attention than others, such as rise in blood pressure (for both sexes), such as hormonal alterations (notably in women), such as Parkinson's and Alzheimer's (that seem to strike randomly, without prior warning) and these are now addressed in turn.

HYPERTENSION The systolic blood pressure of a new-born baby is about 40 (millimetres of mercury), and its diastolic pressure even less. At 10 days the higher figure may have risen to 70, and after a month to 80. It then rises less rapidly, being still under 100 at age 10 years. When full-grown a normal individual has a systolic pressure of 120, with a diastolic pressure of 80, this combination usually written as 120/80. (Systolic is the maximum pressure from the heart, exerted during each pump, and diastolic is the minimum residual pressure before a further pump arrives.) There is nothing abnormal about human blood pressure. Nearly all mammals, whether huge or minute, have maximum pressures between 100 and 200. Humanity lies towards the lower end – at first.

With increasing age, perhaps from the twenty-fifth birthday, there is a tendency for systolic pressure to rise by 0.5 or so a year. By age 60 it may be nearer 140, and by 80 nearer 160. The averages are different between the sexes. Males tend to have higher figures from adolescence, and this average disparity continues into middle age. At old age it may vanish (partly because many high blood pressure males have died, and are therefore no longer included in the totals). African-Americans have higher blood pressure than European-Americans, and the Caribbean-British have higher pressure than the European-British, these differences becoming apparent during the second decade of life. Social class is relevant – the lower the class the higher the pressure – but in the US other ethnic groups, generally of lower class such as Hispanics, Asians, and Native Americans, have pressures similar to US whites. After the age of 30 both male and female blacks have, on average, higher pressure than male whites. Norman M. Kaplan, in the *Lancet*, concluded that 'about a third of the excess mortality of US blacks over non-blacks can be explained by their greater burden of hypertension and other known risk factors'. He thought another third could be explained by 'their lower socio-economic status', leaving the remaining third quite unexplained.

There is still much to be learned. Samburu warriors (of Africa) show no rise in blood pressure with age. Black Jamaicans show a considerable rise. People in west Wales have three times the risk of stroke (not unrelated to hypertension) compared with people in East Anglia. British civil servants (in one Whitehall survey) had an average systolic pressure of 133.7 for the highest grades and 139.9 for the lowest. The relationship between hypertension and risk of heart failure or kidney disease is formidable. The Framingham Heart Study, a follow-up lasting thirty-four years, showed that the risk of congestive heart failure was two to four times higher for individuals in the highest as against the lowest blood pressure group when they entered the study. A fifteen-year survey of over 350,000 individuals (in a so-called Multiple Risk Factor Intervention Trial) showed that a 10 mm higher systolic pressure increased the likelihood of 'end-stage renal disease' by 1.65, after adjustments had been made for 'age, race, cigarette-smoking, serum cholesterol concentration, treatment for diabetes, previous myocardial infarction, and income'. In short a higher blood pressure increases risk of death. On a lighter note there does not seem to be, however frequently alleged, an association between high blood pressure and snoring.

*

What can be done to reduce hypertension? Lifestyle modification is certainly a first approach. Weight should be brought down, if feasible, for those more than 10% above a desirable level. Salt intake should be reduced (to less than 2.3 grams of sodium per day). Alcohol consumption daily should not be more than two glasses of wine/two beers (one pint)/two 'shots' of spirits. Exercise should be undertaken judiciously, says a report from the Dallas medical centre making these strictures. Of these four factors – weight, salt, alcohol, exercise – salt control is considered to be 'especially effective', even if it is tricky for any ordinary human to know if an intake of 2.3 grams is being exceeded. It is also thought that 'lifestyle modification' is not only less feasible but less rewarding for the very old.

Medical treatment of the hypertensive elderly was not supported by sound evidence until the mid-1980s. In the late 1980s and early 1990s evidence surfaced that anti-hypertensive treatment was truly beneficial. A Medical Research Council trial (in the UK) showed 'a significant reduction in stroke and in all cardiovascular disease events' among those being treated. From this and various other trials (notably in Sweden and the US) a consensus of opinion has arisen that it is sensible to treat high blood pressure in the elderly – at least until the age of 84. Facts from beyond that time are still inadequate and uncertain. What is certain is that the target pressure should be amended from young to old. For younger hypertensives the aim should be to get the figures below 140/90. For older individuals the intent should still be to get them down, to less than 160 systolic if it was over 180, for example. It used to be thought that diastolic pressure (in the elderly) should not fall, or be coerced, below 85, but some recent work conflicts with this generality. Current agreement is that systolic pressure is the one to be reduced, however much this may or may not affect the diastolic figure.

Finally, how common is hypertension among the elderly? Too common is one answer, bearing in mind its association with increased risk. The US Third National Health and Nutrition Examination Survey, conducted from 1988 to 1991, showed it was present in 54% of the total population (sexes and races combined) aged 65 to 74. With African-Americans on their own the proportion (of the same age group) was 72%. Even if these proportions stay constant the actual numbers of hypertensives are set to increase along with the general increase of older populations.

PARKINSON'S DISEASE James Parkinson, the English physician who lived from 1755 to 1824, gave his name to this disorder, also known as shaking palsy or paralysis agitans. It is a chronic disease affecting the elderly. Currently there are over 100,000 sufferers in the UK (roughly 1% of those over 65). Tremors are often thought of as the major symptom, coupled with stiff and sluggish muscles, with twitching, and with a distinctive tottering gait; but there are many others. There can be difficulty with speech, and an inability to swallow. The face can be devoid of expression. The feet and legs may even refuse to move, but mental faculties are not affected. As with various other palsies the patient can seem intellectually deficient, mainly though an inability to speak properly, but may be nothing of the sort. Most disturbing fact of all is that, although medication can ameliorate the situation, there is at present no known cure.

Symptoms often start with unwelcome trembling, and stiffness of fingers and hands, or arms, or legs. Greater effort at control, or excitement, can make the situation worse. Tremors and stiffness increase as the disease becomes more marked. Walking can become a business of leaning forward as if instructing the legs to remember their customary function. To begin with they co-operate, as best they can, but then may fail entirely. The annoying tremblings cease only when the sufferer is asleep.

This disease is due to degenerative changes in the basal ganglia. These groups of cells exist in the centre of the brain below the cerebrum. They can also be damaged by a stroke, or poisoning, or injury, but such insults are rare. Unfortunately Parkinson's is less so and, like every other ailment of the elderly, is likely to become more prevalent as the vulnerable elderly become more numerous. Treatment of the disease (as against cure) is possible, but only to a limited extent. With parkinsonism there is a lack of dopamine in the brain's affected part. Giving dopamine to the sufferer does not assist, but a substance known as L-dopa does help because this is then converted to dopamine within the brain. Unfortunately it only relieves some symptoms; it does not banish them. Also it only provides relief to three-quarters of those receiving the treatment. Perhaps two kinds of shaking palsy are therefore involved.

A new and most experimental form of treatment was initiated in the late 1980s. This involved the injection of foetal brain tissue (collected from abortions) into the damaged Parkinson brain. The procedure is expensive (with Americans paying $40,000 for the surgery), complex and controversial. The controversy is largely associated with the ethics of using foetal tissue, receiving permission from the mother, and choosing suitable recipients. (Younger sufferers are favoured. So too those in whom the symptoms are still slight.) As for results, 'we now have good evidence that grafts can survive and have functional effects,' says Olle Lindvall, head of the Lund University team in Sweden which has pioneered much of this work. Niall Quinn, of the Institute of Neurology in London, has said that 'clinical use is still some years away [and] none of us are kidding ourselves that this is going to be a cure'. Some patients have benefited from the operation; some have not. No one yet understands this discrepancy. The operation is also risky, with haemorrhage – and even death – a possibility.

One blessing, announced in 1995, is that fewer foetuses may become necessary. Traditionally about seven are required for each graft, but a team at Cambridge, England, believes it may be able to grow enough tissue from a single embryo to treat 1,000 patients.

It would seem, therefore, as if the disease first properly described by James Parkinson two centuries ago will continue to afflict humanity, notably its chosen victims, for quite a while to come. Among those chosen recently have been Anna Neagle, Kenneth More and A. J. P. Taylor.

ALZHEIMER'S DISEASE   It is currently untreatable. The only truly definitive way of confirming a diagnosis is at autopsy. Four million Americans suffer from it, more than half of them women, and it is the leading cause of persistent dementia in later life, accounting for 50–60% of 'late onset cognitive deterioration' (as the medical world terms the state of going bonkers, being deficient in the top storey, short of a ten-bob note, etc., etc., the phrases used more frequently by the rest of us). Most trenchant fact of all, in any chapter dedicated to Ageing, is that – as one man defined it – 'Alzheimer's is an example par excellence of an age-dependent disorder.'

It is indeed. At age 65 (combining two population studies) less than 1% of individuals suffer from severe dementia. At age 80–85 this proportion has gone up to 15–20%. Not only can half of all people with dementia blame Alzheimer's, but the disease's incidence rises exponentially. If we all lived to 90 one-third of us would remain intact intellectually (according to one estimate), one-third would develop Alzheimer's, and the final third would experience a mixture of what has been called 'benign senescence forgetfulness' plus a variety of diseases producing the dementia syndrome. Risk factors (apart from age) are a family history of the illness, head injury, low intelligence, and Down's syndrome, with family history being the most severe.

Suddenly Woody Allen's wish for achieving immortality through not dying, a sentiment shared by others, can appear less overwhelmingly attractive. Some figures from Australia confirm the increasing preponderance of this disease. That nation's cases have risen by 20% every year since 1981. One in eight people over the age of 65 now develop it, and it is the fourth leading cause of death among the elderly.

Physical hallmarks of this disease include, according to one definition, 'atrophy of brain tissue with loss of specific nerve cells and the presence of neuritic plaques and neurofibrillary tangles'. The atrophy is generally of the outer surface of the cerebral cortex, and is especially pronounced in what is known as the association cortex. The brain undoubtedly loses substance, but it is the selective loss of neurons that is most distinct (and damaging). The loss in weight, although a feature, is hard to quantify, partly because human brains vary so much in weight whether or not from normal or Alzheimer individuals. (Frequently quoted are the brains of Ivan Turgenev and Anatole France, both intellectual giants, but the brain of the former was precisely twice the weight of the latter.) There are three known genetic causes for Alzheimer's, these being some genes on chromosomes 14, 21 and 19. The last is the most frequently encountered. Twin studies have supported the genetic basis of much of Alzheimer's disease (with about half the patients having a positive family history). A Finnish twin cohort, with 13,888 twin pairs for study, has shown the disease's incidence to be significantly higher in one-egg than in two-egg twins, concordance being 18.6% for the former and only 4.7% for the latter.

*

Alois Alzheimer was a German psychiatrist whose interest lay in neuro-anatomical changes possibly associated with mental disorder. In 1907 he applied a new dye to the brain of an individual whose death had occurred five years after the onset of a progressive dementia. The plaques and the tangle, so much a feature of the disease, showed up clearly with the dye. He wrote an account of this work and his name, henceforth, has been linked to this unpleasant disease.

Once initiated it tends to progress rapidly. Before the end of a single year it is usually rampant. Absent-mindedness becomes gross (and not a matter for jest, as with ordinary humans whose brains so frequently let them down). Speechlessness can occur. Reasoning can vanish, along with an awareness of surroundings. An impaired ability to learn new information has been called the cardinal deficit. A decline in language function involves not only difficulty with names but comprehension. Calculation, abstraction and judgement also suffer. Finally, 'sans everything' (to remember Jaques again), almost all cognitive faculties are lost. 'Advanced patients', as one report stated, 'can no longer move about normally and sphincter control is lost so that [they] become bed-bound and incontinent'. Eventually even basic functions, like breathing, are impaired, and the victims will then die.

Spouses (and the victims themselves) may attempt to disguise the disease, covering up mistakes, laughing off mishaps. The engrained habits of conversation, for those still with speech, can misleadingly imply that all is well. 'How are you?' 'Fine, thanks. And you?' 'Fine, thanks.' 'And the wife?' 'She's great.' 'That's good.' 'Yup.' So much of life is also virtual automation – dressing, washing, eating, talking – that a demented individual can get by, for a while, until he leaves home without his trousers, without knowing why he is leaving or where he is going. All us have bad days – forgetting a wallet, keys, names, faces – and all can produce conversational patter without much thought, but an Alzheimer sufferer, still seemingly within the standard range of absent-mindedness, will suddenly go too far, being plainly in a different category. The early days then yield, rapidly, to later days when the disease's presence becomes all too obvious.

An intriguing tale, bolstering the possibility that earlier intelligence might reduce (as already mentioned) 'neuropathological morbid changes', was published in March 1996. During the 1930s, before taking their vows, School Sisters of Notre Dame in Milwaukee wrote brief autobiographies. Some writings were of course better than others and all were recently assessed for grammatical complexity, for low or high 'idea density' and the like. One sister had written 'I prefer teaching music to any other profession', and another: 'Now I am wandering about in "Dove's Lane" waiting, yet only three more weeks, to follow in the footprints of my Spouse, bound to Him by the Holy Vows of Poverty, Chastity, and Obedience.' The first sister died of Alzheimer's and the second lived longer with no cognitive impairment. This is hardly game, set and match for any hypothesis, but ninety-three nuns were involved. The frequency of 'low-idea density' coupled with subsequent Alzheimer's was strong. This does not confirm the theory but David Sharp, in writing of the nuns' tales, considered that 'linguistic skill in early life, rather than education as such, seems to be a better mirror of a person's ability to avoid cognitive decline'.

It was generally acknowledged as courageous when an exceptionally famous victim of Alzheimer's publicly acknowledged the fact. Ronald Reagan, former US president, had often been mocked for apparent lack of intellect, even when in office. More latterly, after welcoming Princess Diana as Princess David and addressing Neil Kinnock by another Labour leader's name, it became clear that not all was well. Ten years earlier a psychologist, having observed Reagan's delayed responses to familiar faces and questions, and having noticed his vocabulary was shrinking, had also concluded that the changes were not those of normal ageing. The Alzheimer's Disease Society (of Britain) has called Reagan's action 'a brave decision'. It welcomed the publicity this former president had aroused, hailing it as 'a great boost to the campaign to remove the stigma from mental illness'.

The same society had earlier wished for a similar boost to be given by a former prime minister of Britain. There was never any official confirmation that Harold Wilson had also been a sufferer, but the media in general presumed this diagnosis. Wilson's abrupt resignation from office shortly before his 60th birthday helped with its promotion. So too the fact that illness (of some kind) dogged his final fifteen years. The ADS, founded three years after that resignation, approached the Wilson family on several occasions in the hope of acknowledgement, but was rebuffed. Hence the society's enthusiasm for Reagan's disclosure. This was made in 1994, long after 1976 when Wilson had resigned. As Dennis J. Selkoe wrote in *Nature* (in June 1995), Alzheimer's disease only 'emerged from relative obscurity two decades ago'. Wilson (whether he had the affliction or not) was therefore at the tail-end of that obscurity. Reagan was in quite a different era of acceptability when admitting Alzheimer's.

Down's syndrome is possibly the best known other form of mental deficiency (being first described in 1886), but not until 1948 was a link first established (by George A. Jervis) between it and Alzheimer's. His suggestion was initially disregarded, mainly because many consider Alzheimer's a normal function of ageing which merely expresses itself earlier in some victims than in others. (The acceleration of its onset with increasing age among the general population had seemed to imply that we would all become victims if sufficiently old. That view has now been discredited.) It seems that senile dementia is better considered as age-related – occurring most within a definite age range – rather than ageing-related, and caused by the ageing process itself. Very elderly people may even be at a diminishing risk from dementia as they become older still.)

The Down's/Alzheimer's association involves a protein known as amyloid beta which is deposited in the plaques that are such an identifying feature of Alzheimer's. In Down's this amyloid is laid down some fifty years before it is observed in the general population. The fact that Down's sufferers have three No. 21 chromosomes (as against two for normal individuals) caused researchers to examine their genetic assortment closely, and to find if it perhaps contained a precursor gene for the amyloid, but these are early days in this work. Nevertheless it is intriguing that Down's (the major congenital cause of severe mental subnormality) and Alzheimer's (the commonest dementia) are associated, with one existing at the start of life and the other appearing, logarithmically, at its other end.

The 4,000,000 Americans, the 800,000 Britons, and many more around the globe with Alzheimer's may or may not be pleased to learn that a simple eye test, if the tests prove positive, could have informed them of their disability even in its early stages. (As 9 million Americans are expected to be suffering from this disease in 2040 they too may not wish to be informed.) Similarly those not yet diagnosed might, or might not, wish to discover – via this test – that they have the disease, even if no other symptoms have yet occurred. The eye tests have proved positive up to a year before other signs of the disease have become apparent. This early diagnosis involves a very dilute solution of the acetylcholine-blocking drug, tropicamide. When this is dropped on the eye the pupil visibly dilates (enlarges) to a marked degree only in those already with the disease *or* about to develop its symptoms. 'Marvellous,' said Leon Thal, Alzheimer researcher at the University of California. The dilaters themselves might be less ecstatic to learn so early of their fate.

But Thal's enthusiasm has cause. Alzheimer's is currently untreatable, but if effective treatments become available it will probably be better to use them before the disease makes itself conspicuous. The drug used for the eye test was chosen because of Alzheimer's link with Down's. Down's patients are hypersensitive to drugs that block the effects of the neuro-transmitter acetylcholine. Therefore the suggestion arose that Alzheimer patients might also be sensitive, and this has proved to be the case. Until this test arose (and its trials are encountering difficulties) the only definitive way of confirming an Alzheimer diagnosis, despite all the manifestations of dementia, was at autopsy.

A possible way to reduce the risk of developing Alzheimer's, a most surprising way, was announced at the American Academy of Neurology meeting held in April 1996. Basically it suggested: 'Keep on taking the tablets.' The health-care habits of 2,000 people living in Baltimore had been closely examined. Particular attention had been paid to the drugs and medication each patient had taken during the previous fourteen years. A 'striking correlation' was found between regular takers of simple pain-killers and a lack of Alzheimer's. The 'more than occasional users' of NSAIDs (non-steroidal anti-inflammatory drugs) were half as likely to develop the disease as non-users.

Scanning the brains of people at risk may also prove to be beneficial. A technique, known as positron emission tomography (PET), measures the quantity of activity in different portions of the brain. This has found lower levels in the parietal cortex of individuals suspected either of impending Alzheimer's or of the disease itself. Once again there is, as yet, no available treatment, but earlier diagnosis will undoubtedly be preferable if some form of cure does arise. When the patient's behaviour does become suspect the 'damage to the brain is far along', said Zaven Khachaturian, of the US National Institute of Aging, Maryland. It will, almost certainly, be easier to prevent damage than to repair it when one of the drugs now being tested shows merit.

Oestrogen, sometimes called the queen of hormones, may also be involved in the Alzheimer story. Production of this substance, as is well known, is reduced in post-menopausal women. Its relative lack thereafter is thought – by some – to put older women at risk from developing Alzheimer's, oestrogen being necessary for the survival of certain neurons. Some women have already been given oestrogen, and some aspects of their dementia have subsequently diminished, but other researchers have been critical of these studies. In short, and once again, these are early days. Women do have higher prevalence rates than men for Alzheimer's, but they also live longer (which may explain the difference). Men may be partially protected because they continue to produce testosterone, the male hormone which can be converted to oestrogen in the brain. It will be intriguing if oestrogen does prove to act beneficially with regard to dementia. Its finger is in, so to speak, so many pies already, such as the one labelled osteoporosis, another concern positively linked with increasing age.

HRT AND OSTEOPOROSIS   One in five British women in their fifties and sixties now takes hormones every day. Three million post-menopausal women in the US were doing so even ten years ago. The purpose of this medication is to alleviate symptoms associated with the menopause, perhaps modify the risk of coronary heart disease, probably reduce bone loss, minimize urogenital complaints, and generally create a better feeling of well-being. On the coin's other side doubts have been expressed about a possible increase in breast cancer, about weight gain, about raised blood pressure and venous thrombosis. What is needed, say many of those involved in this work, is a major trial to discover the safety and effectiveness of HRT. Its long-term effects, says Tom Meade (architect of a plan to recruit thousands of British women in such a trial), are largely unknown.

In May 1996 the Medical Research Council (of the UK) announced its decision to contribute £21 million 'to establish once and for all' the risks and benefits of HRT. In the British part of the study 18,000 women will take the hormones for ten years and will then be followed up for a further ten years. In other countries an additional 16,000 women will be treated similarly. It is hoped the eventual results will demonstrate the benefits of HRT (stronger bones, healthier hearts) over the drawbacks (a slightly increased risk of breast and uterine cancer). Twenty years may seem a long time, particularly to women now approaching the menopause (who want to know now what best to do), but there is no way in which this aspect of the ageing process can be hastened to pacify an urgent need.

What has been known for a long time is that oestrogen supply is reduced at menopause. Normally it is manufactured by each developing follicle. Without an active ovary these egg-bearing follicles do not develop, and neither does their oestrogen. As various aspects of menopause are not entirely welcome, it was likely that a scheme would arise for replacing the missing hormone, thereby negating or reducing unwelcome aspects of 'the change'.

Top of the list in many minds is osteoporosis, the age-related disease that is extremely damaging. In the United States, as was related to a Ciba symposium on this subject in 1988, 'it causes at least 1.2 million fractures (annually) and costs 7 to 10 billion dollars'. Sometimes called 'the silent epidemic', it is yet another complaint to cause increasing trouble in an increasingly elderly population. The three most common fracture sites (for those 1.2 million people) are the vertebrae (538,000), the hip, or proximal femur (227,000), and the distal forearm (172,000). Such vertebral fractures start to increase soon after menopause. The forearm fractures (Colles's) increase until age 65 and then level off. Hip fractures increase only slowly with age until very late in life when the rates become extremely high.

Putting these figures, and frequencies, another way round they mean than one-third of women over 65 will have vertebral fractures, and one-third of women (as against one-sixth of men) will have had a hip fracture by extreme old age. In the US (as of October 1994) some 800,000 hip replacements had been done, with the current annual rate being 120,000. This operation has been described as 'an excellent bargain in terms of quality-adjusted life-years per dollar, relative to other interventions in chronic disabling disease'. Nearly 300,000 elderly Americans suffer hip fractures every year, when the

top of their thigh bones/femurs shear off near the pelvis. As 20% of
the victims die within twelve months, and another 25% spend the
rest of their lives in nursing homes, any form of prevention is
therefore to be welcomed. A lightweight hip pad, filled with polysty-
rene and water, was reported by Harvard University in 1995. The
contents stiffen into a solid on sudden impact. Initial results, published in the *Journal of Biomedical Engineering*, were good and full
clinical trials are planned.

The principal reason for all such fractures is low bone density. This
depends upon the quantity of bone being created and the quantity
being lost. Women will lose about half of their trabecular (internal)
bone and a third of their cortical (external) bone in their lifetimes,
assuming nothing is done to alter the situation. Men also lose bone,
but only two-thirds of the female quantities. (Pierre Delmas, of
INSERM, France's national medical research institute, thinks the male
problem is neglected, even though one-third of all hip fractures
within France occur in men. 'It's time to do something about this,'
he warned in 1995. Bone mass in French men declines by about 4%
each decade, and life expectancy for men is increasing.)

Bone loss begins in both sexes at age 35 or so. It proceeds for
women fairly steadily until menopause when there is accelerated loss
for about ten years. The taking of oestrogen at menopause does
prevent the accelerated phase but not the slow phase. It therefore
does not recreate a youthful situation, but does diminish the likelihood of fracture (by about one half, according to a 1985 survey).
Unfortunately it does not diminish the propensity of the elderly to
fall. Increasing unsteadiness, deteriorating eyesight, more arthritis,
occasional dizziness and the side-effects of certain drugs, as well as
sedatives, all contribute to the difficulty in staying upright.

A further misfortune is that oestrogen, however diligently taken
after the menopause, has little residual effect on bone density among
women of 75 or more, and it is this elderly group which has the
highest risk of fracture. Two recent studies make the point that
chronological age and bone-breaking will always go together – in the
end. One stated that women of average age 79 still had as much hip
fracture whether or not they had taken oestrogen; the other that

oestrogen led to a 63% hip fracture reduction among women aged 65 to 74, but only an 18% reduction among those older than 75. Outsiders (and the elderly) can long for more definite answers, but science (and long-term studies) have not yet obliged. The *New England Journal of Medicine*, in an editorial, stated that oestrogen treatment may have to be started at menopause, and never stopped, to give maximum protection. In this way bone density at age 80 may only fall by 10% after menopause, as against 30% for those who never take the hormone. The article concluded that 'having to take oestrogen for the rest of one's life reduces the appeal of the preventive strategy'.

An alternative strategy, also unsatisfactory in its conclusions, is to consume more calcium. If a person's body is receiving insufficient quantities of this substance for general use it will inevitably resort to the supplies in bone (this containing 99% of all the bodily reserves). As with money, if income is inadequate, take from capital. A recent estimate of calcium requirement (for men and women) considered 800 mg/day to be sufficient. A similarly recent report by the United States Public Health Service affirmed that men in general were consuming sufficient for their needs but women's intake was nearer 550 mg/day. Unfortunately – for simplicity – further work has shown little correlation between calcium intake and bone density. Even 1,500 mg/day given to women shortly after the menopause 'could not substitute for oestrogen' in preventing bone loss from vertebrae.

A third strategy is not to fall. Loose carpets are a hazard. So too slippery floors (as in the bathroom) or highly waxed surfaces (elsewhere). Drugs that impair co-ordination are unhelpful (however helpful with other problems). High and narrow heels can lead to spills; so can darkness in unfamiliar areas. A stick can help (even if its existence admits age to an ageing owner). Fortunately not every fall results in fracture. One famous nursing-home study (by Rodstein in 1964) concluded that only 1.4% of falls resulted in major injury. Of course it is better not to tumble, despite those comforting odds of 70 to 1 against a fracture.

Old people who exercise are less likely to fall, according to a 1995 report in the *Journal of the American Medical Association*. A study sponsored by the National Institute of Aging carried out eight separate investigations with different forms of exercise, ranging from t'ai chi to weight training. Ordinary exercise reduced falling's likelihood by 10%, exercise with balance by 17%, and t'ai chi by 37%. As the Institute reckoned that 30% of people over 65 have a fall each year, and 10–15% of such falls result in injuries, any reduction in such numbers has to be welcomed.

Advice from the Mayo Clinic (of Rochester, Minnesota) listed six measures to reduce bone loss. These were:

- Treat women with premature menopause with oestrogen until age 50.

- Increase physical activity.

- Eliminate tobacco usage.

- Use alcohol in moderation.

- Increase calcium intake (to 1,500 mg/day in adolescence, and 1,000 mg/day in adults).

- Ensure adequate vitamin D intake in the elderly.

Osteoporosis, hormone replacement, oestrogen lack (and replacement), Alzheimer's, Parkinson's, hypertension, centenarian – such words were scarcely used, if ever, when the current crop of elderly individuals were in their infancy. Today they are all familiar, and will become even more entrenched as an old population becomes yet more elderly, in both numbers and percentages. As one man said, on entering a crowded party of acquaintances: 'The only name I can remember these days is Alzheimer's.' So be it – until someone comes up with a cure; but ageing they will (probably) never cure. They may postpone it, perhaps by a handsome margin, but the process is unlikely to vanish absolutely. It, and death, are remorselessly part and parcel of the happening known as life.

SUICIDE

NATIONAL AND LOCAL VARIATION   FINAL EXIT

COUNSELLING   PHYSICIAN-ASSISTANCE   *SEPPUKU*

Father, thank you for the trip to Australia; Mother, thank you for making tasty food; older brother, sorry for being an inconvenience.

NATIONAL AND LOCAL VARIATION Thus the 13-year-old Japanese schoolboy, Kiyoteru Okochi, found hanging from a persimmon tree in the garden of his house shortly after he had written that note in December 1994. The event shocked Japan, partly for the boy's earlier revelations about bullying (being repeatedly dunked until fearing he would drown, and having to give money to his tormentors) but mainly for fear that school pressures and exams were proving excessive. In 1993, according to the Japanese Education Ministry, there were 23,358 reported cases of bullying and 31 of schoolchild suicide – for whatever reason. Japan should not be so concerned about a few self-killings, according to the rest of the world, bearing in mind external prejudice about Japanese enthusiasm for the practice. In fact Japan is not top suicide nation but may be top in its hothouse education system, with tremendous emphasis upon good marks. Hence its concern over childhood death.

Suicide occurs in all nations, whether exam-impelled or not, rich or poor, developed or developing. The WHO's world estimate is 500,000 a year – at least. The actual figure is thought to be nearer a million, with various countries – such as China and the Muslim nations – unwilling to report their truths. All age-groups and all classes are involved, with the highest rates being for middle-aged and elderly men. It is also believed about twenty times as many people attempt suicide as succeed, this tremendous number draining health resources. Whether the attempts are truly determined is irrelevant, as the subsequent care-cost is implemented in any case. No one knows (or hazards) a figure for genuine (as against non-genuine) attempts, or how many might therefore sympathize with Bertrand Russell, Nobel-prize winner and philosopher. He – allegedly – would not commit suicide 'as he would live to regret it'. Others also refrain because they are thus prevented from witnessing the act's aftermath, of tremendous interest to them in particular.

Despite suicide's universality there are substantial rate differences among nations. Western Europe's top country (by a long way) is Hungary. Then follow Denmark, Belgium, France, Luxembourg, Germany, Netherlands, Ireland, Portugal, Spain, United Kingdom, Italy and Greece, with the Greek total one-twelfth of the Hungarian figure. (Greece also appears high in the longevity tables.) For developed nations the suicide rate is (usually) greater than traffic fatalities and much more common than death from cervical cancer. Yet, say many, the concern – and expenditure – on behalf of suicide as against road or cervical cancer deaths is relatively small. (Respective totals for the United Kingdom in 1992 were: cervical cancer 1,532; traffic accidents 3,814; suicide 5,542). In general more men than women kill themselves, being about three to one in Hungary and France, although in Denmark and the Netherlands the proportion is equal. Suicide is not unlawful any longer (in most nations). It is an individual's final wish, and therefore inviolate.

Such clarity is not reflected in public attitudes and legislation should others be involved. In the Netherlands, so often a leader in liberal attitude, a psychiatrist was convicted for assisting in the suicide of a physically healthy patient with a depressive disorder but then, as contradiction, received no punishment. Each patient, in former times, was reckoned to be the guilty one, with suicide itself labelled as a crime. That view is now 'unacceptable', as one psychiatrist reported, with the 'blame now falling on the doctor'. But doctors do not see potential suicides all that often. In *The Health of the Nation*, published by the UK Department of Health in 1992, it was estimated that a general practitioner with a patient list of 6,000 would encounter one patient a year who committed suicide, a frequency rate equivalent to multiple sclerosis, ulcerative colitis, and Crohn's disease. The encounter may be only subsequent (in that the patient has to be deleted from the GP's list), but from half to two-thirds of suicides do see a doctor in the month preceding their death. Unfortunately young adult males, a group increasing in its suicide rate, also form a group less likely to have sought help prior to self-destruction.

Figures collected by an Italian team (from Ferrara) have shown that a circadian rhythm exists for the act of both suicide and attempted suicide. Mere attempts are mostly made in the early evening whereas successes (if that is the correct term) take place during the late morning and early afternoon. Perhaps bodily and hormonal changes are involved. Perhaps, for the attempters, their cry for help is considered more likely to be heard in the early evening. With physiological and environmental factors involved, and with the daily rhythm playing a part, the 'treatment of depressive disorders might therefore be improved by aiming for peak drug concentrations at vulnerable times', as the Italian team suggested.

Individuals who are most vulnerable tend to follow the familiar pattern, and increase their susceptibility with age, but students are particularly prone. So too are British doctors, pharmacists and dentists, who are all twice as likely as other Britons to kill themselves. (Availability of drugs is considered to be relevant, just as those who shoot themselves must first have guns.) Although physiotherapists are part of the medical team they hardly ever kill themselves. Veterinary surgeons top this particular list, being 3.5 times the national average. As for British professions in general, it is Welsh hill farmers who are most suicidal. City-dwellers, longing for a rural idyll and wishing to exchange rat-racing for running sheep, should realize that the hilly sheep country of mid-Wales has the highest suicide rate of the British Isles. Some blame the compulsory dipping and its unwelcome chemicals for depression; others the 365-day, isolated work so fraught with cash concern.

Suicide figures, culled from a variety of sources, are always intriguing but not necessarily revealing. Those Welsh farmer figures are echoed by the figures for Powys as a whole, this lovely (and relatively empty) Welsh county having a suicide rate over twice that for the UK as a whole. The Finno-Ugrian ethnic community, residing largely (and most separately) in Hungary, Finland and by the Urals, tend to have very high suicide rates wherever they live. Australian men kill themselves more frequently shortly after suicide stories have featured in newspapers, but Australian women do not react similarly. Female coffee drinkers (in the US) are less likely to kill themselves than those who abstain. Tea drinking does not appear to offer similar protection. One-third of telephone calls within the UK to the Samaritans (a counselling service) came from men in 1987. By 1995 that proportion had increased to more than half. The male suicide rate in Britain increased slightly (from fifteen to sixteen per 100,000 people) in the decade up to 1995. During that same decade the equivalent female rate dropped from seven to four. (Therefore the male rate became four times that of females.) Men most at risk are those not living with wives, the unskilled, the unemployed and also those excessively employed, such as stressed professionals. Security guards recently began to patrol San Francisco's Golden Gate Bridge hoping to prevent death-leaps (over 1,000 since the bridge's opening in 1937). Prison suicides (in England and Wales) are now running at sixty a year, and 10% of schizophrenics kill themselves.

Young male Britons (aged 15 to 24) have been committing suicide more frequently of late – from 320 in 1982 to 500 in 1992. For their female counterparts the annual toll is about 90, save for 1989 when it leaped to 120. Some observers blame the 'pop scene', partly because many lyrics seem to extol self-destruction – 'One starry starry night, you took your life, as lovers often do' (Don McLean); 'Look on the bright side is suicide' (Nirvana); 'And if a double-decker bus crashes into us, to die by your side, what a heavenly way to die' (The Smiths); 'It's better to burn out than fade away' (Neil Young). Worse still, say some, the leader of the Nirvana group, Kurt Cobain, actually did kill himself in April 1994. Then Richey James, of the Manic Street Preachers, disappeared that same year in a fashion suggesting suicide. Other observers of the pop scene say that drink and drugs are more relevant than role models, with addiction a prime cause of self-destruction.

FINAL EXIT In March 1991, as a considerable step in the liberalization of suicide, the Hemlock Society published a self-help book. *Final Exit: The Practicalities of Self-Deliverance and Assisted Suicide for the Dying* not only hit bookstores but sold extremely well, appearing on the *New York Times* best-seller list for eighteen weeks, notably after its existence had been highlighted by the *Wall Street Journal*. No one knows how many people actually read the 500,000 copies sold in the first few months, and no one knows how many suicides might have been prompted by its appearance, but there was anxiety. According to promotional material the book was 'intended to be read by a mature adult who is suffering from a terminal illness and is considering the option of rational suicide if and when suffering becomes unbearable'. Others believed the do-it-yourself guide would also be read by all sorts of adults, many of whom might then be encouraged to kill themselves, just as all self-help books encourage participation – in anything from Indian cooking and vehicle maintenance to establishing a business or cheating Wall Street.

A group of New York doctors, from Cornell University and the Chief Medical Examiner's office, determined to discover whether 'the number of suicides involving methods recommended in *Final Exit* increased in New York City during the year after its publication'. Their initial, and comforting, finding was that the total number of suicides changed little from the year before publication, but the number of asphyxiations by plastic bag increased from eight to thirty-three. A plastic bag coupled with lethal doses of various prescribed medicines had been recommended in the book as one form of 'self-deliverance'. Copies of the book were found at the scene of nine of the thirty-three suicides by asphyxiation in the first twelve months after publication, and at two of the sixty-six poisonings where recommended agents had been used. On one occasion there was almost a verbatim copy of the suicide note example used in the book.

Overall, according to the New York doctors, 'at least 15 of the 144 people who committed suicide by asphyxiation or poisoning had probably been exposed to the book'. Among that total of fifteen people six had no serious medical disease (either in their records or found at autopsy) but two had cancer, three diabetes, one arthritis, and three a variety of other diseases. 'At least' five of the fifteen had a psychiatric history which included 'a previous suicide attempt, hospitalization, or treatment'. It would therefore seem as if the book had had no adverse effect upon the suicide rate – at least for New York City.

COUNSELLING I once took part in a radio discussion in New York lasting from midnight to 4 a.m. Having exhausted various topics, our chairman abruptly decided to call up the suicide agencies and pretend he was about to kill himself. I was initially appalled, knowing the conversation would be broadcast live, but we looked up relevant numbers and he started dialling. Sensitive concern lessened when an answering machine responded to the first call, asking us to call back during working hours. A second call elicited no response at all. Only the third call received a reply.

The sympathetic woman at the other end did her job splendidly. She listened. She took note. She would have been of great comfort, we all agreed, to anyone in need. What she did not do was accept that our chairman, well known to all who tuned to radio in those night-time hours, was indeed the person calling her. He tried to exaggerate himself, making his voice more so, but failed, and she was left wondering why her caller had pretended to be a radio personality. All of us, including our chairman, were left with admiration that, in the darkened world of New York City, there were helping ears ready and willing to listen to those in need – provided the callers were persistent with their dialling.

The sickness known as suicide does receive scant attention relative to other illnesses of similar gravity. Friendly, voluntary help at the end of a line, or procrastinating answering machines and night-time silence, are modest response for an affliction that kills tens of thousands annually (in the United States alone) and very nearly kills many times that number. The Ebola virus which erupted in Zaïre during the summer of 1995 was global front-page news for days. It killed 164 of the 211 people who were infected, and the place even received a visit from the WHO's Director General. The suicide figure, to say it again, may be 1 million deaths each year, or the Ebola death total every hour and a half.

PHYSICIAN-ASSISTANCE 'Whensoever any affliction assails me, methinks I have the keyes of my prison in mine owne hand, and no remedy presents it selfe so soone to my heart, as mine own sword,' wrote John Donne early in the seventeenth century. Friedrich Nietzsche echoed the same sentiment at the end of the nineteenth – 'The thought of suicide is a great consolation; with the help of it one has got through many a bad night.' This conundrum, of being in the midst of life, and yet contemplating the deliberate ending of that life, is not only strange but pervasive. If the world suicide total is an annual 1 million (or merely 500,000), and if the attempts are twenty times as numerous, it is likely that the thought, so consoling to Donne and Nietzsche, has occurred to very many more, perhaps to hundreds of millions in any year. On one side is life, usually enjoyable. On another is its absence, an alternative without reversal to the only life we know. And yet the first is often forsaken for the second, again and again and again.

The medical profession is not reconciled to deliberate self-killing. Its obligation to save life is countered by an equal obligation not to infringe a person's autonomy. In 'Physician-assisted suicide' by George J. Annas (in the *New England Journal of Medicine*) doctors were said to be 'as frightened and bewildered by the act of death as everyone else'. Lewis Thomas was quoted in the article – 'Death is shocking, dismaying, even terrifying' and a 'dying patient is a kind of freak ... an offense against nature itself'. As a consequence self-caused death, or even self-desired death, is doubly troublesome. The American Medical Association, which ought – say many – to be clarifying the difficult issue, seems 'unable or unwilling to distinguish physician-assisted suicide from killing by a physician', according to Annas in his report. Doctors often use life-threatening procedures, but the general intent is to benefit the patient. Procedures aimed to kill may be acceding to the patient's wishes, but are not beneficial in the ordinary understanding of that word. Should this extra form of service, so contrary to Hippocratic teaching, be added to the medical curriculum?

The most powerful argument against an increasing power to assist patients in suicide, concluded Annas, 'is the danger that this greater latitude will result in abuses disproportionately affecting especially vulnerable populations – the poor, the elderly, women, and minorities'. In a letter to the same *New England Journal* Richard E. Gingrich added: 'There always has been a feeling among the elderly that ... it would probably be best for them to pass on. Legalizing assisted suicide ... could increase this feeling ... not so much out of an overwhelming desire, but out of a sense of duty to their loved ones or to society itself.' Whether or not there is novel legislation, as another correspondent pointed out, physicians are already making life-and-death decisions with patients. Eighty per cent of the deaths in the United States occur in health care institutions (6,000 per day); 70% of these deaths in hospitals 'involve less than the most aggressive care'.

Whatever legislation states, and whatever the medical profession accepts as guideline, the people themselves – the patients – will continue to decide the issue, as best they can. The ten American suicides per year per 100,000 citizens has remained remarkably constant since 1900. The British figure fell from the 1950s to the 1970s but then rose again and shows no signs of lessening. The actual process of dying, as Annas reported, is more feared than death itself. 'Americans say they want to die at home, quickly, painlessly, and in the company of friends and family. Most, however, die in hospitals, slowly, often in pain, and surrounded by strangers.' Suicide can circumvent that issue, neatly and effectively, as it brooks no argument.

SEPPUKU A final word. Many a Roman fell upon his sword but Japanese have achieved more fame for self-destruction. *Seppuku*, or stomach cutting, started at the end of the twelfth century. (Its ideograms can also indicate *harakiri*, but *seppuku* is allegedly more correct.) Minamoto no Yoshitsune, when surrounded by the enemy (in 1189), was first to cut open his abdomen, this region being regarded as the seat of life. He had to endure the resulting painful death, with this agony adding to the act's appeal in the eyes of Japan's warriors. In time *seppuku* became ritualized. A ten-inch dagger was used, and its cut was followed by more immediately lethal cuts to the carotid or heart. Suicide was later assisted by an accomplice who executed the victim. According to Maurice Pinguet, a French author who researched this subject, 'the grand style was to take one's entrails in both hands and give them a vigorous throw in the enemy's direction'.

Such suicides were not always voluntary. *Seppuku* as punishment was banned in 1873, but the Second World War witnessed a resurgence of the act, not compulsorily but in desperation as defeat grew more probable. Kamikaze pilots, named after the 'divine wind' that allegedly hindered the Mongols during their attempted invasion of Japan in 1281, launched some 5,000 attacks during the war. Roughly one-eighth of these hit a target, sinking or seriously damaging 230 ships, and the airborne courage enforced the United States' opinion that an invasion of Japan would meet terrifying opposition. *Tokkotai*, as Kamikaze groups called themselves, may even have assisted in the US decision to use atomic bombs. *Seppuku* still occurs, but rarely. The last famous case occurred when Yukio Mishima, novelist and revolutionary, failed in his wish to make the army revolt. According to Pinguet: Mishima's 'penchant for living in the public eye was so strong that he determined to die as a tourist would imagine a real Japanese ought to do it'.

EUTHANASIA

DELIBERATE DEATH   LEGALITY   THE NETHERLANDS

ADVANCE DIRECTIVES   ASSISTED SUICIDE

> Only a physician can commit homicide with impunity.
> Pliny the Elder

DELIBERATE DEATH Some say that physicians have been committing it ever since. What is less clear is their right to do so, morally, legally. In theory, virtually everywhere, euthanasia (*eu thanatos*, a good death) is not allowed if it incurs a deliberate act, such as the administration of poison. In practice it does occur, whether positively (as with such administration), less aggressively (as with leaving lethal pills by the patient) or passively (as with not following some remedial course of action). Even switching off a life-support machine is said, by some, to be passive, this act being a withholding of care rather than a cause of death.

Some recent cases have, by their fame, highlighted the issues involved. Ramon Sanpedro, of Spain, was so damaged by a swimming accident that he could only move his head. He has expressed a wish to die, but Spanish courts have refused consent. Sue Rodriguez, of Canada, asked in 1992 for death. An incurable disease was attacking her brain, impairing walking, speaking and breathing. The Canadian Supreme Court voted 5–4 against her wish, and she died in 1994 'helped' by an anonymous doctor. Tony Bland, English football fan, suffered crowd-caused brain injury in April 1989, leaving him in a vegetative condition with no hope of recovery. The House of Lords, Britain's ultimate judicial authority, gave permission for the life-giving tubes to be removed. Bland died twenty days later, without anyone assisting his death (and many arguing that a lethal injection would have been more compassionate). 'Diane', attended by Timothy Quill, a US hospice director, was suffering from leukaemia in 1991, having already had vaginal cancer, and wished for death. Quill placed barbiturates by her bed and gave advice about the lethal quantity.

After her death (from those pills) Quill openly declared what he (and she) had done. A grand jury refused to indict him, but the medical establishment – in general – remained opposed to the practice of such assisted death. Jack Kevorkian, former American pathologist, widely known as 'Dr Death', has received little medical support for his admission that he has 'assisted' in twenty Michigan suicides. Nevertheless juries have not convicted him.

As one further example highlighting patients' longings versus established principles, a British rheumatologist was convicted of attempted murder in 1992. Nigel Cox gave potassium cyanide to a woman, tortured by pain, at her request. Had he given a lethal quantity of pain-killers he would, so it was said, never have been brought to trial. (There would also be no trial if the life being terminated had not yet been born. Within the UK, for example, a foetus can be aborted up to term on the grounds that it is likely to be seriously handicapped. Once it is born, however hopeless the general outlook, the legal consensus is that life, however negative or negligible its quality, is preferable to no life.)

*

A poll of 273 British doctors (as reported in the *British Medical Journal* of 21 May 1994) revealed considerable dichotomy, both in experiences with patients and in principles. About half of the physicians admitted they had been asked to hasten certain patients' deaths, and a third had done so. The great majority would be 'prepared to withdraw or withhold a course of treatment from a terminally ill patient, knowing the treatment might otherwise prolong the patient's life'. There was a final question to this poll: 'If a terminally ill patient asked you to bring an end to his or her life would you consider doing so if it were legal?' A total of 20 'strongly agreed', 71 'agreed', 42 were 'undecided', 45 'disagreed', and 21 'strongly disagreed'. In short, 91 were in favour of turning upside-down the traditional Hippocratic code of above-all-do-no-harm (*primum non nocere*), 66 were against, and 42 were neutral.

Nurses also have a point of view, and can act on it. A report in the *New England Journal of Medicine* (published during May 1996) revealed that one in five nurses – of the 852 intensive care nurses who were polled – had hastened the death of a terminally ill patient without the knowledge of the patient's doctor. In general they had either given an overdose of pain-killing opiates or had withheld prescribed treatment. Many said they had acted because doctors were ignoring wishes of the patients and their families or were spending so little time with their patients that they did not comprehend the extent of their suffering. Although only one in five of the nurses had actually hastened death, two out of five said they would help a patient to die were it not for the fear of being caught.

Ordinary people, the patients who might or might not have their lives cut short, are even more clear-cut in their opinion, and increasingly so in modern times. In 1969 a Mass Observation poll found 50% in favour of legalizing voluntary euthanasia, a figure that rose to 69% by 1976, to 72% by 1985, and 79% by 1994. Some polls, as in France and Canada, have put the approving percentage even higher. Ludovic Kennedy, British broadcaster and strong advocate for the cause, recently wrote that the Voluntary Euthanasia Society (of the UK) had initially been a shoestring organization with a 'handful' of members, but it had 16,000 by 1995 and 'half a million (£s) in the bank', thanks to some useful legacies. The World Right to Die Society, not even created until 1980, had thirty-one branches in twenty-one countries by 1995. (The book *A Good Death*, published in 1995 by Elizabeth Lee, was warmly reviewed. One such medical reviewer, although full of praise, wondered how to recommend it 'sensitively' to patients. Plainly that could be difficult. It would therefore seem that death should be contemplated, and discussed, and planned for, when in the midst of life.)

Besides, death is around the corner (or several corners) for all of us. We are all terminal. So when does this word begin to apply? As a leading article in the *Lancet* (of August 1995) asked: 'Are bed-ridden patients with advanced metastatic cancer automatically "terminal" when they may live for several years without any technological intervention?'

The Japanese, leading the world in delaying death, are also leaders in addressing it, or so it can seem. Their physicians take greater care

(than in many other regions of the world) to inform the families fully
of the patient's prognosis. Death is very much a family business,
possibly bolstered by the fact that 50% of elderly people in Japan live
with their children (as against 15% in the US and 10% in England).
Health check-ups are very popular, thus confronting lethal possibilities as well as circumvention. (Such general examinations are
conducted on 20 million annually – out of a population of 123
million.) Family ties remain strong even after death, with ceremonies
held – by Buddhist custom for a dead person – one, three, seven,
thirteen, seventeen, twenty-three, thirty-three, and fifty years later. (A
glass may be raised in occidental nations, with toasts drunk 'to absent
friends' and tears shed on anniversaries, but ceremonies are rarely
undertaken.)

LEGALITY Despite the surge of interest, and a majority of both
patients and doctors favouring change, Britain has not legalized
euthanasia. The British Medical Association, rarely a leader, did not
initially announce its approval. The *Lancet* (in 1960) had echoed that
distaste: 'I hope that we shall hear no more of legislation to enable
men to die in peace, but that doctors will be ready to go on carrying
their responsibilities in this as in so many other matters of life and
death.' In the BMA's pre-1995 opinion only 1 in 5 doctors wanted
euthanasia legalized, but various other polls have not borne out that
assertion. A 1994 survey of 2,000 family practitioners by *Doctor*, a
weekly newspaper for GPs, found its readership divided '44:54 against
legalizing active euthanasia', with 'almost 8 out of 10' believing that
passive euthanasia is 'accepted medical practice'. The Pope, and
ardent Catholics, have expressed strong disapproval, with no retrac-
tion of a statement expressed from the Vatican in 1980: 'Suffering
during the last moments of life has a special place in God's saving
plan.' Ludovic Kennedy admitted the remark might 'say something to
believers' but to him it was a view 'medieval in its thinking and
barbaric in its lack of compassion'.

THE NETHERLANDS has been foremost in searching for better ways of ending life. One Dutch television film, first shown in its country of origin (and then elsewhere, as in the UK), portrayed a doctor actively assisting a terminally ill former restaurant-owner to die. Wilfred van Oijen first injected his motor neurone diseased, 62-year-old patient, Cees van Wendel, to put him to sleep (on the patient's birthday and watched by the patient's wife, Anthionette van Wendel). Then he gave a second, and lethal, injection. It was all very swift – and painless. IKON, the religious broadcasting company which made *Death on Request*, reported after transmission in the Netherlands that twenty telephone counselling lines received more than 200 calls, of which 90% praised the film. In Britain over 100 members of parliament signed a motion condemning the BBC after it had shown the film, and even the Dutch ambassador to Britain sent a formal protest to the UK Foreign Office. The Motor Neurone Disease Society said it was 'a horrible programme'.

A. P. M. Heintz, a Dutch professor of obstetrics (writing in the *British Medical Journal*), considered that the Dutch 'discuss euthanasia with their doctors more than [individuals] in any other country', and that it makes no sense to be 'for or against this form of terminal care'. The basic question is 'whether we accept the right of human beings to decide for themselves how their lives will end'. No doctor needs to be involved against his or her personal conviction. If the patient's views 'conflict with those of their doctor they need the right to be referred to another doctor'. A study by P. J. van der Maas (and others), published in the *Lancet*, affirmed that doctors work hard to find alternatives, and euthanasia is never the easy option. This report states that three times as many patients request this form of ending as eventually receive it. In only 5% of cases is pain the most important reason given; the avoidance both of futile suffering and humiliation were mentioned much more frequently. So too the wish not to be a burden to their family, and to consume cash that might be used more profitably than extending dying. In Heintz's opinion 'euthanasia has always been practised, but requests for it were rare'. People now want 'to have a say where their own well-being is concerned ... The time has come to give this care a solid, legal basis, for the benefit of patients and doctors alike.'

A Dutch nationwide study, published in February 1996, revealed that there had been 'medical decisions' concerning the end of life in 38% of all deaths, and 2.1% of all the deaths 'were brought about by euthanasia or physician-assisted suicide – PAS'. In particular the study had examined homosexual AIDS victims. Of the 131 men in this separate category 22% had died via euthanasia or PAS, and another 13% had 'made medical decisions' concerning their deaths. Both these proportions were much higher with older patients and with those whose AIDS had been protracted.

ADVANCE DIRECTIVES The Galton Institute (formerly the Eugenics Society) of the UK reported in a newsletter that 'the last half century has been a time of greatly increased personal choice [but] strangely the choice of how and when we wish to die is still restricted and unclear'. Suicide ceased to be a crime in 1961, but aiding and abetting someone trying to commit suicide is still criminal. In Germany this is not so. Anyone there in possession of suitable drugs may give them to a would-be suicide. Assisted suicide – where the intention to end life is explicit – is banned in most countries and in forty-four of America's states. In Britain, and in 1985, an amendment to the Suicide Act which would have allowed such aid was presented by Lord Allen to the House of Lords, but it was not enacted.

As yet even the principle of an 'advance directive' (or living will) has not incurred the necessary legislation within the UK. Most states in the US accept that such advanced statements are legally binding, permitting the cessation of life-prolonging 'care' even though the patient may no longer be capable of expressing an opinion. It is expected that some form of similar legislation to Lord Allen's Bill (of 1993) will soon be passed. This will enable 'persons to give directions (or arrange for them to be given) to their physicians regarding the withholding or withdrawal of life-sustaining treatment in a terminal condition, and for related purposes'. By June 1995 almost 200,000 British people had asked for advance directives to help clarify their personal intent, according to the Voluntary Euthanasia Society. The Law Commission has stated that, even without Lord Allen's Bill, adults do already have the right to refuse treatment in advance. In fact treating a patient despite a refusal by that patient is a civil wrong, and may even constitute a crime. What happened to the Dutch restaurateur is not covered by advance directives. These do not ask doctors and nurses to do anything illegal – such as administering a lethal drug. They simply request a cessation of what the Pro-Euthanasia Society calls 'futile, life-prolonging treatment'. In short, the prolongation of life must not become the mere prolongation of dying.

The British Medical Association changed its tune in April 1995. In collaboration with the royal colleges (of nursing, of physicians, and of general practitioners) its new code affirmed that: 'Competent, informed adults have an established legal right to refuse medical procedures in advance. An unambiguous and informed advance refusal is as valid as a contemporaneous decision. Health professionals are bound to comply...' In one sense that is no advance, in that there was already a legal right. In another it is a major shift from the form of paternalism previously recommended, with doctors knowing best. The ball is not yet firmly in each individual's court, but is much nearer.

If the individual happens to be inarticulate, or too young to have and express an opinion, the problems (and ethics) of euthanasia take an extra twist. Two recent cases have highlighted this dilemma. A Dutch doctor gave lethal injections to a 3-day-old girl with spina bifida, hydrocephalus, 'severe' brain atrophy, and lower limb deformities. This gynaecologist did so after he and others had agreed her prognosis was poor, that she would never walk, and would probably live in a vegetative state for perhaps one year. This case was not one of euthanasia, in that the infant was too young to hold a wish, and court hearings were initiated. These were held primarily to 'develop jurisprudence' and not to criticize or admonish the gynaecologist. The college of attorneys-general had already decided he had acted with due care.

The British legislature (in particular three Law Lords) was confronted in August 1995 with a similar dilemma. A child had been born with severe brain damage and would never become mentally competent. It also had to be fed via a tube. After one year the parents requested that the child's life should be terminated (presumably via starvation). This was partly to reduce their own distress but also because their offspring (as *Nature* described it) had 'no future as a person'. Others may argue that everyone, whether with a measurable IQ or not, is a person, and parents have no right to destroy that individual. The Law Lords decided that the child should be made a 'ward of court'. In this fashion the responsibilities of the parent passed to the legal system. Therefore any decision to be made about life or death had become a judicial process which, argued *Nature*'s editorial, was 'right and proper'. It then wondered whether a similar procedure should not be followed in cases of voluntary euthanasia. The person would express a wish to die (and this would be properly confirmed). The medical issues would be outlined (and also properly assessed). The courts would then decide (in public) upon the issue, thus absolving physicians from their former responsibility of shouldering the burden individually.

ASSISTED SUICIDE Despite a general move towards consent for terminating life, Catholic bishops in the US argue that euthanasia and assisted suicide must never be allowed at the 1,200 Catholic hospitals in the US (the largest health care system). The National Right to Life Committee has been vigorous in opposing funding for anything associated with assisted suicide. 'Most Americans do not believe federal tax dollars should be used to promote and support [euthanasia],' said Burke Balch, head of the committee's ethical department. (By April 1996, following the overturning of certain banning orders, observers considered that prospects for a US Supreme Court debate over the right of terminally ill patients to choose physician-assisted suicide had grown brighter.)

Nevertheless the pendulum still swings back and forth. A medical director, writing to the *BMJ*, stated: 'To imply that something written six months previously should dictate what we do today is simply foolish.' A doctor from Connecticut responded to a leading article in *The Economist*: 'Your otherwise balanced view ... was terminally injured by your final point that [legalized euthanasia] would "lighten the burden on doctors". Issues of life and death cannot be decided on the basis of lightening a doctor's load.' In Australia representatives of Aborigines have said that the new (euthanasia) law is 'culturally inappropriate', and elderly Aborigines will now avoid seeking medical help. (Australia's Northern Territory passed in February 1996 the world's first law giving terminally ill patients the right to end their lives. Individuals of sound mind with only a year to live can ask doctors to permit that right. Meanwhile, down in Victoria, seven medical practitioners were being investigated by police, following their admission in a letter to *The Age* that they had 'aided suicide'. Victorian law can impose a sentence of fourteen years on anyone who assists another's suicide.)

The Dutch caused some backtracking by euthanasia supporters when a government report revealed that, in a recent year, there had been 2,300 cases of 'active termination of life' by a doctor at the request of a patient, 400 cases of suicide assisted by a doctor, and more than 1,000 cases of 'life terminating acts without explicit request'. The director of studies at a hospice subsequently reacted: 'When voluntary euthanasia becomes ethically acceptable (to the extent that it is in Holland) non-voluntary euthanasia is an inevitable accompaniment ... Those 1,000 cases occurred when someone decided that another person's life was not worth living. So much for autonomy.'

Considerable hegemony still lies with the doctors. In a television programme one was asked for his attitude to advance directives. 'It depends,' he said, 'if what is written fits in with what we think is right. If it is, then the answer is yes. If not, they are ignored.' The medical profession, in general, does not like the word directive (preferring a word like request). People, in general, who draw up these living wills see little point in them unless they wield authority. Just whose life, they say, is involved?

As *The Economist* concluded, in its article with a balanced view: 'Reverence for life is part of what it is to be human. But no less are freedom and dignity part of what gives life meaning. Let reform of laws on euthanasia and assisted suicide, now long overdue, be guided by that idea.' Or, as Winston Churchill wrote sixty-five years before his own end: 'It often happens that, when men are convinced they have to die, a desire to bear themselves well and to leave life's stage with dignity conquers all other sensations.' Currently a great many lives are ended with only modest dignity. It must be difficult to bear oneself well when plugged with tubes in an alien environment and abjectly subjected to intensive care. There is much to be said for being in control, even when that being is giving up its ghost.

DEATH

THE RIGHT TO DIE   AUTOPSIES   KILLING

HOME v. HOSPITAL   THE CORONER'S OFFICE

DEATH TABLES   SECRECY

To see how others have taken that final journey is the only
help we have when ourselves we enter upon it.
Somerset Maugham

Picasso dead! That's not like him.
A friend on hearing the news

There is no cure for life and death save to enjoy the interval.
George Santayana

If I had any decency I'd be dead. Most of my friends are.
Dorothy Parker, who was to die at 73

THE RIGHT TO DIE It may be inevitable, but that has not stopped
humanity from attempting to control, by either delay or acceleration,
this ultimate event. The US, as one European journalist wrote for
readers back home, is 'the place that pretends death doesn't happen'.
The nation's instinct is 'to treat it as a disease – with no cure yet, but
who knows?' A book review, in the *New England Journal of Medicine*,
pondered 'the spreading belief in "successful aging" – that living
properly with the help of modern medicine will prevent decline and
disease and, by extension, death'. Modern medicine (via its practitioners) can hold an opposing view, as it blames patients for their
'unsuccessful' ageing, their illnesses and death. In the old days, as a
chaplain wrote to the *British Medical Journal*, 'people gave up the
ghost and that was it'. But now, as this writer continued, 'the
professionals have an armamentarium of treatments to keep the very
old and infirm alive a few days longer'. Indeed they have. He
questioned whether the old Hippocratic oath of 'Above all do no
harm' has become 'Always treat until death intervenes'.

Living with mortality was less of an affront in the past. Victorians promoted death vigorously. Queen Victoria, dressed in black for her final forty years, was a steadfast reminder. Since then, with such weeds vanishing, with arm bands and black ties exceptions rather than the rule, and many individuals living whole lives without ever seeing death, a visitor to the planet might presume its absence. US television refers to fatalities rather than killings. The military speak of missing and casualties, as if wounded, dead and captured are similar (which, in warfare, they may be). Ordinary mortals can use all manner of euphemisms rather than the word itself or even stress it unduly, as with the 'dreaded deadeds' (of Britain's radio 'Goons') and comrades being 'very seriously dead' (as in the Second World War). The fact that life equals death is, for most of us for most of the time, not an easy thought.

But, as with the prevention or enhancement of fertility, the control or obliteration of disease, the alleviation of ageing or the tinkering with our genes, human attitudes to death are changing. Some recent books could not have been written a short time ago, such as (the already mentioned) *Final Exit*; *A Good Death: taking more control at the end of your life*; *The Right to Die: policy innovation and its consequences*; *Death and Dignity: making choices and taking charge*; or *The Troubled Dream of Life: living with mortality*. In that final volume Daniel Callahan argues that the US (his nation) is obsessed with trying to control and defeat the 'chaotic forces of nature'. He suggests 'nothing less than a reconstructed view of the self' which incorporates a social and personal acceptance of death's inevitability. This new view, in his opinion, promotes the moral and practical value of a peaceful death.

There was little in common between Jacqueline Kennedy Onassis and Richard Milhous Nixon, save temporary occupancy of the White House and in their dying. It so happened that the final treatment for both of them was in the same hospital. They died within a month of each other, but both succumbed at home. Each had left a 'living will', so named ever since California passed a law – the Natural Death Act – in 1976, permitting people not to have their lives extended by artificial means if terminally ill. Jackie Onassis died twenty-four hours after leaving hospital, and Nixon four days after suffering a stroke. Choice in Dying, a New York organization promoting living wills (now permitted in forty-eight states), had been receiving 100 inquiries a day. When Nixon died they rose to 500, and after America's most famous widow had died they reached 2,000. These two famous exits had moved death, as journalist Rupert Cornwell phrased it, 'from taboo to topic of the moment'. (Britain, somewhat laggardly in this regard, stared to 'consider' living wills in 1995. Its government then set up a working party to examine 'the legal status of advance directives'.)

Some medical cases have also altered opinion about the right to die. (There are so many alleged rights in life, the right to life itself, the right to bear children, not to bear children, to be vaccinated, not to be vaccinated, that a right to die was almost bound to surface – sometime.) Karen Ann Quinlan was among the earliest and most famous of such cases, her parents wishing an artificial life to end when doctors were reluctant to help. Since then the similar cases of Saikewicz, Perlmutter, Spring, Severns, and Eichner have hit headlines, with courts subsequently permitting or aiding the right to die. Father Eichner had not written a will, but had affirmed to his religious group that preservation of life would be unwelcome if he was ever unconscious and on a respirator. At the age of 83 he had a heart attack, suffered brain damage, and was kept alive by the hospital. His fellow brethren thought it not amoral to transform such brain-death into death itself, particularly if not against the wish of a man unable to express that wish. New York's highest court then told the hospital to let this victim die.

The same court acted contrarily 'in the matter of John Storar'. He was 52, had terminal bladder cancer, was a life-long inmate of a mental hospital, and could not express an opinion. His 77-year-old mother gave consent for treatment, but then refused it on her retarded, cancerous, dying son's behalf. The court did not like postulating what the man's opinion might be, and resented another person, even a mother, deciding the issue. Therefore it gave consent for further treatment (and he lived for a few more weeks). A living will would have settled the matter, but Storar could not have written one.

Now that such wills are permissible, their absence can cause problems. Joey Fiori, of Philadelphia, did not write one partly because his motorcycle accident occurred in 1971 before they had become relevant. For most of the time since then he has been in a vegetative state, unable to swallow, to move, to feel, to think – at a cost of $150,000 a year. His mother wants him to be disconnected. The authorities say there is no clear and convincing evidence that the man, aged 45 in 1995, would welcome such a severance. 'I pray that God will take him,' says his mother. No one else seems able to help him on his way, but by no means is he alone in his predicament. About 14,000 Americans are believed to be in a persistent vegetative state. If they are all costing $150,000 the total bill is 2 billion dollars a year.

Two researchers into medical ethics listed, within the *British Medical Journal* (in 1994), what they considered to be justifiable conditions for the non-treatment of incompetent patients. These included:

- irreversible closeness to death;

- neurological damage leading to the permanent destruction of self-awareness and intentional action;

- little self-awareness accompanied by such muscular disability that sustained intentional actions become impossible;

- destruction of memory so that the person who used to exist no longer does, and no other person can evolve instead;

- severely limited understanding by the patient of distressing and marginally effective life-saving treatment that leads to a demonstrably awful life.

The patients themselves may have different ideas about justifiable conditions. An active man suddenly bereft of both legs may wish to die. An elderly woman who breaks her hip may not relish the thought either of operations or of immobility. A creeping disease, whose prognosis is well known (and feared), may well deter its victim from wishing to experience every portion of that downward path. What constitutes a 'demonstrably awful life' can be different for those whose life it is than for medical practitioners who, however sympathetically, are only observers. (More on this subject may be found in the earlier chapter on euthanasia.)

Anti-abortionists, aware of losing ground in the Supreme Court about their principal concern, have been switching attention to the other end of life. They see parallels between young foetuses – incapable, mute, vulnerable – and vegetative individuals, no less incapable, mute or vulnerable. Certain existing laws have proved most convenient for these advocates, such as the Americans with Disabilities Act. This was designed to prevent unfair discrimination, but can be interpreted to support pro-life argument. Ronald Cranford, a neurologist from Minneapolis, who was involved with a family's request to remove the feeding tube from a car-crash victim seventeen years after the accident, is particularly incensed by the legislative manipulations. 'It is these well-meaning laws that now are being turned around and misapplied in the courts by the "pro-lifers".' In his opinion such patients did not have interests to protect since they had no working cerebral cortex.

AUTOPSIES Post-mortems, or necropsies (or autopsies), have been declining, and there is concern. In brief, relatives do not like them. Consent is often difficult to obtain. The dead person may earlier have expressed dislike. General practitioners, having known the affected family for years, may resent requesting such an operation. They may also be unhappy that their diagnosis (and cause of death) might prove to be wrong, with such an error increasing the chance of legislation. Certain religions forbid autopsy and any 'disfigurement of the dead', and some governments have created legislation reducing the need for such examination. Autopsy rates vary nationally around the world – from 49% to 0%.

The concern arises from many in the medical profession. They stress that autopsy is rich with benefit. It can establish precisely the cause of death. It may define other pathological processes which might have been treatable (and were relevant to that death). It assesses the accuracy (and value) of modern investigative techniques. It is important in teaching and it highlights clinical fallibility. The aged are often disregarded as possible candidates for autopsy, but the ageing process and its relationship with disease needs to be better understood. (The elderly, so neglected for autopsy, do lead to the greatest number of deaths, a truism often disregarded when autopsy benefits are being paraded.) The discrepancy between clinical and autopsy diagnoses (concerning cause of death) is around 10%, a proportion – according to one major report – that did not alter between 1910 and 1980.

'Discrepancies of major therapeutic relevance' (according to a 1993 report) can be discovered at post-mortem even with patients who die in intensive care. Where malignant cancer had been diagnosed as cause of death, so said another autopsy survey, this was only accurate on 75% of occasions, and the primary site was correctly postulated in only half of them. One more study revealed 'unsuspected major diagnoses' in 30% of autopsies. Such medical errors may or may not have harmed the patient, by hastening or aggravating death, but they certainly harm the statistics. Death certificates, and their 'Cause of Death', form the source material for many statements about disease. It would seem that clinical diagnosis is not always what it should be, and the wish for more autopsies does have cause.

Even within the same establishment there can be wide variation in the rate of autopsy. One London hospital oscillated between 9% and 30% in a four-year period, but with no discernible long-term trend. Hungary proved to be top country in a twenty-seven-nation survey conducted by the WHO in 1992. Its rate was 49%, mainly because all patients dying in hospital had to undergo autopsy – unless there were acceptable grounds for objection. Australia, Israel, Denmark and Sweden have recently enacted laws which have much reduced their previous quantities. East Germany has also changed downwards since reunification. (One community – Goerlitz – had achieved a 97% rate in 1987.)

Religions are relevant. Hinduism, Buddhism, Shinto, and the major Chinese and Christian religions all permit autopsy, but Islam forbids both the removal of organs and damage to the body. The Muslim corpse should also be buried promptly, but S. A. Geller, in an article on religious attitudes, writes that autopsy of such individuals may be allowed if legally required. Orthodox Judaism forbids it unless the act could save a life. (Medical authorities might argue that the major purpose of their calling is to save lives, with autopsies one tool of their trade.) Israel is more lenient, permitting it in cases of inherited disease or when civil law requires it or when the cause of death needs to be ascertained.

In one British hospital survey, aimed at discovering why more autopsies did not take place, the major reason given was that they were not thought necessary. A lesser reason was refusal by relatives. It is not obligatory that next-of-kin must state why permission is withheld but, in this survey, eight said 'the patient would not have wanted it', five that 'other relatives were unwilling', five that 'they didn't want the funeral delayed', three that 'they didn't want the patient mutilated', two that the autopsy 'wouldn't help the patient now' and one that 'he had suffered enough'. It is almost more surprising that anyone gives consent. Do relatives want to know if the clinical diagnosis was 100% correct? Do they want their loved one, carefully coerced after death into a peaceful position, to be removed and opened up? Do they even care if he/she died from something else? The death is grief enough.

It is easy to suspect that few relations, such as those who give consent, would wish to witness an autopsy. Standard procedure is to cut open the body from its sternum to the base of its abdomen. Organs then exposed are all removed, such as the entire digestive system (from oesophagus to anus), the liver, pancreas, spleen, kidneys, lungs, heart. These are then sliced and examined. If particular attention is required, perhaps to the lungs in cases of possible litigation concerning some industrial pollutant, the organs are dispatched to specialist centres. Arteries are also sliced. So too, if thought necessary, is the brain (a particularly disturbing extra for those who knew the patient). Head skin is first cut and then pulled over the face. A saw is then used to cut through the cranium, permitting the whole brain to be extracted. Finally, after examination of everything has been completed, everything is then returned, as far as possible, whence it came. The half-cranium is replaced and the scalp is folded back before being sewn in place. Something similar is effected with the abdominal incision. All this sewing is intended to make the body look, more or less, as it was before the autopsists began their work.

People do differ from their practitioners in their enthusiasm for autopsy, but not as might be expected. Clinicians, according to a 1994 article in the *Journal of the Royal Society of Medicine*, commonly hold the view that modern methods improve diagnostic accuracy to the extent that autopsy is superfluous. Contrarily, according to a report it also quotes, 91% of families permitting autopsy and 83% of those denying permission considered autopsy beneficial. They rated advancement of medical knowledge as a higher benefit than knowing whether the dead relation had been diagnosed and treated correctly.

KILLING If you want a car immobilized, get a mechanic. If you want it immobilized swiftly, simply and cheaply, get a good mechanic. Similarly, if you want someone executed, as painlessly and pro-fessionally as possible, call in those most familiar with the human body and with death. In short, call in a physician; but that suggestion has caused problems. In the USA thirty-six states now have the death penalty as a possible option, the Supreme Court having overturned an earlier federal ban in 1976. Of that number twenty-eight require a physician's presence at each execution. The medical presence has become favoured – by the state legislatures, partly because preferred methods of execution have become more medical. Instead of the crudity of the electric chair or hanging (or, in older times, rifle fire or even an axe) lethal injections are now widespread, with twenty-five states permitting them. They are less expensive, more acceptable to the witnesses, and are thought to be more humane.

However, as with ending vegetative lives or hastening a terminal situation by withholding care, the healing profession resents its role as life-saver becoming life-destroyer. Veterinary surgeons kill pets, being the best people for the job. Doctors may be best at ending human lives, but are unwilling to assume that role however skilled they would be in doing what has to be done. (It does have to be done as a majority of Americans are now in favour of capital punishment.)

Joseph Ignace Guillotin wanted a system (late in the eighteenth century) that would be swift, humane, cheap and efficient as a means of execution. Being a physician perhaps helped him towards his solution, and his proposed method undoubtedly filled at least three of those requirements. The humane aspect of his guillotine was possibly no worse than any other system the revolutionaries might have favoured. It is not recorded whether he was ever on hand to assist the procedure but, in today's world, the American College of Physicians (representing 80,000 internists) and various other national bodies have taken a strong stand against the participation of health professionals in executions. The American Medical Association, in consequence of protests, has drawn up a set of guidelines. These forbid actions by a physician that –

- cause the death of a condemned person (as in giving an injection);

- help another to cause death (as in prescribing lethal drugs);

- are part of the preliminary process leading to death (as in the prescribing of psychotropic drugs);

- involve monitoring vital signs;

- involve witnessing an execution (as a physician and not as some other form of witness).

What is not forbidden is the certification of death after someone else has declared the condemned person dead. It is all very different from a veterinary surgeon's role. That individual prescribes the drug, prepares it, injects it, and watches for loss of vital signs before pronouncing the victim dead and ready for burial.

Of thirty-six states with the death penalty, according to *Breach of Trust* (a document prepared by the ACP and others), there are ten which have policies against any participation by a doctor. A further eighteen have obeyed the AMA's guidelines, but some actually have laws demanding 'assistance' by a doctor in executions. As these states can also have laws forbidding a doctor from hastening death the physician can be caught two ways (both in causing death and in refusing to cause it). In fact, although state medical licensing boards can discipline doctors (for wrongful acts), none has so far been disciplined for taking part in an execution. 'Execution is not a medical procedure,' bluntly states *Breach of Trust*. The thin line between killing terminal patients, by pulling the plug on so-called vegetables, and efficiently ending the lives of those condemned by society to die can seem, at times, to be very thin. The Hippocratic tradition of never doing harm is nothing like so straightforward as it used to be.

*

Nor, as a slight digression around this topic, is the current concern, notably in the US, about a rising murder rate. Former times were simple. If a murderer was caught he (usually) or she (less often) was condemned to die. In the United States there are now 24,000 murders a year (a number showing no signs of diminishing). Homicide is the eleventh leading cause of death among all age groups. It is the second leading cause among those aged 15 to 24, and the leading cause among male African-Americans aged 15 to 34. Washington DC is not only the nation's capital but reputedly the murder capital (with, in 1993, 467 people killed out of a population of 590,000, or 0.08% in that single year).

New York City, with murders well above the national average, resisted the death penalty until 1994, partly because this ultimate deterrent does not seem to be effective. Texas, responsible for one-third of all executions since 1976, now has almost 400 prisoners waiting on 'death row' (namely one-eighth of all Americans in this unhappy position). Louisiana, leading state for murders, is showing no benefit from the reintroduction of executions. The peak year (in modern times) for US executions was 1935 when 199 were killed. It is generally expected, particularly with New York entering the lists, that that number will soon be exceeded. As nine individuals, who committed their crimes when minors (under 18) have been executed since 1976, a human rights group asserts that the US is world leader in this regard, only beaten by the probable superiority of Iraq and Iran. (The US may also lead the world in its prison population. This has doubled since 1980 to reach 1,350,000, and one in every 189 US residents is therefore currently in gaol. There are also 671,000 on parole and 2,800,000 on probation which means that one in every fifty-two Americans is under some form of correctional supervision.)

It is sometimes alleged that executions not only remove guilty individuals but are more economical than lengthy residences in gaol. However, calculations for Texas, taking into account the legal wranglings beforehand, show a price-tag of $2.3 million for any execution. For Florida the figure is said to be $3 million. That kind of sum can keep a prisoner in a high security establishment for 120 years. The ethical, financial, moral and instructional arguments concerning execution will undoubtedly continue, with or without unhappiness from the medical profession concerning its involvement. In the first week of December 1994 six criminals were killed, the largest weekly total since 1976. In that same week a man was sentenced to death in the electric chair, his crime having been the murder of an abortionist. The motive was revenge (or deterrence) for the killing of unborn children. In short the issues involved in the taking of life are as complex as could be.

HOME v. HOSPITAL Most terminally ill patients (in Britain) wish to die at home. Most do not do so. Almost 70% die in some form of institution, with over 60% in a general hospital. Medical care in hospitals is concentrated on attacking the disease, with an overriding wish for success and eventual recovery. The organizations are less good, as a major report from Scotland indicated (published in the *British Medical Journal*), at palliative care when this should take over from curative care. Hospices, which cater only for the dying after all hope is lost, are better in this regard. Their function is more clear-cut. They do not consider death a failure. They wish only to make it as comfortable as possible.

Care, as the report stressed, 'extends beyond attention to physical needs'. Patients also have emotional, social and spiritual problems. Time therefore should be spent with the terminally ill, listening to them, learning of their fears, their wishes. A commitment to patients is, according to this study, 'rarely observed'. Indeed 'distancing and isolation of patients were evident, the isolation increasing as death approached'. Most of the consultants 'concentrated on the disorder rather than the person'. Senior nurses 'mimicked this behaviour by concentrating solely on the recording of vital signs'. The few senior and nursing staff who did take time were able to 'identify all the needs of the person who was dying, explore various ways to give comfort, and initiate medical and nursing measures to provide relief'.

It is easy to understand why the scarcely mobile, moribund, silent body, with eyes closed and mouth open, in bed No. 24, should receive less attention than the alert, talkative individuals, happily on the mend on either side. Sometimes the dying are even wheeled into separate rooms and therefore further isolation. The Scottish report was also critical of actual medical attention, after collating studies from many sources. These 'consistently show that terminally ill patients suffer severe and unrelieved symptoms increasing in severity as death approaches'. Hospices are more able in alleviating pain, considering it possible to achieve good control of it in about 95% of patients. Nursing care should go hand in hand with medical care, such as cleansing of the skin, care of the mouth, the provision of adequate fluids, and appropriate well-presented food.

One case history (after observation from a distance) can underline these points. She was 41, and dying from liver carcinoma. A supper tray was brought. She was lifted to the side of the bed, and placed unsupported. She soon fell back. Later a cup of tea was brought. She struggled to reach it, but failed. The nurse asked about liquid intake. No reply was forthcoming and '120 cc tea' was written down. The tray was then removed with the food untouched. Thirty minutes later some water was placed on the locker. Visitors then attended for one hour. Afterwards she was asked about tea or coffee. 'I've got water,' said the dying woman. Unfortunately she could not reach it. One hour later she attracted a nurse, asked for water, and was given it (with nine seconds of contact being recorded by the observer). This giving of water had been 'inadequate assistance' as she could not raise her head to drink it. At this point the observer intervened.

Different peoples have different attitudes to death. Muslims, for example, prefer to die in their own homes (and there are 1.5 million of them in the UK). Ideally, the face of the person who has died should be turned towards Mecca. In practice, and in hospital, it is turned to the right. Arms and legs should be straightened. Any clothes should be removed by someone of the same sex and the body then covered by a sheet. The body must be ritually bathed before burial. There is also an obligation that an ill person, whether acquaintance or stranger, be visited. This devotion continues after death and, according to a *BMJ* article, '200–300 people may visit the home of the deceased, often from long distances'. Hospitals can therefore wish, very speedily, that a Muslim death had taken place at home. Haste is also important for the burial, with custom (originating from hot places) demanding a swift interment. Post-mortems are not allowed, the body being sacred and belonging to God. If local law demands such an investigation the relations must acquiesce, but they press for all possible haste.

General practitioners might also wish for more deaths to occur at home. In that way they would at least learn, more swiftly, of a former patient's demise. About twenty of those on a GP's list (of 2,000 or so) die each year, with some fifteen dying in a hospital. Unless a coroner's autopsy is involved, or the hospital bothers to telephone the doctor, days or weeks may pass before notification arrives. Consequently, as a team pointed out in the *British Journal of General Practice*, the earlier 'lack of information may prevent prompt care of the bereaved'. In its survey the team learned that a patient's relatives were the informants in 46% of cases, newspaper obituaries were in 20%, and hospital phone calls in 9%. The GPs themselves were also somewhat lax, with only two-thirds of them recording information about their patients' deaths, and only 17% using the facts for 'audit or research'.

THE CORONER'S OFFICE The coroner's office (in Britain) became 800 years old in 1994. This department is associated, in most minds, with suspicious death, but it originated to protect the Crown's financial interests from corrupt sheriffs. Hence its involvement with treasure trove, wrecks, 'royal fish' (whales, sturgeons), and the valuables of felons and outlaws. A subsequent interest in death was initially also financial. Unless the coroner could ensure that a murdered person was English the victim was presumed to be Norman. Such deaths led to fines for all communities where the bodies were found. Later the coroner's office was broadened to embrace the accidental or sudden death of Normans.

A few centuries later Burke and Hare, the Scottish grave robbers (selling corpses to medical men), caused significantly different legislation. Thenceforth no burial could take place unless it was partnered either by a registrar's certificate or a coroner's order. Further laws followed, also caused by public concern that murders were being concealed. Finally, with the arrival of additional legislation that weakened the coroner's duties, it is even being proposed that the office, so allied with death, should itself be allowed to die. A span of 800 years is quite long enough for any government department, first occupied by three knights and one clerk in every English county.

DEATH TABLES Do take care to get born well, said Bernard Shaw. Childhood will then be more secure. Education will be better. General health is likely to be good. And age at death will probably be greater than for those born less well. US black men between 35 and 54 are 1.8 times more likely to die than similarly aged whites. Black women in this grouping are three times more likely to die than their white counterparts. If blacks and whites in the US are considered as separate nations, and the relevant figures are separated, white America ranks twelfth 'in age-adjusted mortality rates' (near Italy and Australia) whereas black America ranks thirty-third (near Romania). A report (by Pappas in the 1980s) showed that Americans with an annual income less than $9,000 had a death rate three to seven times higher (the disparity being due to race and sex) than those earning $25,000 or more. Americans who never graduated had a death rate two to three times higher than graduates.

All socio-economic status is important but, as Marcia Angell wrote in the *New England Journal of Medicine*, 'no one knows quite how it operates ... We also do not know ... whether there is a threshold above which increasing wealth and education no longer matter.' The fact of unemployment may be relevant to the figures. A British survey showed that men who experienced unemployment in the five years after acceptance (into the trial) were twice as likely to die during the following five and a half years than men who remained continuously employed. Retirement is pertinent. 'Men who retired early for reasons other than illness, and who appeared to be relatively advantaged and healthy, had a significantly increased risk of mortality compared with men who remained continuously employed.' The increased mortality, concluded the report, was non-specific, involving both cancer and cardiovascular disease.

Britain has not been doing well recently in the long life stakes. In 1970 the country ranked twelfth, but it slipped to seventeenth by 1990. During the 1980s incomes of the richest fifth in Britain increased to four to six times the incomes of the poorest fifth. That gap between rich and poor is thought to be influential, with those countries reducing the gap by the greatest quantity showing the greatest improvement in life expectancy. Wars are hardly a blessing, but life expectancy in England and Wales, according to *Scientific American*, 'increased most dramatically in the decades of the two world wars because of the expansion of health care services and guaranteed food rationing for all citizens'. Saudi Arabia is a nation of tremendous wealth, but has a lower life expectancy than the impoverished Indian state of Kerala.

The WHO publicizes a league table showing the countries top in delaying death. In some recent years these nations have been as follows (with the equivalent and always different ordering for women in brackets):

Year	Men	Women
1984	Japan, Greece, Sweden	(Japan, Norway, Netherlands)
1987	Japan, Sweden, Switzerland	(Japan, Switzerland, France)
1990	Japan, Greece, Switzerland	(Japan, Switzerland, France)
1993	Japan, Israel, Sweden	(Japan, France, Switzerland)

The actual ages in 1993 were – for men: Japan 76.3 years, Israel 75.1, Sweden 74.9, Australia 74.8, and Greece 74.7. The top five for women were: Japan 83 years, France 82, Switzerland 81.7, Canada 81, and Australia 80.8. Perhaps Greece is the most surprising inclusion among the top nations, not being significantly wealthy, but some wealthy nations are much lower in the lists, such as Denmark, New Zealand, Germany, Austria, Belgium, the United Kingdom, and the United States.

At life's start the mortality pattern is not tremendously dissimilar. Infant death rates per 1,000 live births were (for 1993): Japan 4.5, Finland 5.2, Singapore 5.4, and Norway, Canada and Switzerland 6.4 each. Highest rates were Romania 23.3, Argentina 24.7, Kazakhstan 26.7, Kirghistan 30.2, and southern Brazil 32.5. Take care to get born well, but also take care where to be born. Not only have birth and infancy to be survived but what comes thereafter. In eight weeks (of 1994) as many people were killed in Rwanda as Britain lost during the First World War. (This was a far greater proportion because Rwanda had, at the time, a population one-sixth of the British number.) Rwanda had been leading the world with its fertility rates, this being an average 8.5 births per woman during her reproductive years (causing population pressure to be thought relevant to all the violence).

Some serious attempts have been made to discover when death intervened in the ancient world. On average, without doubt, people then lived shorter lives. Figures such as thirty-five years (for ancient Greece) and thirty-two (for classical Rome) have often been quoted, but these are misleading. They imply, however much it is understood that one childhood death savagely negates one vintage death, that people in their thirties were dying extensively. The averages do not indicate the expected span, assuming no gross mishap intervened, such as a failed harvest, a terrible plague, a war.

One investigator of this topic recorded all the reliable birth and death dates that were in the *Oxford Classical Dictionary*. A total of 397 individuals were thus collated. Of this number 99 had died from assassination, enforced suicide, or battle. These were therefore excluded, such deaths being irrelevant to the natural span. The range of the remaining 298 was from 19 to 107, with a median of 70 years. When these individuals were arranged chronologically a significant drop in longevity became conspicuous after the second century BC (the use of lead plumbing being suggested as a possible cause of shortened lives). It would seem, from this and similar studies, that humanity had to wait until the second half of the twentieth century before it lived again as long as Romans and Greeks more than two millennia beforehand – provided, of course, all were lucky enough to escape warfare, murder and compulsory death.

SECRECY Death, the final happening, is possibly more secret today than ever before. Everyone knows of road deaths – 50,000 a year within the European Union, 50,000 a year within the United States – but blankets quickly cover all unfortunates before they are whisked to hospital. Warfare was generally a distant matter, with reports of the glorious dead, until pictures increasingly portrayed the truth. Enthusiasm for the Vietnam war waned dramatically when television intervened (along with the body bags).

Medically there can be problems arising from society's unfamiliarity with the ultimate event. The need for resuscitation is a case in point, with relatives often hammering desperately with fists at the very individuals trying to save a loved one's life. Correspondence columns of the *BMJ* were recently opened to this topic: 'I am sure that seeing defibrillation or a pericardial drainage would be unreasonably distressing for most people'; 'A mother, appalled at the sight of cardiac massage, tried to drag the doctor off her daughter'; 'The atmosphere of a resuscitation unit ... is far from being a calm, controlled environment, it often comprises five scared junior members of staff...'; 'Relatives or close friends have occasionally fainted or become hysterical...'; 'There is no time to handle spectators as well as patients when drastic measures are involved.'

Finally, in a world increasingly of rights, 'relatives have the right to be present'. As for the patients, a group all possessing incurable and advanced malignancy was once asked, should their hearts stop beating, if they wished to be resuscitated. The majority said 'Yes', the minority 'No', and a couple – perhaps after a lifetime fraught with indecision – said 'Don't know'.

*

It is difficult conducting the routines of ordinary life if too aware of its brevity. Only 3,650 weeks pass by before a newborn reaches three score years and ten. This time may be long when compared with other creatures, and may seem long when in one's infancy as the future stretches ahead without apparent end, but that number of weeks has been reduced to 2,700 when adult height has been achieved (at age 18), a mere 622 months. (The Chinese speak of 1,000 moons when describing a lengthy life.) Whether considered long or short, reflectively by each of us when somewhere along that path, it is the span that evolution has contrived, giving us opportunity to donate our genes, to nurture their upbringing and then to vanish absolutely, having fulfilled our role.

THE FUTURE

I never gave any written or verbal permission for my children
to be conceived
Miles Kington

Niels Bohr has already been quoted on prophecy, but his remark is well worth repeating: 'Prediction is very difficult, especially about the future.' Perhaps, being a physicist, he knew of some hopeless forecasts by fellow physicists. 'X-rays will prove to be a hoax' and 'Radio has no future,' said the brilliant physicist Lord Kelvin. 'Anyone who looks for a source of power in the transformation of the atom is talking absolute moonshine,' said the equally famous physicist Lord Rutherford. Even Albert Einstein, who lived well into the atomic age, once declared: 'There is not the slightest indication that energy will ever be obtainable (from the atom).' Arthur C. Clarke, a much better exponent of crystal-gazing who has watched (some of his) science fiction becoming science fact, summed up the situation neatly: 'When a distinguished but elderly scientist states that something is possible, he is almost certainly right. When he states that something is impossible, he is almost certainly wrong.'

Various predictions have been made in this book. Some of them may have been troubling. Comfort can therefore be taken from mistaken forecasts of earlier years. Thomas Edison, seeming inventor of practically everything, was halfway through his life when asserting that high tension electricity and alternating current made no sense, either scientifically or commercially. It was also 'apparent' to him (in 1895) 'that the possibilities of the aeroplane . . . have been exhausted'. (By living until 1931 he could observe both statements thoroughly annihilated.) So what price genetic engineering, surrogacy, egg and sperm donation, embryo preservation, elongation of life, adjustment of the menopause, pre-procreation counselling, and all the rest?

One certain answer is that humankind will continue to tinker wherever possible. Long before the arrival of modern times that tendency has been its way. Was there ever a human society which did not mutilate itself, by deliberate incision, tattooing, severing, blunting, piercing? Has any society ever existed which did not seek out medicines, aphrodisiacs, hallucinogenic potions? In more recent years, with invasive surgery possible, with the pharmacopeia multiplied ten-thousandfold, with science still accelerating and with possible options out of all proportion to earlier times, our ancestors would be amazed, intrigued and probably delighted by the current plethora. Besides, we are still in effect our ancestors, no longer hunting and gathering or dying like flies when ill-fortune chose to strike, but our genes are the same (as near as dammit); so too our aspirations and desires, our wish for self-promotion, our courage and timidity. We are them, but in fancy dress. They took drugs and so do we. There is no reason why a passion for alteration will vanish in the years to come.

'The chief characteristic which distinguishes man from other animals is a desire to take medicine,' said William Osler (at Montreal in 1894). And so said Bernard Shaw a few years later. And so say all of us today, in differing fashion, again and again. Every nation could readily spend all its gross national revenue on health, there being so many remedies, procedures, and advances on which to spend the cash. (Unfortunately there is also defence. A recent Save the Children Fund report revealed that Sudan spends $27 per person on defence and $1.3 on health, Zaïre $9.7 on defence and $0.4 on health, and Tanzania $105.3 on defence and $0.7 on health.) Pressure to dedicate more on health (and perhaps less on armaments) will surely become greater in the future as remedies and procedures continue to multiply (and guns and warplanes grow ever more expensive). Human repro-duction, the basis of this book, will certainly not escape this lust for amendment. Indeed there is every likelihood that adjustment of everything adjustable in the creation of subsequent generations will be pursued with even greater vigour.

The procreation of humans is, after all, the key event. It led to us, and it will lead to the next assortment. To look at a photograph of any Victorian scene is to know that everyone in it, however old or young, is dead. So too with us a century from now. The future we are busily creating is for a set of people as yet unknown. We do know we are handing them a planet, more crowded, more polluted and more exploited than ever before; but what are we handing them by way of human reproduction?

It is certainly not the relatively simple tale which we ourselves received. Perhaps infertility will become the rule, with sperm and eggs stored at adolescence, and then mated and implanted only when partnerships have been formed and the time is ripe. Perhaps sex and all its antics will become entirely disconnected from the need for procreation – it being well along that path today. And, almost certainly, some of what we now think of as permissible will seem anathema when inspected from the future, much as we now view child labour, slavery, enforced transportation, or racist persecution.

Conversely much of what we currently consider impermissible may become acceptable (within guidelines) as the years progress. In vitro fertilization was frequently viewed with distaste in its earliest days, but is now (by and large) welcomed. Gender choice of offspring is not now (in general) thought desirable, but neither is the production of undesired children for whatever reason – including, some say, their sex. As for cloning, and the creation of new offspring from the genetic material of adult cells, the world is both alarmed and appalled as well as intrigued and even flippant about its possibilities.

When news broke (in February 1997) from a research unit in Edinburgh that a 6-month-old sheep had been fashioned from a single cell of epithelial tissue the reaction varied from alarmist and terrified to light-hearted and jolly. On the one hand was talk about Hitlerian regiments and Huxleyian epsilons; on the other were happy headline jests inspired by Dolly, the sheep in question. 'Will there ever be another ewe?' and 'Clone, clone on the range' were both preceded by 'There's no such thing as baa-aa-aa-d publicity', this statement arising from the animal's eponym, Dolly Parton (her individual fame not created by mammary tissue, as the sheep had been, but heavily dependent upon it).

As with IVF there will assuredly be emotional cases in the future which will alter our opinion on the merits of such a form of reproduction. A child (lovely, eager, famous) may desperately need bone marrow when no normal form of donor can be found. A couple (each lovely and doubly famous) may wish for a replication of their only child, blown to (genetically usable) bits. A musician, of unparalleled promise, suffers deafness/epilepsy/assassination at age 22 (and millions are donated to reproduce her form). Cloned material would solve all such problems in a most straightforward style.

Similarly there may be situations (and individuals) so repugnant that we, in an act of global unity, attempt to push this particular genie back inside its bottle. A dictator of extreme wealth creates battalions of single-minded automatons quite heedless of death (theirs, or anyone else's). Some narcissus, wishing no dilution of such genetic excellence, evades the usual form of procreation by opting for cloning as supreme alternative. A leader, wishing to maintain power (and stay alive), grows several alter egos to serve as transplantation stores, much as owners of ancient vehicles cannibalize one to keep others on the road.

Genies cannot be returned whence they came, but their powers can be trimmed – by legislation, by universal disapproval. Cloning may even become acceptable, in some form or other, to some or most of us, and be no more than another twist in the matter of human reproduction, or it may quietly be forgotten. Humanity has known, ever since the controlled breeding of crops and animals began, that desirable characteristics can be reinforced by selecting and mating appropriately. It is perhaps strange that humanity, although often in favour of strength, swiftness, beauty or intelligence, has rarely (if ever) applied this principle to itself. That may continue to be the rule, despite superior genetic knowledge and ability. We may wish to continue – whenever possible – with the casual, chancy, happy-go-lucky style of reproduction which we in general employ today, as A meets B, loves B and then makes C (anathema to every racehorse owner, shepherd, dog fancier, or beef and milk producer, save for their own and personal breeding back at home).

*

'Life', said Blaise Pascal 340 years ago, 'is a maze in which we have taken the wrong turning before we have learned to walk.' There have been plenty of wrong turnings since his *Pensées* were published, with the only current difference being that we can now make them much more speedily. Human reproduction is a case in point, perhaps the strongest case. With the genome soon to be available, and with the embryo a marketable and refinable resource, more turnings are now being made concerning the start of life than in all the earlier centuries of humans reproducing. Even more turnings are about to be made – for good or ill (the age-old phrase already used within the introduction to this book). 'If we could first know where we are and whither we are tending,' wrote Abraham Lincoln, 'we could better judge what to do and how to do it.' We could indeed, with 'where we are' and 'whither we are tending' having been the grandiose purpose of this book.

*

To read a newspaper of the day when we were born, notably for those of us longer (and more worn) in the tooth, is to encounter absolutely nothing about human reproduction. The current trouble zones are all there – Balkans, religious feuding, political intrigue, corruption, Middle East – but not a word about divorce, family, abortion, fertility or contraception, let alone such modern extras as surrogacy, test-tube conception, sperm donation or foetal rights. It can seem on occasion today as if little is happening (except in the Balkans etc.) save football, infidelity, royal mismanagement – or yet another new-found twist to the business of creating human beings.

There are many such recent stories from which to choose. A woman wants to become pregnant with her dead husband's sperm. She is refused permission (by the HFEA) because she does not possess his written consent (as he died suddenly from meningitis). His sperm are in storage, having been acquired by electro-ejaculation and then immersed in liquid nitrogen (both such acts being unlawful at the time). A British appeals court overturns the ban, enabling her to travel to another country (also unlawful) for the insemination and for her husband's sperm to be exported. Britain's Human Fertilisation and Embryology Authority had been correct in interpreting the law – written consents are necessary for all manner of subsequent acts – but had been taken aback both by the situation and its considerable publicity. Plainly there will have to be new laws, with these extremely difficult to write, bearing human ingenuity – and scientific advance – in mind.

Simultaneously another woman, following fertility treatment and a boosting of her egg production, conceived eight embryos. Within the previous five years she had had one child, one abortion and one miscarriage; therefore eyebrows were raised that she had been allocated such treatment. They were also raised by her partner, who had not been consulted (being against a pregnancy with her), and by her doctors, who had advised against sexual intercourse because of the excessive egg production. The woman had gone ahead with intercourse, did not tell her partner of the possible outcome, and soon learned she had conceived eight embryos. Having absorbed this amazing news her partner then changed his mind. A public relations consultant quickly arranged a 6-figure deal with a tabloid newspaper eager for news (and the sole rights) about such a massive pregnancy. The partner was promptly lampooned on *Private Eye*'s cover, suggesting he had had no difficulty in choosing names for his eight future offspring – 'Audi, Mercedes, Alfa Romeo . . .'.

The gynaecologist in charge recommended a reduction of the eight to two, a form of selective abortion improving the chances both of those two survivors and of the mother's general welfare. Her partner reacted promptly: 'It's too horrible to contemplate ... It would be nothing more than a selective cull.' The operation was never performed and, in due course, as the medical men had predicted (for no one has ever produced eight offspring from a single pregnancy), she lost all eight. The final media involvement – also not to everyone's liking – was of a well-photographed funeral with eight little coffins, each with bouquets, solemnly borne to the grave by eight black-suited attendants.

A few months beforehand, in total contrast both to a corpse's sperm and an 8-fold tragedy, Louise Brown had celebrated her coming of age. She – in 1978 – had been the world's first test-tube baby, conceived outside the uterus and then implanted. Her birth had paved the way for the successful arrival of 150,000 others who have been similarly initiated, including her 3-year-younger sister Natalie. Unlike the considerable trumpeting on her original birthday Louise Brown planned to celebrate her eighteenth anniversary quietly with her family. Afterwards she would carry on learning to become a nurse, preferably to work with the very young. 'I really love babies. I'm concentrating on my studies now but one thing is sure. I want to have my own children – whatever it takes.'

It will certainly take more time before stories concerning embryos move away from the front pages. In mid-1996, when Louise was picking up her books again, a first batch of human embryos was permitted (by a further ruling of the same HFEA) to be disposed of/ taken from the freezer/sacrificed/murdered/flushed down the toilet – according to publications recording this event. The embryos had formerly existed as fail-safe back-up following in vitro fertilization. Ovaries had been encouraged to produce several eggs rather than one, and several of each personal batch had been fertilized, rather than one. Most of the respective mothers had wanted only one offspring; therefore the other developing forms had been frozen in case of later need. (There was always a possibility that the first egg/ embryo might not implant successfully. It was good to have others in readiness without the need for returning to square one and stimulating ovaries all over again. A plethora of multiple pregnancies in the early days of IVF had resulted from fail-safe being carried right into the womb causing, on occasion, several implantations and many more babies than were desired.)

No one knew whether lengthy storage of human embryos might be harmful, although work with animal embryos indicated otherwise, but the HFEA initially thought five years to be a reasonable span. At the end of that time, if the couples involved had no immediate or likely need, the embryos were to be removed from their tanks of liquid nitrogen. Hence subsequent and scientific talk about the cessation of preservation on the one hand, and the destruction/ murder/slaughter of 'tiny frozen babies' on the other.

As many such embryos (or possibly most) of those which are

conceived naturally following normal intercourse subsequently fail to implant, and therefore die/decay/wither without anyone much wiser about this commonplace event, the scientists were unimpressed by argument from those who declare – vociferously – that every conception should always proceed to term. In theory every couple responsible for each of the extra and unwanted embryos should have given their consent before the removal from cold storage but, in practice, many of these individuals could no longer be located and some of them kept silent, perhaps having no wish to be linked to the lethal procedure.

Legislation involving human embryos is repeatedly being altered, partly as a consequence of changing scientific abilities and also for changing public attitudes. In November 1996 the Council of Europe (to which 40 states belong) permitted the use of embryos for research but forbade their creation for such a purpose. Britain, although a council member, does permit the creation of embryos for research. In practice (at most of the 20 British centres using human embryos) the researchers work with spare embryos from IVF. These are no longer being created in such fail-safe numbers but the HFEA recently doubled their permitted storage time from five to ten years, believing their viability would still be satisfactory and, incidentally, would swell the number available for research. Even so, and without doubt, embryos, storage, implantation, the HFEA and all research in this most emotional area will continue to be a hot potato – in newspaper terms – for a long time to come.

Neither AIDS nor HIV has been far from the headlines since their first invasion in 1981, with good news and bad news always alternating. There is still neither a cure nor an effective vaccine, but research funding continues to be considerable. (A 1996 report stated that the US National Institutes of Health spent $110 on relevant research per death from AIDS as against $1.85 per death from stroke.) Many treatments are prolonging the survival of HIV-infected people, but at great expense. There is wide resentment that research progress has apparently been slow (when the March of Dimes seemed to cure polio in a trice) but cancer, rheumatoid arthritis and multiple sclerosis, for example, are still awaiting effective remedies. The HIV epidemic is far from over but in Europe it has stabilized. (Spain is Europe's most AIDS-afflicted country.) The US toll from AIDS fell (for the first time) between 1995 and 1996 as deaths dropped from 24,900 for January-July to 22,000 in the same period one year later. Elsewhere the disease is proving more successful. According to WHO. there are 6,000 new HIV infections in the world every day (or four a minute), with the most rapid increases occurring in southern and central Africa and in Asia. Nearly half of the 2.7 million adults newly infected in 1996 were women. Of the 830,000 children currently living with HIV almost half contracted it in 1996.

A belief still exists, notably in the US, that AIDS is predominantly a problem for homosexual males. A major survey (undertaken in the 96 largest metropolitan areas of America) recently concluded that half of the new infections arose through the injection of drugs, a quarter through heterosexual intercourse, and the final quarter through homosexual intercourse. Approximately 1 in 300 of US citizens is now HIV-positive. Infection is highest among the African-Americans and Hispanics, with 3% and 2% respectively of such men in their 30s now living with HIV. Another belief still lingers, also in the US, that HIV is a particularly American problem, but Asia (with its 3.5 billion people) is currently facing the most rapidly growing epidemic. India now has the largest burden of HIV infection of any nation.

As for the immediate future there will be 40 million HIV-infected people in the world by the year 2000, with almost 18 million in Asia, and 90% of the new infections will be in developing countries. In that same millennium year there will be 6 million pregnant women and 5–10 million children infected with HIV. If drugs do become available which greatly hinder or even conquer HIV this news will be both good and bad. Effective drugs, however small the price tag, will probably be a solution only for the developed world. Moreover their arrival will take much pressure off the need for an effective vaccine, and only through such a vaccine will the virus known as HIV eventually be eliminated in all the world.

The plague, as some call it, known as abortion shows no signs of being eliminated. The procedure will only vanish when contraception is 100% effective *and* when foetuses are no longer deliberately expelled for medical reasons. In the meantime there are some 50 million abortions a year, with about two-fifths being illegal. Many unskilled practitioners are involved, helping to cause between 50,000 and 100,000 abortion-related maternal deaths a year. Overall, within the world in general, it has become easier to obtain an abortion in the recent past but some of the debate on this topic has become even more acrimonious.

The United States has experienced both ends of the spectrum to a marked degree. 'Pro-life' advocates have been countered by the 'pro-choice' majority, both operating in a nation where 30 million abortions have taken place since the 1973 Supreme Court declaration that such expulsions are a constitutional right. The rate has recently been declining (slightly) but three out of ten US conceptions still end in abortion. Half of all American women are likely to have a pregnancy terminated, and one in five of such patients are evangelical or born-again Christians. Despite the Pope's condemnation of the practice as a 'culture of death' Catholic women in the US actually have a higher abortion rate than their representation in the country as a whole would suppose.

Conventional opinion states that 'most' abortion is on behalf of the very young, the very careless and the very poor. Conversely the Alan Guttmacher Institute reported (in August 1996) that 57.5% of women experiencing abortions *had* been using contraception during the relevant month, that 78.5% of abortion patients were over 20, that 45.6% were over 24, and more than half of all these patients had annual family incomes greater than $30,000.

The American dilemma, seen at its most extreme in the shooting of abortionists, also comes to a head in the so-called Mexico City policy. This has forbidden US aid to any organization promoting or advocating abortion. As a consequence there has been a 'negative impact', as President Clinton said (in February 1997), 'on the proper functioning' of international family planning programmes. What is in the balance, he added, 'are the lives and well-being of many thousands of women and children, and American credibility as the leader in family planning programs around the world'. Such statements have led to charges that he is on a 'crusade to legalize and promote abortion'.

As for the world beyond America there has been further liberalization. In November 1996 Poland overthrew a senate veto and passed a law permitting abortion before the 12th week (provided there had been counselling and a period of reflection). During that same month South Africa discarded an old law permitting abortion only under strictly limited criteria (rape, incest, psychiatric illness) and introduced another which is virtually abortion on demand. During the first 12 weeks a woman need only 'request' termination. From 13 to 20 weeks she should both request it and satisfy four fairly undemanding criteria, such as a pregnancy's harmful effect upon her social and economic circumstances. One further change, in that same month of November 1996, affected the tiny Channel Island of Guernsey. It became the last place in the British Isles to legalize abortion – by 34 votes to 20 in the island's parliament.

These amendments, and others, prompted William Rees-Mogg in the London *Times* to produce some harrowing and entirely relevant statistics. The American abortion rate, proportionately twice that of Britain, is a 'veritable genocide ... comparable in number to the deaths for which Stalin, Hitler or Mao were responsible'. 'Five times as many American babies have died (as a result of the Supreme Court's decision) as Jews were killed in the Holocaust.' As for Britain its abortion deaths, 3 million in the past 30 years, 'have exceeded the battlefield deaths of both world wars. They are the missing generation.'

Two recent stories can sum up this contradictory aspect of our lives. A 43-year-old woman had to undergo a hysterectomy. During the operation her gynaecologist realised she had been 8 weeks pregnant. He was acquitted of performing an illegal abortion, and she later received a five-figure out-of-court settlement from the relevant hospital for the loss of her baby, an event which allegedly contributed to the breakdown of her marriage. The second story followed from a professor's remark that 'the fundamental ethical concern in all fertility treatment must be for the welfare of the offspring'. Why then, wrote a newspaper's reader, 'does this priority not apply in the instance of abortion?' One answer has to be that the newcomer's rights are in short supply until its birth and even then, with severely malformed offspring often receiving scant attention, not overall.

As for eugenics, and 'the right to improve the quality of the newborn population', as China's new law puts it, the world is more united in its opinions. A major genetics meeting, to be held in Beijing in 1998, prompted the *New Scientist* to speak for many when it declared: 'Exporting genetic know-how to a regime that sanctions eugenics is about as morally wholesome as selling Semtex to countries that sanction terrorism.' The Chinese government argues that its new law only permits practices which are already common in the West. Others are not so sure. Chinese doctors are required to recommend an abortion whenever a prenatal test detects 'a serious genetic disease', but the degree of severity is open to debate (and there will assuredly be plenty in 1998). China's population problem, say some, permits China to take extraordinary measures, and it is laudable that China's formidable millions will become fewer than India's formidable millions before so very long, but there are means as well as ends to be considered. Currently China is everyone's pariah in the matter of eugenics, a fact which probably worries China not a scrap.

India's population is expected to exceed one billion by March 2001. It already totals 950 million, and is not expected to stabilize before 2026. China is expected to stabilize somewhat later, but India's relatively enthusiastic fertility will ensure – if United Nations predictions are correct – that India will contain 1.6 billion people by 2050, and will therefore be the winner in this unwelcome race.

The good news is that growth in general seems to be diminishing. This fact is a long way short of global population beginning to diminish, but at least the rate of expansion is lessening. At the start of the 1990s the preferred figure was 90 million more people every year. In December 1996 this quantity was officially reduced to 80 million more a year. The percentage growth rate had been 1.57% a year in 1990 and it became 1.48% in 1995. Women worldwide were having an average of 5 children in 1950 over their reproductive life, and in 1995 this average had dropped to 2.9. In Bangladesh – always a prolific nation – the present rate is 3.5 as against 7 only 20 years ago. As a result of such impressive alterations it is even thought, by those who study demography, that the world's population will never double again. It is now 5.8 billion and will come to a halt at 11 billion. For which, as 11 billion people are sure to say, much thanks.

Europe has already called a halt, but such cessation is not universal. Even Africa's nation of Nigeria is expected – by itself – to overtake western Europe when it reaches 350 million by 2050. The European Union's birth rate in 1995 was the lowest in any peacetime year this century, with its 370 million population producing only 290,000 more births than deaths (as against an expansion of 1.6 million in America's population now standing at 260 million). Within Europe the French are still big breeders (producing two-thirds of the EU's increase); so too the Scandinavians. Italians and Spaniards are the least active in this regard, with Portuguese and Greeks almost as slovenly. If Italians carry on in similar fashion their population will have shrunk by 20 million when the time comes for India to overtake China. Then only one person in 25 will be a European, and only 1 in 175 a Briton (from that same nation which used to rule such a major part of the planet's population). There is still immigration into Europe, trebling the EU's natural growth via babies, but this inflow is being reduced. It peaked in 1992 at 1.3 million but dropped to 800,000 by 1995. Crowded Europe might therefore become less crowded in years to come, and perhaps an even better place to live.

What will certainly grow, notably in other areas, are cities. In general their populations are expanding at double the human growth rate. As nine (or perhaps eight) babies are being born globally every second this means that the planet's conurbations are swelling by 18 (or 16) extra individuals in each same second. In 1950 only 31 of the world's cities possessed more than a million inhabitants. There are now over 200 thus endowed and more than 25 'megacities', these each being home to more than 8 million people. ('Put your money into land, son; they're making no more of that,' runs the old adage. 'Put your money into urban housing, son, they're needing more and more of that,' could be current paternal advice.)

Figures from OECD, the Organization for Economic Cooperation and Development, showed (in October 1996) that the US had become the meanest member of its donor nations, giving only 0.2% of its gross domestic product to international aid. According to Malcolm Potts, steadfast promoter of a global need for contraception, the portion of that modest international contribution allocated to population control 'is equivalent to the cost of one hamburger (with cheese) per US citizen'.

Despite niggardliness, and cost, and papal pronouncements, there is no doubt that contraception is more widely available than ever before. Seemingly contrarily, save in the matter of personal choice, current fertility procedures are permitting more of those who wish for children to have them. And it is equally true that fewer children are dying and that lives are getting longer, but the further fact that more and more of the world's population will be old in future years can seem to be less of a blessing.

Today (according to 1996 figures from the WHO.) there are 540 million people over 60. By 2020 there will be more than a billion, with almost three-quarters of them in developing countries. This increase in ageing will assuredly be a burden upon the younger population, and yet more so in the developing nations where the change is being so abrupt. In such poor communities the average life expectancy at birth rose from 46 years in the early 1950s (only four decades ago) to 64 years by 1990. This figure is expected to reach 72 years by 2020, the sort of life span now enjoyed in the developed world. The suddenness of this change will undoubtedly provoke greater hardship in the poorer places. Instead of many young people having to care for a few old ones there will be relatively few youngsters having to look after a tremendous quantity of elderly individuals.

Britain, by contrast, has already experienced its period of rapid growth of retired people, although there will be another bulge in 20 years' time when the post-Second World War babies reach old age. Its National Health Service, with a brave promise of health and care for all whatever their circumstance and need, has had a difficult time concerning nursing (and beds) for the elderly. Even within the past decade there has been a 25% increase in the over-75-year-olds and a 50% increase in the over-85s. Currently 1 in 20 individuals aged 75–84 is in some form of residential/nursing home; so too 1 in 5 of the over-85s. For the people who have to pay, and whose money is consumed before the state will contribute, the cost can be an awesome £20,000 a year. So what will happen when the next bulge comes along in 2017? In theory the politicians should be making preparations now – an increased levy on something, an old age insurance scheme, and it does seem from some 1997 pronouncements that such thinking is getting nearer the agenda.

Perhaps DNR will become a positive policy and more than another set of medical initials to perplex the rest of us. Just as DNA – for every practitioner – means 'Did Not Attend' (at some appointment, and has nothing to do with chromosomes), so do medical staff know that DNR is 'Do Not Resuscitate'. In theory each patient's condition and likely outcome ought to be the sole guideline whether resuscitation should be attempted. In practice the patient's age is highly pertinent. According to a study published in 1996 of '6,802 seriously ill' patients at five US medical centres the DNR letters were more frequently written for people older than 75 'regardless of the patients' prognosis or treatment wishes'. In short, according to the report, physicians 'may be relying on age to indicate that less aggressive care is appropriate'. Prognosis, it points out, 'is not necessarily tied to a patient's age'.

DNR could become official policy, a form of ageism. As we can say (bluntly) when someone 'on the wrong side' of 80 dies, 'Oh well, they had a good innings', it can therefore seem fair that someone on the right side, irrespective of their condition, should receive more treatment. The Inuit are alleged (by the rest of us, if not by the Inuit) to put their grandmothers under the ice when times are hard. The British, according to a cartoon in *New Scientist*, put them out at bus-stops, the draughty, chilling, unfriendly, seat-less, litter-infested, semi-enclosures which surely hasten death (if a trifle less urgently than ice). Future governments, so beset by aged people and with so few young to pay for them, may well encourage either policy. (And ICE, to be flippant in this regard, may then stand for 'Initiate Chilling End'.)

Euthanasia, assisted suicide, mercy killing and hastened death are all unlikely to vanish from the headlines. With 51 million people dying in the world each year, and with methods for postponing death increasing steadfastly, the current problems are bound to increase. They may even come more into the open. One in five intensive care nurses in the US has intentionally hastened the death of a terminally ill patient without the knowledge of the patient's doctor, according to the *New England Journal of Medicine*. They did so because, in their opinion, the doctors were ignoring the wishes both of the patients and their attendant families. Another US report stated that two-thirds of questioned families believed their dying relatives had suffered 'intolerable' symptoms at the end. Although 59% of patients had asked for treatment which focused on comfort 10% of them had 'received care contrary to this preference'. The study's principal author concluded that the US health care had a long way to go to 'improve its care of people near death'.

And so says much of the world, with current legislation entirely confusing. In Holland, despite some famous recent cases, euthanasia is still illegal but is accepted in practice (provided certain conditions have been met). In Switzerland active voluntary euthanasia and mercy killing are punishable, but passive euthanasia and assisted suicide are not. (Lawyers must have a field day deciding whether lethal pills carelessly/carefully left nearby, with or without helpful labelling, come into one or other category.) The US Supreme Court is currently trying to decide whether a ban on physician-assisted suicide infringes personal liberty. It will probably conclude that medical treatment may be withheld but positive action will not be allowed. The high degree of 'intolerable' symptoms is likely to make ordinary individuals think, and act, otherwise. 'If you want a reason for why the public is looking to suicide,' wrote one report's author, 'there you go.'

Australia's Northern Territory has, however unlikely this may seem to outsiders, been leading the world concerning euthanasia. Its 'Rights of the Terminally Ill Act' came into force in July 1996 and, by February 1997, three people had died under its provisions. Nevertheless the new legislation has experienced considerable criticism, both from within and without the Territory. An Australian Senate inquiry is examining the issue, before deciding whether the law should be overridden. A record number of 12,500 public submissions has been received by those in charge of this investigation. Euthanasia supporters, in favour of the act, are also critical, thinking its requirements are too onerous. Three doctors must be involved, one of whom, a specialist, must be a Territory resident. Apparently the second person to die 'spent weeks' searching for a Territory specialist willing to assist.

There will certainly be further wrangling in Australia – and in every other nation – but Japan experienced a landmark case back in 1962, long before the world in general thought euthanasia might receive positive legislation. In that year a son helped to kill his terminally ill father. The High Court in Nagoya, having debated the case, subsequently laid down six conditions which should all be satisfied before euthanasia could be legally carried out.

These were:

– the inevitability of death despite all medical attempts;

– the suffering of those close to the patient;

– the need to attempt to save the patient from suffering;

– a clear expression of a desire to die from the patient;

– the method of killing should be appropriate;

– the procedure must be performed by a doctor.

The most important point, being the one to dominate later cases, was the patient's consent rather than the family's. In a famous Japanese trial (of 1991) a doctor allowed a terminally ill and unconscious man to die at the request of the family. Following the Nagoya stipulations he was judged to have acted illegally, but was not prosecuted as he had acted under considerable pressure from the family. Similarly the family was not prosecuted as it did not know that the doctor had injected a lethal dose.

*

Human population is expanding. The number of people now dying every year, 51 million, will increase. So will medical expertise in the business of life prolongation. And so, probably, will the 'intolerable' symptoms at the end of life. Therefore what happens in northern Australia, or The Netherlands, or Switzerland or the US, is of considerable concern to those of us alive today. Our time will come, all too assuredly, and it would be very welcome if we could have a say, if so desired, in the matter of how and when we reach our individual ends. We had no choice in our conception. At least we could have influence over our finale, with the manner and style (if not the time) of that conclusion for each of us to choose – should we so wish.

INDEX